The Pictorial Guide to the Living Primates

The Pictorial Guide to the Living Primates
Noel Rowe

Foreword by

Jane Goodall

Introduction by

Russell A. Mittermeier

Pogonias Press   Charlestown, Rhode Island

Library of Congress Card Number: 95-072713

Hardbound: ISBN 0-9648825-0-7
Softbound: ISBN 0-9648825-1-5
Second Printing November, 1999

Pogonias Press
1411 Shannock Road
Charlestown, Rhode Island 02813-3726 USA
401-364-7140
Fax: 401-364-6785
email: abarber@primate.org
This book can be ordered by mail from the publisher.
Please include $4.95 for postage.
Toll-free order number 1-800-296-6310

Editorial Advisers: Louis Budd Myers, Rosemary Sheffield
Layout, design and maps: M. M. Myers
Graphic artists: Stephen Nash, Diana Zadarla

Printed in Hong Kong

Cover photos: Noel Rowe

To the memory of Debbie, who loved me and the birds
& bees and chimpanzees

Front cover: Chimpanzee, *Pan troglodytes*
Spine: White-handed gibbon, *Hylobates lar*
Half title page: Hanuman langurs, *Semnopithecus entellus*
Title page: Assamese macaques, *Macaca assamensis*

Foreword

When I began my study of the chimpanzees of Tanzania's Gombe Stream in 1960, scientists believed there were about 180 species of prosimians, monkeys, and apes, and our entire knowledge of the natural behavior of primates was based on a few short-term studies. In the 35 years since I began my research, primatologists and taxonomists have discovered or reclassified more than 50 additional species. Hundreds of long- and short-term field studies have been carried out in tropical forests around the world. But although the expansion of our knowledge about primates has increased exponentially in my lifetime, there are so many more questions to be answered about the forests and their inhabitants.

E. REGAN - CAMERA 5

Primates, if followed long enough by nonthreatening researchers, will accept us and allow us to study their day-to-day behavior. We have learned not only what they eat and where they sleep but also with whom they fight, with whom they mate, and who their friends are. We know them as individuals with a personality, a life history, a place in their community, and a future. They are just like us in these ways.

The Pictorial Guide to the Living Primates is the first book to show us in color what each of the 234 species looks like. It gives us details of their lifestyles and some of their interesting behaviors, as well as range maps and their conservation status. It is sad to realize just how many of these species are in danger of disappearing from the earth.

There are two central themes in this lavishly illustrated book. First, it shows us how beautiful and diverse the primate order is. It shows us our place in the family tree. We are very different from the leaping bush babies, which can see in the dark, and the prehensile-tailed spider monkeys, which swing through the canopy. Our closeness to the great apes is undeniable; we share 98% of our genetic material with chimpanzees, for example. Like us, great apes can use and make tools, are capable of rational thought, and experience emotions that are probably similar to those we describe as joy and sadness, contentment and despair. They have a sense of empathy.

Second, the book serves to remind us that without our help many of these monkeys, apes, and prosimians will vanish. Our fellow primates are the link we humans have to the diversity of our evolutionary past, and our future will be one of profound impoverishment without them. We are a problem-solving species—the challenge we face for the immediate future is how to leave room for the other 99.99% of the species with which we share this planet.

Noel Rowe, the author, has visited primate field study sites and conservation projects on three continents to take many of the photographs for this book. In the course of this work he came to realize how little we know about so many primates and how little protection they receive as their forest homes are being invaded by human hunters, loggers, and farmers. In 1992 he founded Primate Conservation Inc., a nonprofit organization that gives seed money for field research and conservation projects on the least-known and most-endangered primates. This is good news for eager young field researchers anxious to set out, as I did in 1960, to make exciting discoveries about little-known species.

As we turn the pages of this book, we are caught up in the author's love of his subjects, his sense of wonder and delight. It will surely help motivate us to protect our closest relatives in their natural habitats. Everyone who acquires the book will use it extensively and treasure it forever.

Jane Goodall, Ph.D.

Jane Goodall Institute
P.O. Box 14890
Silver Spring MD 20911
Phone 1 800 592 5263

Contents

v **Foreword** Jane Goodall, Ph.D.

1 **Introduction** R. A. Mittermeier, Ph.D.

3 **Overview**
   3 How to Use This Book
   10 Protecting and Conserving Primates

13 **PROSIMIANS**—Prosimii
  13 **Lorises, Pottos, and Bush Babies**—Loroidea
    14 Lorises and Pottos—Loridae

    18 Bush Babies—Galagonidae

  27 **Lemurs**—Lemuroidea
    28 Dwarf Lemurs—Cheirogaleidae

    34 Sportive Lemurs—Megaladapidae

38 Lemurs and Bamboo Lemurs—Lemuridae

47 Indrids—Indridae

51 Aye-ayes—Daubentoniidae

52 **Tarsiers**—Tarsioidea
  53 Tarsiers—Tarsiidae

57 **HIGHER PRIMATES**—Anthropoidea
  59   **Neotropical Primates**—Platyrrhini/Ceboidea
    59   **Marmosets and Tamarins**—Callitrichidae

  81   **Cebids**—Cebidae
    83   Night Monkeys—Aotinae

    85   Titi Monkeys—Callicebinae

    93   Capuchins and Squirrel Monkeys—Cebinae

    101   Sakis and Uacaris—Pitheciinae

    107   Howler Monkeys—Alouattinae

    112   Spider Monkeys and
           Woolly Monkeys—Atelinae

119 **Old World Monkeys**—Catarrhini/Cercopithecoidea
  119 **Macaques, Baboons, Guenons, and Colobines**—Cercopithecidae
    120 **Cheek Pouch Monkeys**—Cercopithecinae

    153 **Guenons**—*Cercopithecus*

    169 **Leaf-eating Monkeys**—Colobinae

207 **Apes**—Catarrhini/Hominoidea

  208 **Lesser Apes (Gibbons)**—Hylobatidae

  219 **Great Apes**—Pongidae and Hominidae
  220 Orangutans—Pongidae

  224 Gorillas, Chimpanzees, and Humans—Hominidae

234 **Acknowledgments**
235 **About the Photographs**
236 **Biographical Information**
236 **Glossary**
239 **Resources and Bibliography**
257 **Index**

Introduction

All of our planet's biodiversity is of great importance, but clearly certain groups of species and a small number of particularly rich ecosystems, mainly in the tropics, deserve special attention. Among these are the nonhuman primates—the monkeys, apes, lemurs, lorises, galagos, and tarsiers that are our closest living relatives and occupy a special position in the imagination of our own primate species, *Homo sapiens*. As research over the past four decades has shown, these animals can teach us a great deal about ourselves and our evolution, and they have played an important role in many different kinds of research aimed at a better understanding of our health, our behavior, our language, and our unique position in the animal kingdom.

Aside from the ways in which primates can enlighten us about ourselves, research on them has an intrinsic value as well and is very relevant to research on tropical ecology. About 90% of all primates are found in the tropical rain forests of the world, the richest and most diverse terrestrial ecosystems. Primates occupy an important role in these habitats as seed dispersers, seed predators, and even pollinators. Our knowledge of how important they are in tropical forests grows every year.

Unfortunately, wild populations of nonhuman primates are in trouble in all of the 92 countries in which they occur, with some of the most serious problems being in those nations richest in primate species, among them Brazil, Colombia, Madagascar, Indonesia, China, and Viet Nam. Primates are threatened by destruction of forests and other natural habitats, by hunting for food, and by live capture for export, although the impact of the last of these has diminished considerably in recent years.

Almost half of the 250 primate species are considered to be of conservation concern by the Primate Specialist Group of the Species Survival Commission (SSC) of the World Conservation Union (IUCN), and one in five is in either the endangered or the critically endangered category, meaning that without better protection they will be extinct in the next two decades. Although we have not yet lost a single primate taxon in this century (an excellent record compared with that of many other vertebrate groups), we will enter the next millennium with a large portion of the order Primates on the edge.

In order to prevent the extinction of a significant percentage of our primate relatives, a variety of actions are needed. Foremost among those actions is to obtain more and better protection for primates and better information about them. We still are amazingly ignorant about basic issues of taxonomy and geographic distribution of most primate species, and new taxa continue to be determined either through taxonomic revisions of museum specimens or by actual discovery in the wild. The most striking examples of new discoveries over the past decade have been in Madagascar and Brazil. In Madagascar two distinctive new species, *Hapalemur aureus* and *Propithecus tattersalli,* were discovered and described in the latter half of the 1980s, and a third species, *Microcebus myoxinus,* described over 100 years ago and forgotten, was "rediscovered" in 1993 and is now recognized as the smallest living primate. In Brazil the record is even more striking, with five new species being discovered and described since 1990 *(Callithrix mauesi, C. nigriceps, C. saterei, Leontopithecus caissara,* and *Cebus kaapori).* In addition to these spectacular discoveries, almost every new expedition obtains significant new information about distribution, ecology, and conservation status. Consequently, in many of the most basic ways our research on primates is still in its infancy.

One of the great gaps in primatology has been the lack of easy-to-use guides for both the specialist and the layperson. Good field guides and illustrative material about birds have long been available, but someone interested in primates needed to assemble a library of bulky monographs and a large collection of reprints to get some feel for these animals. To fill this gap, several new publications are being produced, among them Conservation International's tropical field guides and now the first truly comprehensive illustrated guide of all primate species by Noel Rowe, which I am very pleased to introduce here.

With this book, Noel Rowe has made a major contribution to our understanding of primates. In addition to his own photographs, obtained in many remote corners of the planet, he has also gathered together a series of top-quality photos by other specialists that cumulatively gives us a real feel for the gestalt and the great diversity of this unique mammalian order. This publication should be of considerable use to professional primatologists and, I think, should do a great deal to stimulate interest among the nature-loving public as well. I encourage all of you to read this book and use it in the field or in your local zoo to learn more of these wonderful creatures and get more involved in conservation efforts to ensure their survival.

Russell A. Mittermeier
Chairman, IUCN/SSC Primate Specialist Group
President, Conservation International

Extinct Lemurs

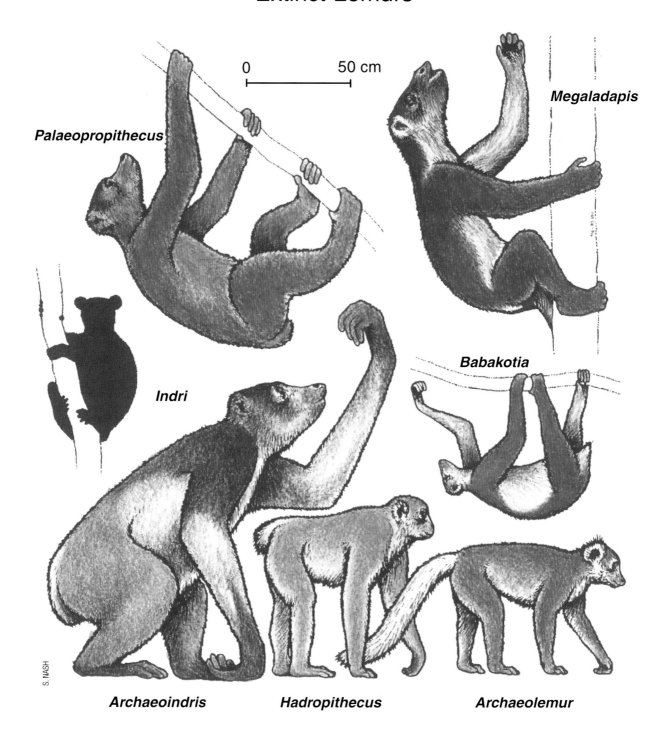

Figure 1. The indri is the largest living lemur. It is placed here for size comparison.

Overview

In 1992, after taking several courses in primatology, I was curious about what the primates that were being discussed looked like. I went to the library to find a book with photographs so I could fit the names to the faces. I found several good books, but most of the species I looked for were not illustrated and were lumped together with related but more common species. I wanted a book that showed me the diversity of primates and told me some basic information about each species, like the many guides to birds or amphibians. The book you are holding is the outgrowth of that desire.

My hope is that this book will give you a greater understanding of the beauty and diversity of primates and will motivate you to help protect these species from the fate of the lemurs illustrated in Figure 1. All of those lemurs were large, well-adapted species that lived on Madagascar when the first humans arrived 1500 years ago. They are now all extinct and known to us only as subfossils. The evidence strongly suggests that humans were the cause of their demise.[168] Without your help, many of the species living today will be lost forever.

How to Use This Book

The Pictorial Guide to the Living Primates provides a photograph or an illustration of each of the living prosimians, monkeys, and apes of the world. A map shows the range of each species, and a color bar indicates the species' current conservation status. The text accompanying each species is divided into nine categories to make it easy to compare information about different species. For some species the information in some categories is not available (**NA**). Usually, "NA" means that the information is completely unknown or has not been published. To a lesser extent, it means that the information was not available to me, because of the difficulty of researching the entire primate order. Rather than gloss over the lack of information about some primates, I have used the "NA" designation in the hope of stimulating people to study those species and fill in the gaps.

Throughout the book the word "primates" is used to refer to all of the species of primates except *Homo sapiens.* This usage is not in line with most of the literature, which refers to all but one of the 234 species of primates as nonhuman primates.

The Pictorial Guide is meant to be useful to students, primatologists, and anyone interested in primates. The text is not all-encompassing. It is meant to introduce you to some of the interesting behaviors of primates on a species-by-species basis. References are provided for those who would like to find more detailed information.

The book is divided into sections, with each superfamily, family, subfamily, or genus that represents an adaptive radiation receiving an introductory page. A scale drawing is included so that the reader can compare the relative size of the primate with a five-foot-six-inch human primate. When the largest and the smallest species of each group are illustrated there are two silhouettes. Within each genus the species are arranged alphabetically by Latin name. The book begins with the suborder **Prosimii,** whose name means "before apes," and continues with the suborder **Anthropoidea,** to which humans belong.[234] Of the two primate suborders, the prosimians are morphologically more similar to the ancestral primates found in the fossil record from 40 to 50 million years ago. Although prosimians are often referred to as "primitive," they have had much longer to evolve and adapt to their environment than the apes, for which the first fossils date back only half as far.[234] The prosimian suborder is divided into two

infraorders: Lemuriformes and Tarsiiformes. **Lemuriformes** is divided into two superfamilies: **Loroidea** (the lorises, which live in Africa and Asia) and **Lemuroidea** (the lemurs, found only on the island of Madagascar). **Tarsiiformes** has only one superfamily, **Tarsioidea** (the tarsiers, which survive only on islands in Southeast Asia). An alternate system of classification, which takes into account derived anatomical features, has different names for the two suborders: Strepsirrhini and Haplorrhini in place of Prosimii and Anthropoidea. This alternate system places the Lemuriformes in Strepsirrhini and places the Tarsiiformes in Haplorrhini. The suborder named both Anthropoidea and Haplorrhini is divided into two infraorders, Platyrrhini and Catarrhini. These two classification systems coexist in the literature and are two ways of categorizing the same information.

The suborder Anthropoidea comprises the true monkeys and the apes. The anthropoids are made up of three major radiations, each classified as a superfamily. The **Ceboidea,** or cebids, are sometimes referred to by their infraorder name of **Platyrrhini,** meaning "flat noses." These New World monkeys live in the Central and South American Neotropics. The **Cercopithecoidea,** commonly known as the Old World monkeys, live in Africa and Asia. **Hominoidea,** or hominoids, includes the lesser apes and the great apes and is the superfamily to which *Homo sapiens* belongs. The last two superfamilies are also referred to by their infraorder name of **Catarrhini,** meaning "round noses." The catarrhines shared a common ancestor that lived during the Miocene epoch about 20 million years ago.[234]

Maps

The maps are topographic and show the general range of each species in yellow bordered with red. The geographic region or countries in which the primate lives are listed below the map. On a scale this small, the maps are

India, Sri Lanka

Kenya, Somalia, Ethiopia

Madagascar

not exact. Even if they were exact, they would soon be out of date because we humans continue to expand our own range and modify the natural world to the detriment of most other species. The maps do give a relative idea of the location of each species' range and how large or small the range is. Very small ranges are indicated by an arrow and are the most likely to be the homes of endangered or critically endangered species. Small ranges indicate either that the species are very specialized or that the areas contain the only natural habitat left for them. Even species with broad ranges may be found only where suitable habitat still remains and hunting is not extensive.

Please consult the world map inside the covers for the entire distribution of all nonhuman primates. If you wish to see more detailed maps, consult the following authors: Corbet and Hill,[130] Emmons,[198] Groves,[312] Hershkovitz,[355,356,357] Mittermeier et al.,[580] Nash et al.,[608] Niemitz,[942] Oates et al.,[635] Rylands et al.,[953] and Wolfheim.[898]

Taxonomy

Readers familiar with primates may notice common names and scientific names that are unfamiliar and species that were formerly considered to be subspecies. This text primarily follows the taxonomy of Groves (1993),[315] found in *The Mammal Species of the World: A Taxonomic and Geographic Reference*. Groves has revised the entire order, elevating some subspecies to species and vice versa. There will always be differences of opinion about phylogenies, species, and subspecies.[422] Taxonomy is based on available information and is subject to personal interpretation. As more field studies are undertaken, more anatomical features measured, and more fossils uncovered, our understanding of the many species will change and so will some of the taxonomy. Nevertheless, no matter what we humans label them or how we classify them, these are the animals with which we share the earth and a common ancestor, and the photographs in this book show us what they look like.

The taxonomy category in the text includes the number of subspecies within each species. This is interesting because it reflects the amount of variation within a species. Usually the more subspecies there are, the more diverse and robust the species is within its range. The fewer subspecies there are, the less diversity is found. Some taxonomic groups are **monotypic,** meaning they have have only one representative. A case in point is the aye-aye (*Daubentonia madagascariensis*, page 51), a species that is literally one of a kind. It is the only species in the only genus in its family, and it has no living subspecies. If there is controversy about a species' Latin name or about whether a primate is a true species or a subspecies of another species, the word **disputed** appears at the beginning of the taxonomy section.

Distinguishing Characteristics

A brief description of what each species looks like is included because the identifying features are not always visible in the pictures. Photographs show us only individuals. They cannot show the whole range of variation within a population or the variation from one population to another. So the individuals you see in the forest may look different from the photos, and in captivity there may be hybrids.

Physical Characteristics

The physical characteristics section provides five categories of measurements for each species, in both metric and U.S. units. The average **head and body length** (combined), **tail length,** and body **weight** for the species are listed, as well as the ranges of those measurements for males and females when available. The smallest primate is the pygmy mouse lemur (*Microcebus myoxinus*, page 32), with a weight of 30.6 grams (1.1 ounces);[580] the largest is the male gorilla, which weighs 169 kilograms (373 pounds).[234] Individual body size determines how likely it is that an organism will become prey to another species. Predation is a real worry for a small primate. In a study of the contents of one owl's nest in Madagascar, the remains of 400 mouse lemurs were found during one month.[204] Alternately, the predators of gorillas and other large primates are humans and large cats.

The **intermembral index** is a number derived from the ratio of the length of the forelimbs to the hind limbs. This information is important to paleontologists, who can only make inferences about the behavior of the fossil animals they find. All primates are on a scale from 50 to 150. A given intermembral index indicates a general type of locomotion. For instance, the ratio for the northern lesser bush baby (*Galago senegalensis*, page 21) is 52, which means its hind legs are longer than its forelimbs and it is therefore better adapted to leaping than to walking or swinging. Generally an intermembral index of 50–69 indicates a species that is a vertical clinger and leaper; 69–105 indicates a species that is quadrupedal (walks on all fours) on medium-sized forelimbs and hind limbs; and 105–150 indicates a species with long arms and shorter legs that are used for suspensory locomotion. The one exceptional intermembral index is 70, which is seen in humans. Even though other primates occasionally walk or stand bipedally, *Homo sapiens* is the only one that habitually walks on two legs.

The **adult brain weight** is listed for many species, because we humans like to pride ourselves on our enlarged brain. A ratio of body size to brain weight allows us to compare ourselves with other primates.

Habitat

The habitat is the specific environment in which a primate is found within its range. If you went to look for a gray squirrel, which is common throughout North America, you would look for it in or near trees, not in an open field. You would probably have the best luck finding it in large nut-bearing trees rather than in bushes.

Most primates live in the tropical regions of the world. Arboreal primates inhabit forests. There are many types of forests, depending on climate, altitude, soils, and forest age. **Primary forests** have not been disturbed in several hundred years. Their structure and composition of tree species are very different from those of **secondary forests.** The latter are forests that are regenerating from disturbance caused either naturally by fire or hurricane or

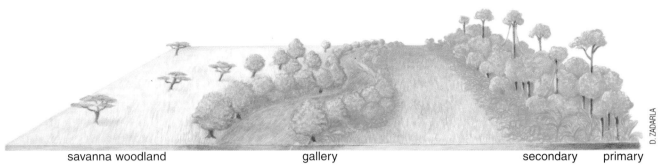

savanna woodland   gallery   secondary   primary

Figure 2. Some common primate habitats.

by human logging and clearing. Some primates can survive only in undisturbed forest, whereas others prefer the regenerating forest. Many can live in forest that has been selectively logged, but very few arboreal primates can live without forest. Those that do will be exterminated by humans for raiding crops, the only food left to them.

There are at least 40 types of forest mentioned in the literature. A few of the most common types are illustrated in Figure 2. For definitions of other terms, refer to the **glossary**.

Primates may prefer different levels in the forest. Some species are found on the ground, some in the **understory** (the middle), and more high in the **canopy** and the tops of the emergent trees (Figure 3). Some primates, such as rhesus macaques (*Macaca mulatta,* page 126), are very adaptable to different habitats within their range. Other primates are extremely restricted. For example, the pygmy marmoset (*Callithrix pygmaea,* page 67) lives only on the edges of rivers and in swamps that contain the main four or five species of trees on which it depends for edible saps.

Diet

What a primate eats is determined by its size, its teeth, and the anatomy and physiology of its stomach. Small primates need high-energy foods in easily digestible packages, so they eat fruits and insects and avoid leaves, which are plentiful but low-energy foodstuffs. Alternately, big-bodied primates would soon starve if their diet were restricted to relatively rare, high-energy foods. Instead, they eat more leaves, which, though low in energy value and hard to digest, are usually more abundant.

When the diet is known in detail for a species, the foods types are listed according to the percentage of time the primate spends obtaining them, from most time spent to least time spent. These percentages are specific to the population studied and may be different from other studies. The general categories that are commonly assigned to these species—frugivore (fruit eater) and folivore (leaf eater)—have not been used, so that you can better appreciate the diversity of diets that most primates have. Some of the seasonal changes in the diet are noted. Whenever possible, the number of plant species that are utilized has been included to show how important the biodiversity of habitats is to the species living in them.

Figure 3. Levels of the primary tropical rain forest.

Life History

Every individual has a life history. **Weaning**, as defined in this book, denotes the amount of time an **infant** is dependent on its mother for milk. After an individual is weaned, it is generally considered a **juvenile** until it is sexually mature. **Sexual maturity** is thus an estimate of adulthood in primates. Many primates, such as gibbons (*Hylobates* spp., pages 208–217), have a **subadult** phase, comparable to human adolescence, in which they are capable of reproduction but need to find a mate, a territory, or a more elevated position in the social hierarchy. For females, **age at first birth** indicates full adulthood. The **estrus cycle,** the length of the **gestation,** and the **birth interval** are all indications of how often a female is likely to reproduce. A species' **life span** is given in years and measures the maximum amount of time between birth and death rather than an average. Some primates, like Pennant's red colobus (*Procolobus pennantii tephrosceles,* page 174), mate and give birth in all months of the year but may have a **birth peak,** when several females give birth in the same few months. Other primates, such as lemurs, have a strict **mating season** and **birth season** and mate or give birth only during a one or two-month period each year, even if taken into captivity. Most of the larger primates have only one **offspring,** but some of the prosimians and tamarins (Callitrichidae) often have litters of two or more; this information is noted when it is known. A great deal of the life history information has been learned from captive studies and does not necessarily correspond to primates in their natural habitat.

A species' life history is important because it indicates how many offspring a given primate can have in a lifetime, which is invaluable information when considering the primate's conservation status. For instance, it is possible for a female gray mouse lemur (*Microcebus murinus,* page 31) to produce more than 30 offspring in a lifetime, whereas an orangutan female would be lucky to have 5 babies.

Locomotion

There are four basic types of primate locomotion: **quadrupedal, bipedal, vertical clinging and leaping,** and **brachiation.** A primate chiefly uses one of the four types but may use all types at least some of the time. The types of locomotion a species employs indicates the kind of supports it uses in its habitat. For instance, Verreaux's sifakas (*Propithecus verreauxi,* page 49) are vertical clingers and leapers. They prefer to cling to small tree trunks and leap distances of up to 10 meters (33 feet) from one trunk to another, but if necessary they can hop bipedally across the ground. Japanese macaques (*Macaca fuscata,* pages 124–125) are a good example of a quadrupedal animal. They use all four limbs, palms flat, to climb or walk, as do most primarily ground-dwelling primates. Chimpanzees and gorillas knuckle-walk, a specialized version of quadrupedalism.

Some larger primates traveling through trees use suspensory locomotion. Hanging below the branches, they spread their weight among several small supports.[234] This form of locomotion has major advantages; the alternative—balancing above a single large branch—would limit movement to a much smaller proportion of the tree. The gibbons (*Hylobates* spp., pages 208–217) use an exceptional form of suspensory locomotion called brachiation (Figure 4). They can swing from branch to branch in the canopy of mature forests.

Social Structure

The social structure of a species is based on how an individual relates to other members of its species. Primates are all social animals, but they have different types of social organizations. Many prosimians and the orangutan are listed in this book as **solitary foragers.** Although they do not live in large gregarious groups, they are not as solitary as was first assumed. They may forage alone, but female bush babies, for example, relate to their offspring and may sleep in groups of up to six individuals. The males have ranges that are larger and overlap the ranges of several females. This is often referred to in the literature as a **noyau** social system.

The **one male–one female** group can be a monogamous pair that defends a common territory, like titi monkeys (*Callicebus* spp., pages 85–92) and gibbons (*Hylobates* spp., pages 208–217). In many cases the male defends the female from other males, and the female excludes other females from the male. Many tamarins and marmosets (*Saguinus* spp., pages 68–76, and *Callithrix* spp., pages 60–67) found in South America live in **multi-male–one female** groups. Cultural anthropologists would call this a polyandrous social structure rather than a polygynous or harem social structure of one male with multiple females. *The Pictorial Guide* uses the simple descriptions of social structure rather than the latter terms. If more than one social structure has been reported, the word **variable** begins the category. Tamarin and marmoset groups often vary in composition, because they raise their offspring communally. All the members of the group share food with and help carry and care for the infants. The group may have multiple females, but only the dominant female reproduces. The subordinates are hormonally suppressed and do not ovulate until they leave the group or the breeding female dies.[201]

Many leaf monkeys have **one male–multifemale** groups in which all the females are related or bonded. The females remain in the group and defend its territory and its resources. The one adult male is peripheral to the group

Figure 4. Composite photo of white-handed gibbon *(Hylobates lar)* demonstrating brachiation.

and defends the females from other males. Young males, which leave the group before adulthood when a new male takes over the troop, may remain solitary or join an all-male bachelor troop.

There are several types of **multimale–multifemale** groups, which have complex intragroup (within-the-group) relationships. Japanese macaques live in groups organized around a strict female (matrilineal) hierarchy. Female macaques remain in the troop in which they are born (natal troop) and inherit the same rank in the hierarchy as their mother. Male macaques, though bigger and dominant to the females, leave their natal troops at maturity and often emigrate every four years to a new troop to gain more reproductive opportunities and fortuitously avoid inbreeding.

Hamadryas baboons (*Papio hamadryas,* page 136) live in multimale–multifemale troops of up to 400 individuals, but within the large troop, which congregates on rock cliffs to sleep, are three other levels of organization. The basic foraging and reproductive unit is a one male–multifemale group. Interestingly, each small group is made up of unrelated females whom the male has adopted as immatures or coerced into following him with a ritual neck bite. In a troop composed of unrelated females, the male is the center of attention. A few such one-male groups make up a clan in which the dominant males are related. Several clans make a herd in which the clans know each other. The two or three herds, which share the cliff, do not intermingle, and there is little emigration between herds.

The multimale–multifemale social structure with the most fluidity is called the **fission-fusion community.** It is seen in spider monkeys (*Ateles* spp., pages 112–115) and chimpanzees (*Pan troglodytes,* page 230). The related males forage with and groom each other and often patrol their common territory. Unrelated females are found foraging alone with their offspring. Female members emigrate to new troops at maturity. Occasionally most of the community gathers at a rich feeding site, but usually they are in dispersed subgroups.

Most of these social structures are not necessarily fixed. They are in constant flux as individuals grow, reproduce, and join or leave the group by emigrating or dying. Changes in resources, intragroup and intergroup (between-group) competition, and predation all affect the social structure. The gender and age (if known) of the individuals that leave the group are listed under **emigration**.

One of the most important theoretical questions currently being discussed by primatologists is, Why do primates live in groups? Is it to defend limited resources, to avoid predation, or both? These seemingly simple questions are the basis of the hypotheses that researchers test in order to understand how natural selection works.

The classification of primate social structures is a by-product of trying to understand the selective factors that have increased the reproductive success of the primates that maintain these social groups. For instance, a male Hanuman langur (*Semnopithecus entellus,* page 184) that can exclude all other males from a group of females will have much greater reproductive success for his genes than if he mated with only one female. The male langur is said to have a **reproductive strategy,** which has evolved from physiological and behavioral adaptations that have maximized his offspring's survival.

A male's strategy may not be the same as a female's. A langur male, for example, has an average tenure of 27 months in a group of females. Females have an average birth interval of 17 months, and an infant is weaned in 13–20 months.[389] When a new male becomes the leader of a group, females with a newborn infant cannot give birth to his infant for 17 months, and his infant will not be weaned by the time that he is forced out. The male langur's strategy is to kill the unweaned infants in the group. This **infanticide** has been seen in at least 19 other primates and in many other animals, including lions. By killing the infant, the male forces the female to come into **estrus** (start ovulating) within a month. She becomes pregnant with his offspring, which will be two years old and weaned by the time he is ousted as leader by the new male.

The reproductive strategy for the langur female is to wean her infant as early as possible. Failing early weaning, she mates with the male and has his infant even though he has killed her former infant.[389] Langur females that are pregnant when a new male takes over the troop often have a pseudoestrus in which they mate with the new male in order to make him believe that the offspring is his when it is born.

A less violent male reproductive strategy has been found in the woolly spider monkey (*Brachyteles arachnoides,* page 117). In this species, males show little aggression toward each other while all of the males in a group mate with an estrous female.[947] It has been hypothesized that each male competes for reproductive success by producing more sperm more often. This strategy has been labeled sperm competition, and researchers can judge if a species uses this strategy by measuring the ratio of the weight of the testes to the weight of the body. The larger the ratio, the more important sperm competition is to the species.

Group size includes an average and/or a range of how many individuals per group have been observed. Often the numbers from different studies vary greatly for the same species. The size of a troop depends on local environment, climate, predation, and group dynamics. A large group may split into two smaller troops that divide their former territory.

The **home range** is the approximate amount of land used by a group throughout the year. This figure is not exact. A home range may vary a great deal from year to year because of changes in weather, resource availability, competition from other groups of the same species, and human interference, such as hunting, logging, and agricultural expansion. The **day range** is the amount of distance traveled in an average day by a group; for nocturnal species the **night range** is given. The core area is the amount of space used exclusively by a group within its home range.

For some species, like titi monkeys (*Callicebus* spp., pages 85–92), home ranges are exclusive territories that are defended and maintained for years. Other species have home ranges that overlap to varying degrees, from 10% to 90%. The size of a home range and the size of a group give conservationists an indication of the population density, or how many individuals live in a given habitat. A small home range usually means that a primate will have a larger, denser population than a species with a large home range. It may be easier to find a primate in the wild if it has a small home range.

The metric system is usually used to measure ranges. A **hectare** (abbreviated as "ha") is 10,000 square meters, or about 108,000 square feet. One hectare is roughly equivalent to 2.5 acres. There are 100 hectares in a square kilometer, and about 2.5 square kilometers in a square mile.

Behavior

The behavior category is divided into several sections for well-studied species. For less studied primates, categories for which information were not located are not included. The general behavior is considered first. The activity pattern of the species is listed as **nocturnal** for primates that are active at night and **diurnal** for primates that forage during the day. Primates often rest at midday or midnight, but they are not generally considered to be **crepuscular** (active only at dawn and dusk). Some prosimians, however, are considered to be **cathemeral** in that they are irregularly active throughout both the day and the night. All of the nocturnal primates are prosimians, except for the night monkeys (*Aotus* spp., pages 83–84), which are the only nocturnal anthropoids. Since three-quarters of all prosimians are nocturnal, it has been hypothesized that they are closer to the nocturnal ancestors of all primates.[234] It has also been reasoned that the prosimians not isolated on the island of Madagascar were unable to compete during the day with monkeys and that only the nocturnal prosimians survived.

Most primates are **arboreal** and come to the ground only occasionally. A few species, like gelada baboons (*Theropithecus gelada,* page 143), are terrestrial. They spend all their time on the ground and sleep on cliffs rather than in trees.

The behavior section includes an assortment of information gleaned from the literature about foraging behavior, infant care, interactions between individuals of a troop, and activity budgets. This information is not exhaustive, because of spatial limitations. The references will lead you to more detailed information about most of these species.

Arboreal primates often forage at specific levels of the forest. For instance, spider monkeys (*Ateles* spp., pages 112–115) forage mostly in the canopy, whereas titi monkeys (*Callicebus* spp., pages 85–92) prefer the middle levels to the forest and many marmosets (*Callithrix* spp., pages 60–67) prefer the dense tangle of the understory.

Allomothering, which is another name for baby-sitting, is the care of an infant by another member of the troop besides its mother. This behavior is mentioned for many species because researchers are trying to understand its adaptive value. Does it help the mother, the infant, or the baby-sitter? Why do the females of some species such as langurs allow allomothering and other species never allow it?

Primates have a large complement of behaviors with which to communicate and facilitate interactions with other members of the troop. These communicative behaviors have evolved because primates are social animals. Facial expressions, body postures, and tail positions effectively convey information. Unfortunately, many primate behaviors are misinterpreted by humans who do not know the species or the context of the behavior they are witnessing. For instance, Old World monkeys react to eye-to-eye staring as threatening behavior and so they threaten back. A male proboscis monkey (*Nasalis larvatus,* pages 203–204) will exhibit his erect penis as part of a threat display. In chimpanzees (*Pan troglodytes,* page 230) this same behavior is part of a sexual display meant to invite a receptive female to mate.

One behavior common to many primates is called presenting. A monkey presents his hindquarters to another monkey, which then mounts the presenter. This is not necessarily sexual behavior but may be merely a greeting that affirms the dominance hierarchy for both participants, with the subordinate monkey presenting to and being mounted by the dominant member.

Grooming is one of the most common interactions seen. It serves the purpose of cleaning dirt and ectoparasites from the fur, but it is just as important as a sign of friendship and cooperation. If you have watched monkeys during an extended period of grooming, you know that the experience is obviously as pleasurable for them as having our back scratched or our neck massaged is for us. Researchers can determine a great deal about the relationships within a troop by studying the interactions between grooming partners.

Although fighting is the most attention-getting behavior, researchers such as Frans de Waal are studying reconciliation behaviors in several species.[156] Chimpanzee males (*Pan troglodytes,* page 230) have a special facial expression and posture when making peace after a fight. In this way the tensions of group life are reconciled by the proper gesture.

An **activity budget** is a list of the percentages of time spent in regular daily activities such as rest, feeding, and traveling. It is interesting to compare the amounts of time that different species or genera spend on each activity. For example, one species may rest much more than another.

The **association** category lists which other species are found with the primate being described. Many Old World monkeys (Cercopithecinae, pages 120–167) travel and forage in multispecies associations on a daily basis, similar to mixed-species flocks of birds. The different species probably benefit from this association by being more likely to detect potential predators and improve their foraging.[204] The species that associate most often usually have diets that do not overlap, so they are not competing for the same limited foods, as they would be if they were in larger groups of their own species. The lack of mixed-species groups in Asia may be due to the lack of a large aerial predator like the crowned hawk eagle in Africa or the harpy eagle in South America.[855]

The **mating** category mentions the behavior associated with breeding. Multiple-mount matings are often mentioned because researchers are studying this behavior to understand its adaptive significance. The females of many primates that live in multimale troops visually advertise their reproductive state with prominent genital swellings.[420] Species without swellings often have proceptive behaviors in order to communicate their willingness to mate. In some species, like yellow baboons (*Papio hamadryas cynocephalus,* page 138), the males try to **consort** with a female friend and keep her away from other males. In other species, such as Japanese macaques (*Macaca fuscata,* pages 124–125), the females mate **promiscuously** with all the males. In many species, however, promiscuous mating takes another form. The subadult and subordinate males mate with a female early in her cycle, and only the dominant male mates with her at the peak of her estrus cycle, when she is most likely to conceive.

Scent marking is another form of communication used by many prosimians and Neotropical primates but is less important in Old World monkeys and apes. Although the behavior is observable, the message that is chemically communicated is difficult to decipher but should be an interesting area of future research in primatology.

The **vocalizations** of many species are often better documented than other behaviors, simply because it is easier to hear some primates than to see them.

Researchers use sonograms to analyze different calls. Some primates have calls that are species specific. The bush babies (Galagonidae, pages 18–25) are similar in this way to birds like flycatchers—the different species look similar to us humans, but we can tell them apart by their calls. Understanding this has changed the taxonomy of bush babies. When conventional methods were used before the 1980s to compare museum specimens, only 6 species were identified.[599] *The Pictorial Guide* includes 12 species as well as 2 subspecies that have been proposed for recognition as species. At this point it is not known how many more primate species may finally be documented by classifying them according to the way they identify themselves rather than the way we visually identify them.

The final behavior category is **sleeping site**. Where a primate sleeps is important for two reasons. Researchers may be able to locate a species if they know where it sleeps. For species such as the gray-backed sportive lemur (*Lepilemur dorsalis,* page 34), the lack of large trees for sleeping sites may be a limiting factor in their distribution and could be a cause for their declining numbers.

There is only space enough in this book to give a few interesting behaviors for each species. The superscript numbers provided correspond to the references found at the end of the book, beginning on page 240. You can use these references to find more detailed information about species. Also provided are a list of popular books written for general audiences and a few World Wide Web sites that will link you to more information about primates (page 239).

A separate conservation-oriented category of the **threats** to primates has not been included for every species, because the threats to primates are almost always the same: hunting and habitat destruction. The hunting may vary from subsistence hunting to commercial hunting for food, pets, and medicine, but it is all still hunting. Habitat destruction is a problem for the majority of species as the human population continues to expand and we consume an ever greater proportion of the earth's resources.

Perhaps the most important information listed for each species is its conservation status. The color of the bar in the margin identifies its status (Figure 5). **Critically endangered** is the highest category and is relatively new. A species is designated as critically endangered if it faces a 50% chance of extinction in the wild within the next 10 years or three of its generations. An **endangered** species has at least a 20% chance of extinction in the wild in 20 years or five of its generations. A **threatened** species is likely to become endangered within the foreseeable future throughout all or a significant portion of its range. The **vulnerable** category indicates at least a 10% probability of extinction in the wild within 100 years. A **rare** species has a small world population and is not presently endangered or vulnerable but is considered at risk. If there is not enough information about a species to legally determine its category of endangerment, the term **data deficient** is used rather than the **lower risk** designation. Each color bar lists the organization that has determined the category. The United States Endangered Species Act

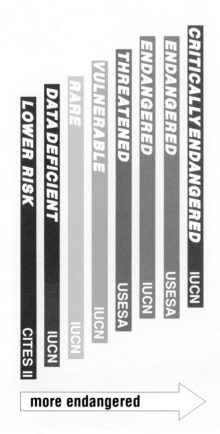

Figure 5. Conservation status.

(**USESA**) has the force of law. The World Conservation Union (**IUCN**) is a group of governments, conservation agencies, research institutions, and nongovernmental agencies from 120 countries. Its mission is to promote the protection and sustainable use of living resources.[337] The IUCN's Red Data Books list the threatened species of the world and recommend that governments protect them. The Convention on International Trade in Endangered Species (**CITES**) is an agreement among more than 100 signatory nations to abide by the guidelines set forth in the treaty to regulate the international trade in endangered species. Trade sanctions can be applied to countries that do not abide by the treaty. We must remember, however, that laws are only a mechanism to protect endangered species. To effectively protect a species, these laws must be actively enforced. Alternatively, each species and its habitat must be valued more as a perpetual life-sustaining system than as a disposable commodity.

These endangerment categories may seem remote, but make no mistake—the extinction of many prosimians, monkeys, and apes will happen unless more people get involved in the protection and conservation of these primates and their habitats.

Protecting and Conserving Primates

Most primates live in tropical, developing countries, and humans compete with them for resources. In many parts of the world, primates are exploited for food, "medicine," and commercial trade. Primates that raid crops are shot or poisoned. The forest habitat that is home to most species is being logged and cleared at an alarming rate by commercial loggers and subsistence farmers for land and firewood. The fate of several primate species will be decided in the next five years. If we humans collectively and as individuals do not act to protect these endangered primates and their habitats, they both will disappear—lost forever! We should not be lulled into complacency by another upbeat report from the media. We must get involved, write letters, educate others, and help the local communities that live near the forest to preserve their wildlife.

Sustainable development is defined as "increasing or maintaining productivity at levels that are economically viable, ecologically sound, and culturally acceptable, through the efficient management of resources with minimum damage to the environment or human health."[22] This is the current theoretical phrase used by governments, international funding agencies, and conservation organizations for their programs to save endangered species and help the local people who will decide the fate of those species. The key word is "sustainable." We must be vigilant to see that what is called sustainable really is sustainable in the long term. What is *not* needed is more big development schemes that exacerbate the problem and lead to further destruction of forest habitats.

"Protection" is the word that must be stressed in regard to the conservation of primates. Both the individuals and their forest home must be protected with laws and enforcement. Monkeys are often a hunter's main target, because they are the largest diurnal mammal that is easy to detect and shoot. Females with infants are the preferred quarry, and their loss hastens population decline. Many endangered primates live only in primary forest, which has the most valuable timber. In many countries, including the United States, primary forest exists only where it has been protected. Primary forest is a productive biological system from which valuable forest products can sustainably be obtained. Most of the nutrients of tropical forests reside in the vegetation, not in the soils. Thus, once the trees are cut, the few nutrients in the soil are depleted by human cultivation in a few seasons. Primary forest will survive only if people who understand its true value educate others and together they defend the forests from the forces of greed.

Captive conservation and release is another avenue that has been tried recently with the golden lion tamarin (*Leontopithecus rosalia*, page 79). The program was a limited success but had an enormous cost for each individual released. Clearly it would be cheaper for us and better for the primates if they are protected in their natural habitat, where they can be viewed as integral members of their habitat rather than as bored captives. If a primate species exists only in captivity, it is no longer a part of a natural evolutionary system but a living specimen in a museum.

The individual primates that are lucky enough to survive and be rescued from illegal poachers and traffickers should be the only source of primates used for captive breeding and display. They must be the ambassadors for their species. From them, we can learn about their species and appreciate their uniqueness—and our own.

Ten Things You Can Do for Primate Conservation
1. Call, write, or fax your U.S. congressional representative (U.S. House of Representatives, Washington DC 20515; 202-224-3121) and senators (U.S. Senate, Washington DC; 202-224-3121) and urge them to support legislation that
   - protects endangered species and their habitats.
   - increases funding for foreign aid programs that specifically address sustainable development and the conservation of global biodiversity.
   - increases enforcement of wildlife trade laws to stop the smuggling of endangered species.
   - curtails the use of primates in inhumane experiments.
   - encourages family planning and discourages human population growth. The earth has finite resources—we must control ourselves.
2. Call, write, or fax the president of the United States (The White House, Washington DC 20500; 202-456-1414) and ask that the United States impose sanc-

tions on countries that do not enforce their endangered species laws or that violate international wildlife treaties.

3. Register to vote, and vote for candidates who support the protection of the environment, biodiversity, and endangered species.

4. Write to the president of the World Bank (1818 H Street NW, Washington DC 20433) and ask that the World Bank make its loans to countries conditional on their protection of their national parks, biodiversity, and endangered species.

5. Ask your local zoo to adopt a park in a developing tropical country and provide it with supplies for educating local people about their endangered wildlife.

6. Get your local school to adopt a park or reserve or a school in a community near a park and share information and ideas about your wildlife and theirs.

7. Join the Peace Corps, which has many environmental and wildlife conservation projects in habitat countries.

8. Take a working vacation with Earthwatch (680 Mount Auburn Street, Box 403, Watertown MA 02272-9104; 800-776-0188; http://www.earthwatch.org) and get actively involved in primate field research in a habitat country.

9. Be an ecotourist rather than a sightseer. Be sure the tour company you choose follows the principles of ecotourism set forth by the Ecotourism Society (802-447-2121). The idea is to ensure that part of the money you spend will benefit the local community.

10. Join and support in any way you can some of the following organizations, which support the conservation of primates.

Bonobo Protection and Conservation Fund
Georgia State University Foundation
University Plaza
Georgia State University
Atlanta GA 30303

Conservation International
1015 18th Street NW, Suite 1000
Washington DC 20036

Dian Fossey Gorilla Fund
800 Cherokee Avenue SE
Atlanta GA 30315-9984

Flora and Fauna Preservation Society
Great Eastern House
Tenison Road
Cambridge England CB1 2DT

International Primate Protection League
P.O. Box 766
Summerville SC 29484

Jane Goodall Institute
P.O. Box 14890
Silver Spring MD 20911

Jane Goodall Institute (UK)
15 Clarendon Park
Lymington, Hants SO41 8AX
United Kingdom

The Nature Conservancy International
1815 North Lynn Street
Arlington VA 22209

Orangutan Foundation
824 South Wellesley Avenue
Los Angeles CA 90049

Primate Conservation, Inc.
1411 Shannock Road
Charlestown, Rhode Island 02813-3726 USA

Wildlife Conservation Society—International
Wildlife Conservation Society—Bronx Zoo
Bronx NY 10460

Wildlife Preservation Trust International
1520 Locust Street Suite 704
Philadelphia PA 19102

World Wildlife Fund
1250 24th Street NW
Washington DC 20037

Slow loris *Nycticebus coucang*

Southern lesser bush baby *Galago moholi*

12

Prosimians

"Prosimians," the common name for the suborder Prosimii, means "before apes."[234] Prosimians are said to be more "primitive" than other primates, because some of their anatomical characteristics are found in some other mammals but not in anthropoid primates. For instance, the tapetum (found in all prosimians except tarsiers) is a layer in the retina of the eye that reflects light and enhances night vision. It is found in cats and other nocturnal mammals whose eyes shine in the dark when a flashlight is pointed in their direction. Prosimians also have a moist rhinarium (also found in dogs), a structure of the nose that enhances the perception of smell.[234] Prosimians are active at night and communicate by marking territories with their special scent glands.

Some aspects of prosimian reproductive anatomy differ from those of anthropoid primates. For example, prosimians have a different-shaped uterus (bicornuate) and a different type of placenta (epitheliochorial).[234] Many prosimians have at least four nipples and produce litters rather than a single infant.

Common tree shrews *(Tupaia glis)* were considered to be primates until the 1980s. They now have their own order, Scandentia, with 1 family and 16 species.

Additionally, the skulls of prosimians have smaller brain cases than those of higher primates and do not have closed eye sockets. The two halves of the lower jaw of prosimians are mobile, not fused together as in anthropoids.[234]

Lorises, Pottos, and Bush Babies *Superfamily:* Loroidea

Lorises and bush babies are the two families in the superfamily Loroidea. The name Loroidea is the correct nomenclatural spelling for what was referred to as the Lorisoidea before 1987.[412] They are small and nocturnal and weigh less than 2.0 kilograms (4.4 pounds). They have a dental formula of

$$\frac{2.1.3.3.}{2.1.3.3.} \times 2 = 36$$

which means they have 2 incisors, 1 canine, 3 premolars, and 3 molars in each half of the upper and lower jaw, for a total of 36 teeth.[234] Teeth are very important to paleontologists who study the evolution of mammals, because teeth are often the only parts of an animal that fossilize and can be recovered after millions of years. The shape of the teeth of each species is unique and indicates its diet.

Lorises and Pottos *Family:* Loridae

The family Loridae[315] is made up of the lorises, which are found in southern Asia, and the pottos, which are found in Africa. Both groups have slow quadrupedal locomotion and do not leap. They move slowly to avoid being detected by predators and to get close enough to their prey to grab it. Lorises and pottos have very short tails, smaller ears and longer bodies than bush babies, and equal-length forelimbs and hind limbs. Their ankle and wrist joints are more mobile than those of other primates, and their thumbs are more opposite to their other digits.[234] These characteristics allow the lorids to have a powerful grasp. Special blood vessel storage channels in their hands and feet enable lorises to contract their muscles and hold tightly to a branch for hours, seemingly without fatigue or pain.[650] In this way they remain hidden from predators during the day. Many nocturnal primates have not been well studied and new discoveries will be made. For example, in 1996 a new genus and species of potto has been proposed, with the name *Pseudopotto martini*.[945]

Bush Babies *Family:* Galagonidae

The bush babies of the family Galagonidae[412] get their common name from the vocalizations some species make, which sound to humans like a crying human baby. Found only in continental Africa, bush babies have slender bodies, long bushy tails, and large mobile ears. Their hind limbs are longer than their forelimbs, this makes them good leapers. Males have a baculum, a slender bone that reinforces the penis. Baculums are found in many prosimians and anthropoid primates[363] and in many other mammals, including the raccoon. Four genera of bush babies are currently recognized. Many of the species look similar, and taxonomists disagree about how many species there are. Recent fieldwork has shown that the vocalizations of bush babies are different and that each species recognizes its own by sound and smell and less by sight.[43] To show the diversity of species-specific advertisement calls, sonograms have been included in the species descriptions that follow.

13

Golden Angwantibo *Arctocebus aureus*

Angwantibos forage on vines for insects.

Taxonomy Disputed. Elevated from a subspecies of *A. calabarensis* in 1980.[315]

Distinguishing Characteristics Golden angwantibos are reddish brown above, with reddish buff underparts. Juveniles are darker than adults.

Physical Characteristics Head and body length: 244mm (230–260) *[9.6in (9.1–10.2)]*.[739] **Tail length:** 15mm *[0.6in]*.[739] **Weight:** 210g (150–270) *[7.4oz (5.3–9.5)]*.[739] **Intermembral index:** NA. **Adult brain weight:** NA.

Habitat Primary and secondary forest. Regenerating vegetation with lianas and dense undergrowth is preferred.[738]

Diet Animal prey, 85%; fruit, 14%.[109] Caterpillars make up 90% of the prey; the rest is crickets, beetles, and ants, which bush babies find unpalatable.[109]

Life History NA. When a mother forages, she leaves her infant suspended under a branch.[109]

Locomotion Slow-climbing angwantibos move easily above or below branches.[109]

Social Structure Solitary foraging. The male's range overlaps the ranges of several females (noyau). There are no matriarchies.[41] **Group size:** 1–2. **Home range:** NA. **Night range:** NA.

Behavior Nocturnal and arboreal.[412] Golden angwantibos have poor eyesight but can smell a caterpillar from a meter *[3ft]* away.[109] Before eating a caterpillar that has irritating hairs, they hold it by the head and wipe their hands down it to remove the hairs.[109] They have not been observed higher than 15 meters *[49ft]*.[109] Their defensive posture is to arch the back, tuck the head under the body, and position their mouth between their front and back legs in order to bite the attacker.[109] **Vocalizations:** 3 main vocalizations. A hoarse growl is a threat, a *weet* call is for distress, and a *tsic* is a contact call.[109] **Sleeping site:** Dense foliage.[109] The mother and infant sleep together.

Cameroon, Congo, Gabon

Angwantibo *Arctocebus calabarensis*

Taxonomy Disputed. Formerly included *A. aureus* as a subspecies.[315] Because of this taxonomic change, some of the life history information for this species may belong to *A. aureus*.

Distinguishing Characteristics Angwantibos have an orange to yellow brown back and white, light gray, or buff underparts.[663] The index finger is reduced.[48]

Physical Characteristics Head and body length: 220–603mm *[8.7–23.7in]*.[739] **Tail length:** 8mm *[0.3in]*.[739] **Weight:** 266–465g *[9.4–16.4oz]*.[739] **Intermembral index:** 89.[425] **Adult brain weight:** 7.7g *[0.3oz]*.[346]

Habitat Primary and secondary forest, natural clearings,[370] and edges, as well as *Gmelina* plantations.[898]

Diet Animal prey.[898]

Life History Weaning: 3.5mo.[464] **Sexual maturity:** 9–10mo.[606] **Estrus cycle:** 36–45d.[848] **Gestation:** 133d.[663] **Age 1st birth:** 9mo.[464]

Birth interval: 4mo.[41] **Life span:** 11y.[41] Births: Year-round. Birth peak: Jan.[663] Offspring: 1. Females have a postpartum estrus. Infants cling to the fur.[41]

Locomotion Slow climbing; no leaping.

Social Structure Solitary foraging.[606] **Group size:** 1–2. **Home range:** NA. **Night range:** NA.

Behavior: Nocturnal and arboreal. Angwantibos hunt in small trees in undergrowth 0–5 meters *[0–16ft]* high.[41] **Mating:** Duration is 4 minutes.[175] **Scent marking:** The male scent-marks a female by "passing over" her back and rubbing her with his scrotal gland.[522]

Cameroon, Nigeria

Slender Loris *Loris tardigradus*

Taxonomy 6 subspecies,[412] some of which may be valid species after further study.[315]

Distinguishing Characteristics Coloration of the slender loris varies with the subspecies: gray, brown, or reddish, with or without a dorsal stripe.[130]

Physical Characteristics Head and body length: 180–260mm *[7.1–10.2in].*[739] **Tail length:** 4–7mm *[0.2–0.3in].*[739] **Weight:** Range of means 102–285g *[3.6–10.1oz].*[739] **Intermembral index:** 93.[412] **Adult brain weight:** 6.7g *[0.2oz].* Baculum length: 14.2mm *[0.6in].*[174]

Habitat Tropical rain forest, dry semideciduous forest, scrub forest, swamps, and montane forest up to 1850m *[6070ft].*[398]

Diet Mostly insects, with young leaves, shoots, hard-rind fruits,[412] flowers,[737] eggs, and small vertebrates.[94] Often insects are detected by smell.[412]

Life History Weaning: 5–6mo.[737] **Sexual maturity:** 10mo.[398] **Estrus cycle:** 29–40d.[398] **Gestation:** 166–169d.[398] **Age 1st birth:** 17–18mo.[398] **Birth interval:** 9.5mo.[398] **Life span:** 15y.[737] Mating seasons: Apr–May, Oct–Nov.[41] Birth seasons: Yes in the wild, but year-round in captivity.[398] Offspring: 1[812] or 2 (22%).[737] Slender lorises have 2 pairs of mammary glands. Estrus is signaled by swelling and reddening of the genitalia.[398] Infants cling to their mother's fur.[41] An infant may be born successfully headfirst or in a breech position. Other than licking, the mother does not assist the infant as it crawls to her nipples to suckle. The placenta is ingested.[812]

Locomotion Quadrupedal climbing.

Social Structure Solitary foraging. Adults may sometimes forage in pairs. **Group size:** Sleep 2–4.[898] **Home range:** 1ha. **Night range:** NA.

Behavior Nocturnal and arboreal. Slender lorises occasionally come to the ground. They can move fast if alarmed.[412] They rest rolled into a ball.[412] In captivity, mothers have cared for infants other than their own. The mother responds to an infant's vocalizations more than to sight.[812] **Mating:** The female urine-marks profusely before mating. There are multiple copulations, 6 in 2 hours.[398] Mating times from 2 to 17 minutes have been reported.[175] During mating, the female

Even relative to other primates, the slender loris has exceptionally mobile joints.

is suspended from a branch by all 4 limbs. **Scent marking:** Territory is scent-marked with urine rubbed on branches. If threatened, slender lorises growl, stare, and produce an unpleasant odor from the brachial organ.[412] **Vocalizations:** 6; no alarm call.[737] The territorial call is a whistle given mostly by males.[737] **Sleeping site:** Tree hollows are not used.[41]

India, Sri Lanka

Slow Loris *Nycticebus coucang*

The coloration and size of the slow loris vary from the northern to the southern part of its range.

Taxonomy Disputed. 4 subspecies, some of which may be valid species after further study.[313]

Distinguishing Characteristics Slow lorises vary in coloration from ash gray to whitish. They have a dark dorsal stripe and dark rings around the eyes.[130]

Physical Characteristics Head and body length: 265–380mm *[10.4–15.0in].*[739] **Tail length:** 13–25mm *[0.5–1.0in].*[654] **Weight:** 230–610g *[8.1–21.5oz]*[654] and ♀ 1195g *[42.1oz]*, ♂ 1207g *[42.6oz].*[436] **Intermembral index:** 88.7.[412] **Adult brain weight:** 10g *[0.4oz].*[346]

Habitat Tropical evergreen rain forest up to 1300m *[4265ft].*[654] Slow lorises prefer forest edges, which have more supports and insect prey.[417]

Diet Fruit, 50%; animal prey, 30%; gums, 10%;[41] shoots, bird eggs, cocoa,[654] and insects that have a repugnant taste and smell.[848]

Life History Weaning: 6mo.[399] **Sexual maturity:** ♀ 17–21mo, ♂ 17–20mo.[871] **Estrus cycle:** 42.3d[848] (29–45).[399] **Gestation:** 191d.[399] **Age 1st birth:** NA. **Birth interval:** 12–18mo.[606] **Life span:** 20y.[650] Estrus is year-round, signaled by reddening and enlargement of external genitalia. Neonates have gray on the body and silvery white limbs and hands with long, glistening hairs that disappear at 11 weeks.[704] Infants are parked while the mother forages.[41]

Locomotion Quadrupedal slow climbing.[704]

Social Structure Solitary foraging.[654] The male loris's range may be larger than the female's.[522] No matriarchies are reported.[41] **Group size:** NA. **Home range:** NA. **Night range:** NA.

Behavior Nocturnal and arboreal. Slow lorises may hang from a branch upside down to eat with both hands.[41] **Association:** Sympatric with pygmy lorises *(N. pygmaeus).* **Mating:** Duration is 3–7 minutes.[175] **Scent marking:** When under stress, the slow loris scent-marks its head and neck with its brachial

organs (glands located on the inside of the upper front arm). In captivity, urine marks are made.

Vocalizations: Slow lorises make a "low buzzing hiss" when disturbed. The greeting call is a single low-pitched rising note. Estrous females give frequent high-pitched whistles. When distressed, infants produce ultrasounds.[412] **Sleeping site:** Tangles. The slow loris uses many sites and sleeps rolled up in a ball, the head buried between the thighs.[412]

Southeast Asia

Slow loris infants cling to their mothers for up to 7 weeks.

Pygmy Loris *Nycticebus pygmaeus*

Taxonomy Disputed. A proposed third *Nycticebus* species, called *N. intermedius,* has not been accepted. It is included in this species.[315]

Distinguishing Characteristics Pygmy lorises have bright orange brown fur, with a faint or absent dorsal stripe.[130]

Physical Characteristics Head and body length: 210–290mm *[8.3–11.4in].*[130] **Tail length:** Vestigial. **Weight:** ♀ 372g *[13.1oz],* ♂ 462g *[16.3oz].*[436] **Intermembral index:** 91.[234] **Adult brain weight:** NA. Males are 23% larger than females.[437]

Habitat Secondary forest.[816]

Diet NA. These lorises have been seen near trees with gouges resembling those made by species that eat exudates.[816]

Life History Weaning: 133d (123–146).[871] **Sexual maturity:** ♀ 9mo, ♂ 17–20mo.[871] **Estrus cycle:** NA. **Gestation:** 188d.[871] **Age 1st birth:** NA. **Birth interval:** 12–18mo.[871] **Life span:** 20y.[436]

Locomotion Quadrupedal running, climbing, and walking.[437]

Social Structure NA. **Group size:** NA. **Home range:** NA. **Night range:** NA.

Behavior Nocturnal and arboreal. Pygmy lorises move more quickly than slow lorises *(N. coucang).*[816] They reportedly are active all night, pausing only to feed.[181]

Association: Sympatric with slow lorises *(N. coucang).*[130]

Viet Nam, Laos, Kampuchea (Cambodia)

Potto *Perodicticus potto*

Taxonomy 3 subspecies,[412] some of which may be valid species with further study.[739]

Distinguishing Characteristics Pottos are brownish overall. Their index finger is vestigial. The neck vertebrae have long spinal processes (apophyseal spines of 3–9 cervical vertebrae) that project above the muscles.

Physical Characteristics Head and body length: 305–390mm *[12.0–15.4in].*[625] **Tail length:** 37–100mm *[1.5–3.9in].* **Weight:** 850–1600g *[30.0–56.4oz].*[625] **Intermembral index:** 87.5.[425] **Adult brain weight:** 14.3g *[0.5oz].*[346] **Baculum length:** 21mm *[0.8in].*[175]

Habitat Primary[464] and secondary forest; plantations and wooded savanna.[629]

Diet Fruit, 65%; gums, 21%; animal prey, 10%.[370] Pottos eat insects that other species find unpalatable, especially ants, which make up 65% of their insect

prey. During wet season, they store fat.[629]

Life History Infant: 4mo.[41] **Weaning:** 4–6mo.[606] **Juvenile:** 4–8mo.[41] **Subadult:** 8–18mo. **Sexual maturity:** ♀ 8mo, ♂ 6mo.[109] **Estrus cycle:** 39d. **Gestation:** 197d (194–205).[138] **Age 1st birth:** NA. **Birth interval:** 12–13mo.[109] **Life span:** 26y. Offspring: 1–2. Infants are parked while the mother forages. At 8 months, female offspring sleep alone.[109]

Locomotion Quadrupedal, slow climbing; no leaping.[464]

Social Structure Solitary foraging, 96%; pairs, 4%.[898] The territory of a female and her young is overlapped by the larger territory of 1 or more males (noyau). There are no matriarchies.[606] **Emigration:** Males leave their mother's home range at 6 months of age. **Group size:** 1–2.[898] **Home range:** ♀ 6–9ha, ♂ 9–40ha.[109] **Night range:** NA.

Behavior Nocturnal and arboreal. Pottos are slow-moving and well camouflaged (cryptic) hunters in the forest canopy (920–940m *[3019–3084ft]*).[109] Their apophyseal spines are said to be used for defense. When threatened, the potto puts its head down and presents an attacker with the spines and a scapular shield.[109] If frightened by a snake, the potto may escape by intentionally falling. In captivity, when one member of a

pair was sick, the other shared food. Infanticide has been observed. **Association:** Sympatric with Matschie's bush baby *(Galago matschiei).* **Mating:** Polygynous.[41] Pottos may have a prolonged intromission, because of their long baculum (21mm *[0.8in]*).[174] **Scent marking:** Pottos define their territory by urine marking large branches.[109] They scent-mark the opposite sex.[412] Apocrine glands in the skin of their genitals produce a secretion that acts as an insect attractant.[138] When a potto scratches these glands during self-grooming, it spreads the scent over its body.[522] These glands also produce a fear scent that warns other pottos of danger.[522] The potto has a strong "curry-like" odor.[109] Females have preclitoral pockets for scent marking.[109] **Vocalizations:** Mother and young maintain contact by means of a high-pitched *tisc* call. Females in estrus emit a whistling call.[625] Pottos have a high-pitched distress call and groan when threatening other pottos.[41] They have no loud call.[41]

Sleeping site: Thin branches in thick foliage.[109] Males sleep alone.[41]

Coastal West Africa, Central Africa

Southern Needle-clawed Bush Baby *Euoticus elegantulus*

Taxonomy Disputed. Formerly included in *Galago*.[315]

Distinguishing Characteristics Needle-clawed bush babies are named for their pointed, clawlike fingernails.[608] They have large eyes, a short muzzle, and a large tooth scraper used to extricate gums.[313] Their body is cinnamon to red above, with silvery white underparts. The tail has thick fur with a white tip.[608] The large cecum is an adaptation for digesting gums.[604]

Physical Characteristics Head and body length: 188mm (106–270) *[7.4in (4.2–10.6)]*.[608] **Tail length:** 243mm (194–337) *[9.6in (7.6–13.3)]*.[608] **Weight:** ♀ 271g *[9.6oz]*, ♂ 300g (270–360) *[10.6oz (9.5–12.7)]*.[436] **Intermembral index:** 64.[234] **Adult brain weight:** 5.8g *[0.2oz]*. Ear length: 30mm (24–40) *[1.2in (0.9–1.6)]*. Baculum length: 16.9mm *[0.7in]*. Males are 7% larger than females.[437]

Habitat Primary and secondary rain forest, riverine forest. These bush babies forage in forest canopy (3–50m *[10–164ft]*).[898]

Gabon, Congo Republic, and Cameroon

Diet Gums (of 5 tree species[109]), 75%; animal prey, 20%; fruit, 5%;[370] birds.[94]

© CNRS ECOTROP - A. R. DEVEZ

Life History Weaning: NA. **Sexual maturity:** 10mo.[606] **Estrus cycle:** NA. **Gestation:** 135d. **Age 1st birth:** NA. **Birth interval:** NA. **Life span:** NA. **Births:** Year-round.[109] Offspring: 1. Infants are carried for 60 days.[608]

Locomotion Leaping, quadrupedal running, climbing. This bush baby can leap up to 5.5 meters *[18ft]*.[109] It lands with all feet and maintains a vertical position during its leap.[109]

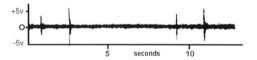

Social Structure Solitary foraging.[109] **Emigration:** Males emigrate and females remain near their natal territory. **Group size:** Sleep 1–7.[109] **Home range:** NA. **Night range:** NA.

Behavior Nocturnal and arboreal. Needle-clawed bush babies detect insect prey by sound, catch prey with their hands, and transfer it to their mouth to kill it.[109] Their clawlike nails allow them to climb smooth-barked trees in search of exudates. They can smell gums and visit up to 100 gum sites per hour.[109] **Association:** Sympatric with Demidoff's bush babies (*Galagoides demidoff*) and Allen's bush babies (*Galago alleni*). **Mating:** This bush baby is believed to have long copulations, because it has a long baculum[175] relative to its body size.[174] **Scent marking:** These bush babies never urine-wash their feet but deposit urine directly on branches. **Vocalizations:** 6 main vocalizations. The long-range contact call is said to be like a "bird call."[662] The advertisement call is a short, high-pitched *tsic* given at dawn.[608] **Sleeping site:** Tree forks hidden by dense foliage, at several sites.[109] Several individuals sleep in a tightly packed ball on the fork of a branch.[109]

Northern Needle-clawed Bush Baby *Euoticus pallidus*

Taxonomy Disputed. Recognized as a species in 1989.[315]

Distinguishing Characteristics Northern needle-clawed bush babies are paler in color than southern needle-clawed bush babies (*E. elegantulus*) and have no white on the tail tip. Their long ears have no extra fold. The eyes are bright orange to brownish gold.[363] The yellowish gray body has a dorsal stripe and white underparts.[313] The penile morphology is distinctly different from that of *E. elegantulus*.[20]

Physical Characteristics Head and body length: Less than 200mm *[7.9in]*.[313] **Tail length:** >243mm *[>9.6in]*.[313] **Weight:** NA. **Intermembral index:** NA. **Adult brain weight:** NA. **Habitat** NA.

Diet NA.
Life History NA.
Locomotion NA.
Social Structure NA. **Group size:** NA. **Home range:** NA. **Night range:** NA.
Behavior Nocturnal and arboreal. No field studies had been published as of 1995.

Nigeria, Cameroon and Bioko (Equatorial Guinea)

S. NASH

Allen's Bush Baby *Galago alleni*

Taxonomy 2 subspecies.

Distinguishing Characteristics
Allen's bush babies have a pointed muzzle and a dark face with circumocular rings.[412] They have a gray back, pale gray underparts, and reddish arms, thighs, and flanks. The tail is dark gray, with or without a white tip.[110] They do not have a sacculated cecum.[412]

Physical Characteristics Head and body length: 199mm (155–240) *[7.8in (6.1–9.4)].*[608] **Tail length:** NA. **Weight:** 314g (200–445) *[11.1oz (7.1–15.7)].*[608] **Intermembral index:** 64.[234] **Adult brain weight:** 6.1g *[0.2oz].* Males are 11% larger than females.[608]

Habitat Primary forest, swamp, moist lowland rain forest,[898] montane forest. This species is rarely found in secondary growth.[898]

Diet Fruit, 73%; animal prey, 25%.[109] This species has been seen eating fallen fruit, invertebrates,[608] and frogs.[94] Its stomach can hold 20 grams *[0.7oz]* of food (8% of body weight).[109]

Life History Weaning: NA. **Sexual maturity:** 8–10mo.[606] **Estrus cycle:** NA. **Gestation:** 133d.[608] **Age 1st birth:** NA. **Birth interval:** 12mo.[41] **Life span:** 12y.[41] **Births:** Year-round.[109] **Birth peak:** Jan.[606] Mothers carry infants for 45 days by mouth.[41]

Locomotion Leaping from vertical supports. The hind feet do not land first.[608]

Social Structure Solitary foraging (86%). Males are solitary, but 2–3 females with offspring may sleep in the same hollow.[109] Matriarchies are present.[606] Males are aggressive to other males. **Group size:** Sleep 1–4.[606] **Home range:** ♀ 8–16ha,[109] ♂ 30–50ha.[109] **Night range:** NA.

Behavior Nocturnal and arboreal. Allen's bush babies hunt primarily below 10 meters [33ft].[109] They prefer undergrowth 1–2 meters *[3.3–6.6ft]* high with vertical supports.[41] **Association:** Sympatric with Demidoff's bush babies *(Galagoides demidoff)* and southern needle-clawed bush babies *(Euoticus elegantulus).*

Bioko (Equatorial Guinea), Gabon, Cameroon, Congo

© CNRS ECOTROP - A. R. DEVEZ

This bush baby is canting (supporting itself on a branch with feet only) in order to catch insects with its hands.

Vocalizations: Nine.[662] Advertisement calls are "loud low-pitched croaks."[608] These bush babies are reported to "mob" (alarm-call) at leopards. The infant distress call, which attracts other individuals, is mimicked by hunters in Gabon.[109] **Sleeping site:** Tree hollows in multiple locations.[109]

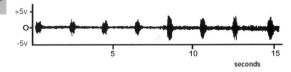

Somali Bush Baby *Galago gallarum*

DIGITALLY COMPOSITED

Taxonomy Disputed. Monotypic.[313] Elevated from a subspecies of *Galago senegalensis* in 1989.[315]

Distinguishing Characteristics
Somali bush babies are buff to sandy brown, with light gray underparts. The flanks are brightly colored. The brown eye rings are incomplete; there is a prominent stripe between the eyes.[608]

Physical Characteristics Head and body length: 167mm (130–200) *[6.6in (5.1–7.9)].*[608] **Tail length:** 252mm (205–293) *[9.9in (8.1–11.5)].*[608] **Weight:** NA. **Intermembral index:** NA. **Ear length:** 35mm (30–40) *[1.4in (1.2–1.6)].*[608] **Adult brain weight:** NA.

Habitat *Acacia* woodland, thorn scrub.[608]

Diet Seeds, animal prey, gums.[608]

Life History NA.

Locomotion NA.

Social Structure NA. **Group size:** NA. **Home range:** NA. **Night range:** NA.

Behavior Nocturnal and arboreal. Somali bush babies come to the ground to feed.[608] **Association:** Sympatric with *G. senegalensis* along parts of range and with Zanzibar bush babies *(Galagoides zanzibaricus)* and Garnett's greater bush babies *(Otolemur garnettii)* in riverine parts of range.[608]

Kenya, Somalia, Ethiopia

Matschie's Bush Baby *Galago matschiei*

Taxonomy Disputed. Name changed from *Euoticus inustus* in 1989.[315]

Distinguishing Characteristics Matschie's bush babies are dark brown, with dark eye rings and a stripe between the eyes. They have pointed clawlike nails, and the tail is not bushy.[608]

Physical Characteristics Head and body length: 166mm (147–184) *[6.5in (5.8–7.2)]*.[608] **Tail length:** 255mm (240–279) *[10.0in (9.4–11.0)]*.[608] **Weight:** 210g (196–225) *[7.4oz (6.9–7.9)]*.[608] **Intermembral index:** 60.[234] **Adult brain weight:** NA. Ear length: 39mm (37–42) *[1.5in (1.5–1.7)]*.[608]

Habitat Primary and secondary tropical forest and edges.[444]

Diet Animal prey (including caterpillars, beetles, and other insects[444]), gums, fruit.[608]

S. BEARDER (NATURAL HABITAT)

Life History NA. Offspring: 1.[608]
Locomotion NA.
Social Structure NA. Group size: NA. **Home range:** NA. **Night range:** NA.

Behavior Nocturnal and arboreal. Matschie's bush babies move from canopy to thick tangles of understory.[608] **Association:** Sympatric with Demidoff's bush babies *(Galagoides demidoff)*, Thomas's bush babies *(G. thomasi)*, and pottos *(Perodicticus potto)*. **Vocalizations:** The alarm call is loud and shrill. This species also produces a "hoarse insect-like churr."[444]

Zaire

Southern Lesser Bush Baby *Galago moholi*

The lesser bush baby lands feet first when it jumps to the ground to catch insects.

Taxonomy Disputed. Elevated from a subspecies of *G. senegalensis* in 1989.[315]

Distinguishing Characteristics Southern lesser bush babies are gray, with yellow-tinged underparts. They have eye rings, a nose stripe, and larger ears than those of *G. senegalensis*.[337]

Physical Characteristics Head and body length: 159mm (149–168) *[6.3in (5.9–6.6)]*.[337] **Tail length:** 228mm (113–279) *[9.0in (4.4–11.0)]*.[608] **Weight:** ♀ 188g (140–229) *[6.6oz (4.9–8.1)]*, ♂ 211g (160–255) *[7.4oz (5.6–9.0)]*.[337] **Intermembral index:** 51.5.[337] **Adult brain weight:** NA. Ear length: 39mm (33–50) *[1.5in (1.3–2.0)]*.[608] Males are 17% larger than females.[436]

Habitat Semiarid *Acacia* woodland, savanna, forest edge, "orchard bush."[337]

Diet Gum and animal prey, including butterflies, moths, and beetles. Fruit eating has not been reported.[337]

Life History Weaning: 61d (70–100).[606] **Sexual maturity:** 6.7mo3 (9–12).[41] **Estrus cycle:** NA. **Gestation:** 123.5d (121–124).[848] **Age 1st birth:** 10–12mo.[398] **Birth interval:** 4–8mo.[41] **Life span:** 16y.[41] Births: Oct–Nov, Jan–Feb.[337] Offspring: 1–2. First pregnancy is a single infant, with twins later.[464] Females have a postpartum estrus[848] and carry their infants by

Central Southern Africa

mouth for 53d.[41]

Locomotion Vertical clinging and leaping; bipedal hopping when foraging on the ground.[337]

Social Structure Solitary foraging. The male's territory overlaps the territories of 1–5 females.[337] Matriarchies are present.[606] **Group size:** 1–3.[464] **Home range:** ♀ 6.7ha (4.4–11.7), ♂ 11ha (9.5–22.9).[337] **Night range:** NA.

Behavior Nocturnal and arboreal. Dominant males are larger, are more vocal, and have an odor; subordinate males are smaller, are quiet, and lack odor.[606] **Association:** Throughout most of

20

Southern Lesser Bush Baby *continued*

range, sympatric with *Otolemur crassicaudatus*; in other parts of range, overlaps with northern lesser bush babies *(G. senegalensis),* Demidoff's bush babies *(Galagoides demidoff),* and Zanzibar bush babies *(G. zanzibaricus).* **Mating:** Polygynous.[41] The male approaches an estrous female with a low clucking vocalization and, while mounting, makes a loud call that ends in a

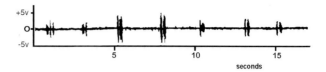

whistle.[848] **Scent marking:** Urine is deposited in a cupped hand and wiped on the foot of the same side. Males mark more than females. **Vocalizations:** A territorial advertisement call is answered back and forth for up to an hour by others. These bush babies also

make sharp tonal calls.[337] **Sleeping site:** Tree hollows.[41]

Northern Lesser Bush Baby *Galago senegalensis*

Taxonomy Disputed. Includes *G. s. granti,* which some consider a full species.[315] Four species of *Galago* were lumped into this species until they were studied in their natural habitat.

Distinguishing Characteristics Northern lesser bush babies are brownish gray or gray, with yellow on the flanks.[608] The head is broad, with a short muzzle, dark eye rings, and a white stripe between the eyes.

Physical Characteristics Head and body length: 165mm (132–210) *[6.5in (5.2–8.3)].*[608] **Tail length:** 216mm (195–303) *[8.5in (7.7–11.9)].*[608] **Weight:** ♀ 193g *[6.8oz],* ♂ 210g *[7.4oz].*[41] **Intermembral index:** 52.[234] **Adult brain weight:** 4.8g *[0.2oz].*[346] **Ear length:** 37mm (21–57) *[1.5in (0.8–2.2)].*[412] **Baculum length:** 16.7mm *[0.7in].*[175] Males weigh 11% more than females.[436]

Habitat *Acacia* woodland, savannas, thorn scrub, gallery, forest edge.[933]

Diet Animal prey, 52%; gums, 48%.[41] These bush babies eat gums from 2 tree species[612] but no vertebrate prey.[41]

Life History Weaning: 70–100d.[606] **Juvenile:** 2.5–12mo.[606] **Sexual maturity:** 11–13mo.[606] **Estrus cycle:** 31.7–39d.[848] **Gestation:** 142d.[608] **Age 1st birth:** 10–12mo.[398] **Birth interval:** 6mo.[398] **Life span:** 16y.[346] Mating seasons: Yes. Birth seasons: 2. Offspring: 1–2.[606] Females have a postpartum estrus;[606] the vagina is closed except during breeding season.[848]

Locomotion Vertical clinging and leaping, bipedal hopping. The hind feet land first.

Social Structure Solitary foraging.[606] **Group size:** NA. **Home range:** NA. **Night range:** NA.

Behavior Nocturnal and arboreal. Northern lesser bush babies prefer *Acacia drepanolobium* gum, whose flavonoids may have estrogenic effects and may be lower in tannins than other gums.[612] During antagonistic interactions these bush babies box, grapple, and bite.[606] In captivity without means of dispersal, there is no incest avoidance; males mount mothers and sisters.[606] **Association:** In East Africa, part of the range overlaps with Somali bush babies *(G. gallarum),* southern lesser bush babies *(G. moholi),* and thick-tailed greater bush babies *(Otolemur crassicaudatus).* **Mating:** Duration is 7–12 minutes.[175] Individuals have been observed to mate 22 times the first day of estrus.[848] **Scent marking:** These bush babies are hypothesized to use urine washing for territorial marking, social communication, and enhancing

R. A. BARNES

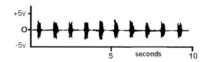

their grip on branches.[606] Both genders urine-wash the same amount.[606] **Vocalizations:** 21.[662] The advertisement call is a low-pitched single call repeated for up to an hour. The male's vocalization during courtship chases of females sounds like a human baby's cry—hence the name "bush baby." The aggressive, defensive, and alarm vocalizations are similar to *G. moholi's,* but the contact-seeking calls are distinctive.[933] **Sleeping site:** A few different sites are used, including "flat leaf nests, tree hollows, or branch forks in thorn trees."[41]

Equatorial Africa

(right margin) **LOWER RISK   CITES II**

Demidoff's Bush Baby *Galagoides demidoff*

R. AUSTING CINCINNATI ZOO

Taxonomy Disputed. This species includes the subspecies *G. d. rondoensis*.[313]

Distinguishing Characteristics
This species is the smallest bush baby. It has a narrow head with a pointed upturned muzzle.[608] The eye rings are variable. The tail is not bushy.[608]

Physical Characteristics Head and body length: 129mm (73–155) *[5.1in (2.9–6.1)]*.[608] **Tail length:** 179mm (110–215) *[7.0in (4.3–8.5)]*.[608] **Weight:** ♀ 69g *[2.4oz]*, ♂ 81g *[2.9oz]*.[436] **Intermembral index:** 67.5.[425] **Adult brain weight:** 2.7g *[0.1oz]*.[346] Ear length: 24mm (14–35) *[0.9in (0.6–1.4)]*. Baculum length: 13.2mm *[0.5in]*.[175]

Habitat Primary and secondary forest, gallery, savanna, montane forest up to 2000m *[6562ft]*.[898] Secondary vegetation (0–10m *[33ft]*) is preferred.

Diet Animal prey, 70%; fruit, 19%; gums, 10%;[109] leaves, buds.[662]

Life History Weaning: 53d.[109] **Juvenile:** 2–6mo.[606] **Sexual maturity:** 8–10mo.[41] **Estrus cycle:** NA. **Gestation:** 111–114d.[109] **Age 1st birth:** NA. **Birth interval:** 12mo.[41] **Life span:** 13y.[41] Births: Year-round.[109] Birth peak: Jan–Apr.[109] Offspring: 1.

Locomotion Quadrupedal running and jumping. After jumps, the forefeet land first.[608]

Social Structure Solitary foraging. Matriarchies are present.[606] Males maintain territories for 1–3 years.[109] Group size: Sleep 2–5, up to 10.[898] Home range: ♀ 0.6–1.4ha, ♂ 0.5–2.7ha.[898] Night range: NA.

Behavior Nocturnal and arboreal. Demidoff's bush babies hunt in tangles,[109] detecting prey by hearing and sight rather than by smell.[370] They catch fast-moving insects, such as moths and grasshoppers.[370] After foraging solitarily, individuals unite at dawn.[109] When interacting with adults, if the young hold their tail in a corkscrew, they are not attacked.[109] **Association:** Sympatric with Allen's bush babies (*Galago alleni*), Matschie's bush babies (*G. matschiei*), southern lesser bush babies (*G. moholi*), southern needle-clawed bush babies (*Euoticus elegantulus*), and, in parts of range, Thomas's bush babies (*Galagoides thomasi*). **Mating:** Mating is prolonged, lasting up to an hour,[174] and takes place while the pair are suspended beneath a branch. They appear to get stuck while mating but do not have a true lock like dogs.[169]

Scent marking: These bush babies scent-mark with urine and by rubbing their hands, cheeks, or chest on substrates.[667] **Vocalizations:** 11.[662] The gathering call carries 100 meters *[328ft]*. The advertisement call is a crescendo that increases in pitch and speed of repetition.[608] Individuals alarm-call when they encounter snakes. During estrus, females produce plaintive squeaks.[109] **Sleeping site:** Spherical leaf nests or dense vegetation tangles.[41] Several females may sleep together in groups of 2–10.[898]

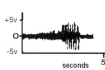

Senegal to Uganda

S. K. BEARDER (NATURAL HABITAT)

The rondo dwarf bush baby was recently discovered in southeast Tanzania and may be a new species.

Thomas's Bush Baby *Galagoides thomasi*

J. WICKINGS

Taxonomy Disputed. Elevated from a subspecies of *G. demidoff* in 1989.[608]

Distinguishing Characteristics
Thomas's bush babies are brownish gray, with gray underparts and a black anal scent gland. Their narrow face has a long muzzle and dark eye rings. The genital morphology is distinct from that of Demidoff's bush babies (*G. demidoff*).[20] The tail is not bushy.

Physical Characteristics Head and body length: 146mm (123–166)

[5.7in (4.8–6.5)].[608] **Tail length:** 261mm (150–233) *[10.3in (5.9–9.2)]*.[608] **Weight:** 99g (55–149) *[3.5oz (1.9–5.3)]*.[608] **Intermembral index:** 67.[234] **Adult brain weight:** NA. Ear length: 27mm (23–33) *[1.1in (0.9–1.3)]*.[608]

Habitat Marshy, gallery, and highland forests.[887]

Diet NA.

Life History NA.

Locomotion Quadrupedal running, leaping, and hopping.

Social Structure NA. **Group size:**

Thomas's Bush Baby *continued*

NA. **Home range**: NA. **Night range**: NA.

Behavior Nocturnal and arboreal. Thomas's bush babies forage at higher levels of forest (above 20m *[66ft]*) than Demidoff's bush babies *(G. demidoff),* which stay in understory (below 10m *[33ft]*) and use the ground.[887] **Association**: Sympatric

Zaire, Angola, Uganda, Cameroon

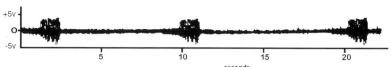

with Matschie's bush babies *(Galago matschiei)* and, throughout most of range, *G. demidoff.*[608]

Vocalizations: The advertisement call is a rolling call that rises to a crescendo and is repeated several times.[887]

Zanzibar Bush Baby *Galagoides zanzibaricus*

L. T. NASH (NATURAL HABITAT)

Taxonomy Disputed. Elevated from a subspecies of *Galago senegalensis* in 1971.[315] Grant's bush baby *(G. z. granti)* is included in this species but has been proposed as a full species.[313]

Distinguishing Characteristics Zanzibar bush babies are brown, with pale underparts, a white streak on the nose, and a dark-tipped bushy tail.[337]

S. K. BEARDER (NATURAL HABITAT)

Grant's bush baby *(G. z. granti)* is emerging from its sleeping site in a tree hollow.

Physical Characteristics Head and body length: 159mm (149–168) *[6.3in (5.9–6.6)].*[337] **Tail length:** 220mm (170–265) *[8.7in (6.7–10.4)].*[337] **Weight:** ♀ 137g (118–155) *[4.8oz (4.2–5.5)],* ♂ 149g (130–183) *[5.3oz (4.6–6.5)].*[337] **Intermembral index:** 58.4.[337] **Adult brain weight:** NA. Ear length: 35mm (27–46) *[1.4in (1.1–1.8)].*

Habitat Secondary lowland rain forest,[603] coastal dry forest, evergreen[608] and montane forest.[337]

Diet Animal prey, 70%; fruit, 30%.[338] This species eats birds and insects,[337] especially beetles and ants.[338] It does not eat *Acacia* gum.[337]

Life History Weaning: NA. **Sexual maturity:** 12 mo.[607] **Estrus cycle:** NA. **Gestation:** 120d.[533] **Age 1st birth:** NA. **Birth interval:** 4–8mo.[533] **Life span:** NA. Mating seasons: 2. Birth seasons: Feb–Mar, Aug–Oct.[41] Offspring: 1.[41] Mothers carry infants by mouth.[41]

Locomotion Quadrupedal running and leaping. After leaps, the front feet land first.[337]

Social Structure Solitary foraging. Females share territories with a daughter or a sister.[610] The male's territory overlaps that of 1–2 females.[610] Smaller males have larger ranges that overlap more females' ranges.[605]

Emigration: Males emigrate and females remain.[610] **Group size:** Sleep 1–6.[41] **Home range:** ♀ 1.6–2.6ha, ♂ 1.9–2.9ha.[605] **Night range:** 1576–1940m *[5171–6365ft].*[610]

Behavior Nocturnal and arboreal. Zanzibar bush babies are recorded below 5 meters *[16ft]* 91% of the time[338] and use branches and tangles less than 5cm *[2in]* in diameter. Males and females have closer associations in this species than any other bush babies that have been studied.[610]

Association: In parts of range, sympatric with thick-tailed greater bush babies *(Otolemur crassicaudatus),* Garnett's greater bush babies *(O. garnettii),* Somali bush babies *(Galago gallarum),* and southern lesser bush babies *(G. moholi).* **Vocalizations:** These bush babies give long-distance calls more often during full-moon peri-

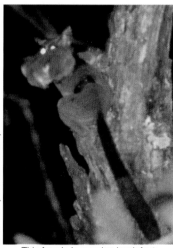

L. T. NASH (NATURAL HABITAT)

This female is carrying her infant in her mouth.

ods and travel more on moonlit nights.[603] Their advertisement call is loud and lasts 3–4 seconds. **Sleeping site:** Tree hollows[533] in a few different sites.[41] Large males may sleep with the females and offspring.[610]

East Coastal Africa from Ethiopia to Mozambique

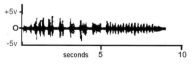

VULNERABLE

IUCN

Thick-tailed Greater Bush Baby *Otolemur crassicaudatus*

Dark form of *O. crassicaudatus*.

Taxonomy Disputed. 4 subspecies.[412] This genus was separated from *Galago* in 1974.[315]

Distinguishing Characteristics This species is the largest bush baby. Its fur varies from gray to black, and it has big ears and a large muzzle. Its tail is thick and bushy.[608]

Physical Characteristics Head and body length: 313mm (255–400) *[12.3in (10.0–15.7)]*.[608] **Tail length:** 419mm (342–495) *[16.5in (13.5–19.5)]*.[412] **Weight:** ♀ 1242g (1122–1497) *[43.8oz (39.6–52.8)]*,[436] ♂ 1495g (1126–1750) *[52.7oz (39.7–61.7)]*.[436] **Intermembral index:** 69.5.[608] **Adult brain weight:** 11.8g *[0.4oz]*.[346] Ear length: 62mm *[2.4in]*.[608] Baculum length: 22.4mm *[0.9in]*. Males are 20% larger than females.[436] The baculum is as long as a baboon's.[174]

Habitat Gallery, coastal forest, savanna woodland,[608] montane forest, bamboo, plantations.[174]

Diet Gums, 62%; fruit, 21–33%; animal prey (invertebrates, reptiles, birds, mammals[94]), 5%.[41;534]

Life History Weaning: 70–134d.[606] **Sexual maturity:** 18–24mo.[41] **Estrus cycle:** NA. **Gestation:** 126[528]–135d.[608] **Age 1st birth:** NA. **Birth interval:** 12mo.[41] **Life span:** 15y.[41] Birth season: Oct–Nov. Offspring: 2–3.[608] Mothers

carry infants by mouth and later on the back for 70 days.[608]

Locomotion Quadrupedal climbing and leaping. These bush babies walk and run along medium and large branches. When leaping, they land forefeet first.[608]

Social Structure The male is a solitary forager. The female and her offspring forage as a cohesive group. A male's range overlaps that of several females (noyau).[42] Matriarchies are present[412] and it is possible that they have complex social networks. **Group size:** 1–6.[42] **Home range:** ♀ 7ha, ♂ 10ha.[41] **Night range:** NA.

Behavior Nocturnal and arboreal. Thick-tailed greater bush babies prefer upper levels of forest (4–12m *[13–39ft]*).[41] **Association:** Sympatric with southern lesser bush babies *(Galago moholi),*[608] Garnett's greater bush babies *(Otolemur garnettii),* and, in parts of range, northern lesser bush babies *(G. senegalensis)* and Zanzibar bush babies

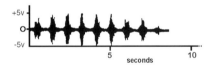

(Galagoides zanzibaricus).

Mating: Mating is prolonged (1–2h).[169] **Scent marking:** A large scent gland on the chest is used to mark branches and communicates an individual's identity.[122]

Vocalizations: The advertisement call is "drawn out loud cries repeated at regular intervals for a duration of 4 seconds."[608] Up to 5 individuals call simultaneously.[533] The repetitive structure of the call enhances its long-distance penetration.[533] Primary forest undergrowth transmits sound best between frequencies of 1000 and 1500 Hz; this call is close to 1500 Hz.[533] The call is given more often during mating season (24.5 calls/h) than at other times (5.4 calls/h). **Sleeping site:** Dense tangles, tree holes,[42] or flat leaf nests.[42] The male sleeps alone. The female sleeps with her offspring (2–6).[41]

Kenya, Tanzania, Rwanda, South Africa, Angola

S. M. ROWE

Light form.

Garnett's Greater Bush Baby *Otolemur garnettii*

Taxonomy Disputed. 4 subspecies.[412] Elevated from a subspecies of *O. crassicaudatus* in 1979.[315]

Distinguishing Characteristics Garnett's bush babies have reddish to grayish brown fur. The tail tip is brown, black, or white.[608]

Physical Characteristics Head and body length: 266mm (230–338) *[10.5in (9.1–13.3)]*.[608] **Tail length:** 364mm (310–400) *[14.3in (12.2–15.7)]*.[412] **Weight:** ♀ 721g *[25.4oz]*, ♂ 822g *[29.0oz]*.[41] **Intermembral index:** 69.[234] **Adult brain weight:** NA. **Ear length:** 45mm (34–55) *[1.8in (1.3–2.2)]*.[608] **Baculum length:** 26mm *[1.0in]*.[175] Males are 18% heavier than females.[437]

Habitat Coastal, riverine, and highland forests.[608]

Diet Fruit, 50%; animal prey (including beetles, ants, termites, snails, and birds[338]), 50%.[41]

Life History Weaning: 140d.[611] **Sexual maturity:** 12–18mo.[608] **Estrus cycle:** NA. **Gestation:** 130d.[528] **Age 1st birth:** NA. **Birth interval:** 12mo.[41] **Life span:** 15y.[41] Birth season: Aug–Nov.[41] Offspring: 1–2. Mothers carry infants by mouth[41] and park them while foraging.[875]

Locomotion Quadrupedal running and climbing, bipedal hopping.[605]

Social Structure The male is a solitary forager. Related females have overlapping territories.[611] Matriarchies are present.[606] Males have larger territories than females. **Emigration:** Males emigrate. Females do not.[611] **Group size:** NA. **Home range:** ♀ 11.6ha, ♂ 17.9ha.[41] **Night range:** 1600–3000m *[5250–9843ft]*.[611]

Behavior Nocturnal and arboreal. Garnett's greater bush babies range above 5 meters *[16ft]* 50% of the time and on branches larger than 5cm *[2in]*. They rarely come to the ground.[608] In a captive behavioral study, head-cocking behavior was used most when a moving stimulus such as insect prey was introduced to the cage.[700] **Association:** Sympatric with thick-tailed greater bush babies *(O. crassicaudatus)*, Zanzibar bush babies *(Galagoides zanzibaricus)*, and, in riverine parts of range, Somali bush babies *(Galago gallarum)*.[608] **Mating:** Promiscuous.[41] The prolonged duration of mating (up to 120min) may be a form of mate guarding.[175] When this species is housed with *O. crassicaudatus* in captivity, the two will mate but rarely produce offspring, perhaps because of the different penile morphology of these closely related species.[20] **Vocalizations:** The most notable call is an advertisement call that lasts more than 4 seconds, longer than that of *O. crassicaudatus*. **Sleeping site:** Vine tangles and rarely tree hollows. Males sleep alone. Females sleep with their offspring.[41]

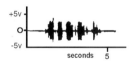

seconds 5

Somalia, Tanzania, Zanzibar

Ring-tailed lemur *Lemur catta*

Lemurs

All members of the superfamily Lemuroidea are found on the island of Madagascar, which has been separated from East Africa for over 100 million years.[580] Just as Darwin's finches evolved into many forms on the isolated Galápagos Islands of Ecuador, lemurs have diversified without competition from other primates from the African continent. Humans arrived in Madagascar about 1500 years ago. Almost one-third of the lemurs that existed then are now extinct. Subfossil remains show that the majority of the extinct lemurs were large, slow moving, and diurnal. The evidence suggests that human hunting and disturbance of the environment caused their demise.[168] Today most of the remaining lemurs face the threat of extinction for the same reasons, and only with concerted human effort to protect them will they survive.

The "flying lemur" (*Cynocephalus* spp.) is a misnomer given to an odd gliding mammal of the order Dermoptera from Asia. It is more closely related to bats than to primates.[625]

The name "lemur" should technically refer only to the genus *Lemur,* but it is often used to refer to all of the primates that inhabit Madagascar. The Lemuroidea are classified as five distinct families: Cheirogaleidae, Megaladapidae, Lemuridae, Indridae, and Daubentoniidae. There is little sexual dimorphism in body size and weight among lemuroids. This superfamily has retained the tapetum (a reflective layer in the eyes), although many lemurs are diurnal. As a rule, in lemur species that live in social groups, females are dominant to males. In most other primates, males are dominant to females. Not all members have retained the ancestral dental formula of $\frac{2.1.3.3.}{2.1.3.3.} \times 2 = 36.$[234]

Dwarf Lemurs   *Family:* Cheirogaleidae

The family of small nocturnal lemurs (Cheirogaleidae) comprises a diverse group of four genera. Although some experts believe that this family may be phylogenetically closer to the African prosimians,[820] it is traditionally placed with other lemurs inhabiting Madagascar. This family includes the smallest living primate, the pygmy mouse lemur *(Microcebus myoxinus),* which weighs a little over an ounce (30.6 grams). The fat-tailed dwarf lemurs (*Cheirogaleus* spp.) are the only primate known to hibernate for long periods (up to six months) during the dry season. The hairy-eared dwarf lemur *(Allocebus trichotis)* was thought to be extinct until 1990, when it was rediscovered in northern Madagascar.

Sportive Lemurs   *Family:* Megaladapidae

The family Megaladapidae, whose name was changed from "Lepilemuridae," contains only the genus *Lepilemur.* The sportive lemurs are leaf-eating, territorial, night creatures. Adult sportive lemurs lose their upper incisors, and their dental formula is $\frac{0.1.3.3.}{2.1.3.3.} \times 2 = 32.$[234]

Lemurs and Bamboo Lemurs   *Family:* Lemuridae

The family Lemuridae is made up of four genera: *Eulemur, Lemur, Varecia,* and *Hapalemur.* The ring-tailed lemur *(Lemur catta)* is the most well-known species of the superfamily Lemuroidea because of its banded tail and its adaptability to life in zoos. Ringtails, which are more terrestrial than other lemurs, live in large groups of up to 32 individuals. Most lemurs are seasonal breeders, but a ringtail female is receptive only one day per year. *Eulemur* is the most common and diverse genus of Lemuridae. It contains five species, most of which are sexually dichromatic—that is, males and females have different patterns of fur coloration. Several of these species are cathemeral (active part of the day and part of the night).[234]

The ruffed lemur is the only species in the genus *Varecia.* Ruffed lemurs are known for their large size and loud alarm barks. They are the only large lemur to produce multiple offspring in a litter. The bamboo lemurs (*Hapalemur* spp.) are so named because all three species, though different sizes, live primarily on a diet of bamboo. The golden bamboo lemur *(Hapalemur aureus),* discovered in 1986, has adapted to digest the high levels of cyanide found in the shoots of a bamboo species upon which it feeds.

Indrids   *Family:* Indridae

The three genera of folivorous primates that make up the family Indridae are strictly arboreal leapers. The smallest and only nocturnal indrid is the so-called woolly lemur *(Avahi laniger).* The indrids that belong to the genus *Propithecus* are called sifakas because their alarm call sounds like that name. One *Propithecus* species lives in the spiny desert of southern Madagascar. The largest of all the living lemurs and the one with the shortest tail is the indri *(Indri indri).* It lives in territorial family groups from which it broadcasts a very loud and haunting spacing call each morning. All indrids have a reduced dental formula of $\frac{2.1.2.3.}{2.0.2.3.} \times 2 = 30.$[234]

Aye-ayes   *Family:* Daubentoniidae

The most morphologically specialized (derived) family of lemurs is Daubentoniidae, which has only one species, the nocturnal aye-aye *(Daubentonia madagascariensis).* Its large brain-to-body ratio is an indication of intelligence. The aye-aye has incisors that continue to grow throughout its life, as do the incisors of a rodent. The aye-aye uses its specialized teeth to gnaw dead wood and uses its elongated third finger to probe for insect larvae. Unlike any other primate, it has a dental formula of $\frac{1.0.0.3.}{1.0.0.3.} \times 2 = 16.$[234]

Hairy-eared Dwarf Lemur *Allocebus trichotis*

B. MEIER

5.3in].[547] **Tail length:** ♀ 150, 165mm *[5.9, 6.5in]*; ♂ 160, 195mm *[6.3, 7.7in]*.[547] **Weight:** ♀ 85g (78–90) *[3.0oz (2.8–3.2)]*, ♂ 92g (75–98) *[3.2oz (2.6–3.5)]*.[436] **Intermembral index:** NA. **Adult brain weight:** NA.

Habitat Primary lowland rain forest.[547]

Diet In the wild—unknown. In captivity—insects, 70%; sweetened rice broth, fruit. The long tongue of these dwarf lemurs suggests that they feed on nectar.[547]

Life History NA. Birth season: Jan–Feb.[547]

Locomotion Quadrupedal running and leaping. Individuals suspend themselves from branches with their hind limbs.[547] They reportedly stand erect on their hind limbs to spot danger.[10]

Social Structure Variable. Solitary or 1 male–1 female.[10] Individuals may forage and feed alone.[580] **Group size:** Sleep 2–6.[580] **Home range:** NA. **Night range:** NA.

Behavior Nocturnal and arboreal.[10] Hairy-eared dwarf lemurs grab insects with both hands and often eat insects while hanging upside down from a branch.[547] They are less active during dry seasons in June to September.[547] They store fat all over the body, not just in the tail, as other dwarf lemurs do.[580] At dusk before leaving the nest, they groom each other intensively.[547] **Vocalizations:** At least 4. The most notable is an "alternating squeal." **Sleeping site:** Holes in dead trees less than 4m *[13ft]* from ground. In captivity, vertical cylindrical cavities with a diameter of 5–7cm *[2–3in]* are preferred. Adults sleep with juveniles.[547]

Madagascar

Taxonomy Disputed. This monotypic genus and species was once included in the genus *Cheirogaleus*.[315] Until 1989 it was known only from 5 museum specimens and was thought to be extinct.[546]

Distinguishing Characteristics Hairy-eared dwarf lemurs are brownish gray, with long hair on their ears and a long tongue.[580]

Physical Characteristics Head and body length: ♀ 130, 145mm *[5.1, 5.7in]*; ♂ 125, 135mm *[4.9,*

Greater Dwarf Lemur *Cheirogaleus major*

Taxonomy 2 subspecies. *C. m. crossleyi* may be a valid species after further study.[315]

Distinguishing Characteristics Greater dwarf lemurs are gray brown to reddish brown, with dark rings around the eyes and a fleshy nose.[580] Swelling of the tail with fat is less noticeable than in fat-tailed dwarf lemurs *(C. medius)*.[580]

Physical Characteristics Head and body length: 250mm *[9.8in]*.[580] **Tail length:** 275mm *[10.8in]*.[580] **Weight:** 235–470g *[8.3–16.6oz]*.[529] **Intermembral index:** 71.5.[425] **Adult brain weight:** 5.9g *[0.2oz]*.[346]

Habitat Rain forest, moist seasonal forest, dry forest that is not deciduous.[580]

Diet Fruit, young leaves, flowers, animal prey (mostly insects).[580]

Life History Weaning: 70d.[606] **Sexual maturity:** NA. **Estrus cycle:** 30d.[848] **Gestation:** 70d.[848] **Age 1st birth:** NA. **Birth interval:** NA. **Life span:** 8.8y.[346] Birth season: Nov–Feb.[580] Offspring: 2–3. During mating season the testes enlarge. Infants are born with closed eyes and are carried in the mother's mouth.[580] By 3 weeks, they can climb vertical limbs, eat fruit, and follow their mother out of the nest while she forages.[848]

Locomotion Quadrupedal.[580] This dwarf lemur leaps less often than *Microcebus*.[580]

Social Structure Solitary foraging. **Group size:** NA. **Home range:** ♀ 4ha.[529] **Night range:** NA.

Behavior Nocturnal and arboreal. Greater dwarf lemurs hibernate from April to September during the cool dry "winter" in Madagascar.[529] They are one of the few primate species that hibernate. **Mating:** Repeated matings of 2–3 minutes consist of the "male holding a receptive female with his hands and licking her flanks

and neck without biting. The male vocalizes, waves his tail, and grips the female's ankles."[848]

Scent marking: Fecal scent marks, 40cm *[16in]* long, are made.[732]

Vocalizations: This species has no alarm call, but its drawn-out, high-pitched whistling noises are probably long-distance territorial signals.[662] **Sleeping site:** Small groups sleep together.[580]

Greater Dwarf Lemur *continued*

Madagascar

Fat-tailed Dwarf Lemur *Cheirogaleus medius*

Taxonomy 2 subspecies.[580]

Distinguishing Characteristics Fat-tailed dwarf lemurs are mostly gray, with lighter undersides visible between the front and back legs. The ears are concealed.[580]

Physical Characteristics Head and body length: 200mm *[7.9in]*.[580] **Tail length:** 200mm *[7.9in]*.[580] **Weight:** 142–217g *[5.0–7.7oz]*.[580] **Intermembral index:** 68.[234] **Adult brain weight:** 2.9g *[0.1oz]*.[346]

Habitat Secondary forest, spiny desert, dry forest, gallery forest.[580]

Diet Fruit, flowers, nectar, leaf buds, gums, and animal prey, especially beetles and chameleons.[370]

Life History Weaning: 61d.[606] **Sexual maturity:** ♂ 10–14mo.[235] **Estrus cycle:** 19.7d (8–23).[898] **Gestation:** 61.6d (61–64).[580] **Age 1st birth:** 12mo. **Birth interval:** 12mo.[398] **Life span:** NA. Mating season: Oct–Nov.[580] Offspring: 1–4, usually 2.[898] The vagina is closed except during breeding season.[898] Adults begin to hibernate in April; the young feed until May before hibernating.[898]

Locomotion Quadrupedal.

Social Structure Solitary foraging.[580]

Madagascar

Group size: Sleep 1–5.[580] **Home range:** 4ha.[368] **Night range:** NA.

Behavior Nocturnal and arboreal.[692] In the 6 months that these lemurs are active, their body weight almost doubles, from 142 to 217 grams *[5.0–7.7oz]*,[580] and their tail volume increases from 200 to 530 cubic cm *[12.2–32.33 cu in]*.[370] They are lethargic for 7–9mo, losing 100 grams *[3.5oz]* of body weight, and emerge from hibernation at the end of November when rainy season starts.[368] Body temperature is 33–38°C in summer but can be as low as 15°C during winter hibernation.[235] **Mating:** Males fight over estrous females.[235] During breeding season 10 copulations in 30 minutes have been observed.[235] **Scent marking:** Fecal marks, 10cm *[3.9in]* long, are made.[732]

Vocalizations: The distress call is a "succession of resonant 'ouii-ouii-ouii'" calls.[662] Both genders make a squeaking copulation call.[235]

Sleeping site: Fat-tailed dwarf lemurs hibernate in hollow tree trunks during the dry season with 3–5 others.[370] Suitable hibernation sites and food are thought to be the limiting factors in the species' distribution.[368]

The fat-tailed dwarf lemur stores fat in its tail during the warm wet season and hibernates during the cool dry season.

ENDANGERED

USESA

Coquerel's Dwarf Lemur *Microcebus coquereli*

the second half, they engage in social activities such as grooming.[647] They feed on insect secretions first, then hunt sometimes at ground level.[647] Unlike fat-tailed dwarf lemurs (*Cheirogaleus medius*), they have no seasonal weight change.[664] **Scent marking:** 4 kinds—anogenital marking, urine marking, salivary marking, and "odor discharge," a volatile odor release that even humans can smell.[647] **Vocalizations:** Infants "purr" when licked by the mother. The low-pitched contact call is the best way to locate these lemurs in the forest. Their alarm call is a *ptiao*. They have a contact-rejection call that is a grunting crescendo, and their distress call consists of "high-pitched short calls repeated at high intensity."[662] **Sleeping site:** The daytime is spent in spherical nests 2–10 meters *[6.6–33ft]* above the ground, in the forks of a large branch. Made of small interlaced lianas, branches, and leaves,[647] the nests are 50cm *[20in]* in diameter. Each lemur uses up to 12 nests.

Madagascar

Taxonomy Disputed. Monotypic.[898] The genus name was changed from *Mirza* in 1985.[315]

Distinguishing Characteristics Coquerel's dwarf lemurs are brown or gray brown, with large ears and a hairy, thin tail.[580]

Physical Characteristics Head and body length: 200mm *[7.9in].*[315] **Tail length:** 330mm *[13.0in].*[315] **Weight:** ♀ 302g *[10.7oz]*, ♂ 307g *[10.8oz].*[436] **Intermembral index:** 70.[234] **Adult brain weight:** NA.

Habitat Dry deciduous forest with dense underbrush;[898] gallery forest; near semipermanent ponds.[370]

Diet Fruit, animal matter (including insects, frogs, bird eggs,[370] and chameleons[580]), flowers, gums, nectar, buds.[370] These dwarf lemurs spend up to 50% of their time in June licking the larval secretions of Flatidae (Homoptera) off the branches where the insects feed.[370]

Life History Weaning: 86d.[234] **Sexual maturity:** 9mo.[692] **Estrus cycle:** NA. **Gestation:** 89d. **Age 1st birth:** NA. **Birth interval:** 12mo. **Life span:** NA. Mating season: Oct.[580] Estrus is signaled by swelling and reddening of the clitoris and the vulva opening.[664] Mothers carry young

infants by mouth and park them while foraging. The young leave the nest at 3 weeks of age.[664] At 5 months, infants are able to forage alone.[647]

Locomotion Quadrupedal.[580]

Social Structure Variable. 1 male–1 female pairs[647] and male territories 4 times the size of a female's.[432] Some nests have only females; other nests have males, females, and young.[580] **Group size:** Sleep 1–3.[580] **Home range:** 2.5–3ha.[692] **Night range:** 1000–1500 sq m *[10,760–16,140 sq ft].*[370]

Behavior Nocturnal and arboreal. Coquerel's dwarf lemurs leave their nest about 6 p.m. and return about 6 a.m. On colder nights they are less active but do not hibernate.[647] The first half of the night they forage by themselves;

Gray Mouse Lemur *Microcebus murinus*

Gray mouse lemurs store fat in their tails and are less active in winter but do not have a true hibernation.

Taxonomy Disputed. 1 subspecies elevated to a full species *(M. myoxinus)*.[735] Researchers may describe other mouse lemur species with further study.[317]

Distinguishing Characteristics
Gray mouse lemurs have a gray back, off-white underparts, and long, fleshy ears.[580]

Physical Characteristics Head and body length: 125mm *[4.9in]*.[580] **Tail length:** 135mm *[5.3in]*.[580] **Weight:** 109g *[3.8oz]*.[436] **Intermembral index:** 72.[234] **Adult brain weight:** 1.8g *[0.1oz]*.[346] Females are 17% larger than males.[436]

Habitat Secondary forest,[464] spiny desert,[580] humid, dry, deciduous, gallery forest, brush—all with dense foliage and lianas.[898]

Diet Fruit, animal matter, flowers, gums, nectar, exudates.[580] Animal matter includes beetles, spiders, occasionally tree frogs, and chameleons, as well as secretions from homopteran larvae.[580]

Life History Weaning: 60d, 90d.[346] **Sexual maturity:** 8.5mo (7–10).[606] **Estrus cycle:** 45–55d.[848] **Gestation:** 60d (59–62).[580] **Age 1st birth:** 9.6–29.3mo.[848] **Birth interval:** 12mo.[398] **Life span:** 15.5y.[346] Mating season: Sep–Mar.[302] Birth season: Nov–Jun. Offspring: Usually 2–4.[464] During mating season the testes increase to 8 times their size in nonbreeding season.[848] After mating season the vagina closes.[848] The female transports her infants by mouth, and though their eyes are closed until day 4, they can cling to a branch while she forages. Infants can crawl at day 10.[464]

Locomotion Quadrupedal; rapid scurrying and agile leaps.[464]

Social Structure Solitary foraging. A male's range may overlap several females' ranges.[580] **Group size:** Sleep 1– 5,[464] up to 15.[580] **Home range:** 0.07–2ha.[692] **Night range:** NA.

Behavior Nocturnal and arboreal. The activity of mouse lemurs decreases during the dry season, but they have no true hibernation.[370] They store fat in their tail, which can expand from 50 to 200mm *[2.0–7.9in]* in captivity and from 20 to 95mm *[0.8–3.7in]* in the wild.[370] Owls prey heavily on this species in May and June; the remains of 58 gray mouse lemurs were found in owl pellets collected at one owl nest site.[302] **Association:** Sympatric with pygmy mouse lemurs *(M. myoxinus)*.[580] **Mating:** A vaginal plug forms after copulation.[848] **Scent marking:** These lemurs have apocrine glands in the cheeks and genitalia. They mark their territory by rubbing their faces on the end of twigs. They also wash their hands and feet in urine or urine-mark directly on branches.[294] Scent glands may contain information about reproductive condition; females mark more often during estrus.[294] Scent marking with urine probably provides information about social status, because dominant animals scent-mark more often.[294] **Vocalizations:** Gray mouse lemurs have high-pitched vocalizations that are partially beyond the range of human hearing.[662] Males produce a trill call during mating season to advertise their presence. Females may be able to "assess the quality of a male" by his call.[934] This species has a specific mating call.[662] **Sleeping site:** Spherical nests or tree holes during the day.[580] Males sleep alone or in pairs.[464] Several females may sleep together with their infants in a nest.[464]

Madagascar

Pygmy Mouse Lemur *Microcebus myoxinus*

The pygmy mouse lemur is the smallest known primate.

R. A. MITTERMEIER

Taxonomy Disputed. First described in 1852, this species was noted again by researchers in the 1960s but not described in detail until 1992.[580] It is now accepted as a full species rather than as a subspecies of *M. murinus*.[317]

Distinguishing Characteristics Pygmy mouse lemurs are the smallest primates in the world.

Their body is orange-tinged rufous brown, with creamy white underparts. A dark median stripe runs down the back, and a white stripe runs from the forehead to the nose.[580]

Physical Characteristics Head and body length: 61mm [2.4in].[580] **Tail length:** 136.2mm [5.4in].[580] **Weight:** 30.6g (24.5–38) [1.1oz

(0.9–1.3)].[580] **Intermembral index:** NA. **Adult brain weight:** NA.

Habitat Deciduous dry forest.[580]

Diet NA.

Life History NA. Testes increase in size when breeding season starts.[735]

Locomotion NA.

Social Structure NA. **Group size:** NA. **Home range:** NA. **Night range:** NA.

Behavior Nocturnal and arboreal. When pygmy mouse lemurs are spotted with a flashlight beam at night, they often freeze rather than run for safety, as other mouse lemurs do.[580] **Association:** Sympatric with gray mouse lemurs *(M. murinus)*.[580]

Madagascar

Brown Mouse Lemur *Microcebus rufus*

Taxonomy Elevated from a subspecies of *M. murinus* in 1977.[315]

Distinguishing Characteristics Brown mouse lemurs have a brown to rufous back. Their underparts are lighter, and their ears smaller, than those of gray mouse lemurs *(M. murinus)*.[580]

Physical Characteristics Head and body length: 125mm [4.9in].[580] **Tail length:** 115mm [4.5in].[580] **Weight:** ♀ 49g (41–63) [1.7oz (1.4–2.2)], ♂ 50g (35–70) [1.8oz (1.2–2.5)].[180] **Intermembral index:** 71.5.[425] **Adult brain weight:** NA.

Habitat Primary and secondary moist forest to drier forest; old plantations; *Eucalyptus* groves.[580]

Diet Fruit (47 types[32]), animal matter, young leaves, flowers.[580]

Life History Weaning: NA. **Sexual maturity:** NA. **Estrus cycle:** NA. **Gestation:** 61d.[180] **Age 1st birth:** NA. **Birth interval:** NA. **Life span:** NA. Mating season: Sep–Oct.[32] Birth season: Nov–Dec;[32] 2 seasons per year in captivity.[606] Offspring: 1–3.[32] In mid-August

the testicles begin to enlarge.[32]

Locomotion Quadrupedal climbing.[711]

Social Structure Solitary foraging. Home ranges overlap extensively. **Group size:** Sleep 1–4.[31] **Home range:** NA. **Night range:** NA.

Behavior Nocturnal and arboreal. These lemurs feed at different heights, from low shrubs to treetops.[580] **Scent marking:** Marks are made with urine and feces.[580] **Sleeping site:** During the day, leaf nests (usually 1.8–10m [5.9–33ft] high), tree holes, and even old bird nests.[580] Individuals sleep alone or females sleep with small infants.[31]

NATURAL HABITAT

Madagascar

Fork-marked Lemur *Phaner furcifer*

Taxonomy Monotypic genus with 4 subspecies.[580]

Distinguishing Characteristics Fork-marked lemurs get their common name from the black dorsal stripe that runs down their brown back, bifurcates on the crown into 2 stripes, and continues to the eyes.[111]

Physical Characteristics Head and body length: 227–285mm *[8.9–11.2in]*.[436] **Tail length:** 285–370mm *[11.2–14.6in]*.[436] **Weight:** 460g (350–600) *[16.2oz (12.3–21.2)]*.[436] **Intermembral index:** 68.[234] **Adult brain weight:** 7.3g *[0.3oz]*.[346]

Habitat Primary and secondary deciduous and gallery forest up to 1000m *[3281ft]*.[898] Secondary forest with a continuous canopy is preferred.

Diet Fruit, animal matter, flowers, gums, buds, exudates.[580] Fork-marked lemurs are specialized gum feeders;[370] the tree genus most often used as a gum source is *Terminalia*. They also eat the secretions of insects (Homoptera).[111]

Life History Weaning: NA. **Sexual maturity:** NA. **Estrus cycle:** 15d. **Gestation:** NA. **Age 1st birth:** NA. **Birth interval:** NA. **Life span:** NA. Mating season: Jun.[580] Birth season: Nov.[580] Offspring: 1. Females have 4–5 estrus cycles per mating season; the vulva is open only 2–3 days per 15-day cycle.[111] Mothers carry infants by mouth.[580]

Locomotion Quadrupedal running and leaping.[649] Fork-tailed lemurs move very rapidly.[580]

Social Structure Variable. These lemurs have been reported to live in pairs or be solitary.[464] Males and females remain in vocal contact during the night and sleep in the same hole. Males maintain their territory by vocalizing throughout it.[111] **Group size:** 1–4.[580] **Home range:** 4ha. **Night range:** NA.

Behavior Nocturnal and arboreal. Fork-marked lemurs are active from "15 minutes before complete dark"[111] through the night.[649] They call repeatedly and visit gum-producing trees. Each individual has "access to 9 gum trees."[111] Females eat first, and the male follows behind the female 2–5 meters *[7–16ft]*.[111] "After 10 p.m. they [all] hunt insect prey, preferring moths and mantids."[111] When these lemurs are spotted with a flashlight, they bob their head.[580] **Scent marking:** Males use their throat gland to mark females.[111] No territorial scent marking has been reported. **Vocalizations:** Fork-marked lemurs have vocal confrontations in which all participants (3–9 individuals) make the *kui* call at once.[662] Other vocalizations include the *hong* sound, a contact call; the male-and-female-duet *tia* call; an alarm and distress call; and a staccato call for contact rejection.[662] Males have special long-distance calls to which other males respond.[662] **Sleeping site:** Tree holes or spherical leaf nests during the day.[649]

R. A. MITTERMEIER

Madagascar

Gray-backed Sportive Lemur *Lepilemur dorsalis*

ENDANGERED

USESA

Taxonomy Disputed. Chromosomal evidence has split what was once considered to be 1 species, *L. mustelinus,* into 7 species.[315]

Distinguishing Characteristics Gray-backed sportive lemurs have a gray to brown face and short ears. The back, tail, and chest are darkish brown.[580]

Physical Characteristics Head and body length: 254mm (250–260) *[10.0in (9.8–10.2)].*[820] **Tail length:** 266mm (260–278) *[10.5in (10.2–10.9)].*[820] **Weight:** <1000g *[<35.3oz].*[580] **Intermembral index:** NA. **Adult brain weight:** NA.

Habitat Gallery forest; bush;[464] humid forest with a short dry season.[580]

Diet Fruit, leaves, bark.[580]

Life History NA. Birth season: Aug–Nov.[580] Offspring: 1.[580]

Locomotion Vertical clinging and leaping.[580]

Social Structure Solitary.[580] Ranges overlap.[464] **Group size:** NA. **Home range:** NA. **Night range:** NA.

Behavior Nocturnal and arboreal. Gray-backed sportive lemurs have been studied only briefly. **Vocalizations:** These lemurs have a variety of calls that vary in intensity and tonal character, ranging from weak squeaks to powerful high-pitched calls. The maternal call sounds like a kiss.[662] **Sleeping site:** Large trees. Logging has seriously reduced such sleeping sites. Researchers have provided nesting boxes in limited areas.

Madagascar

R. A. MITTERMEIER

Milne-Edwards' Sportive Lemur *Lepilemur edwardsi*

ENDANGERED

USESA

R. A. MITTERMEIER (NATURAL HABITAT)

Taxonomy Disputed. Chromosomal evidence has split what was once considered to be 1 species, *L. mustelinus,* into 7 species.[315]

Distinguishing Characteristics Milne-Edwards' sportive lemurs have a grayish brown back and head, a brown to dark gray face, and a light brown tail. They have a median stripe down the back.[580]

Physical Characteristics Head and body length: 280mm (272–292) *[11.0in (10.7–11.5)].*[820] **Tail length:** 277mm (269–290) *[10.9in (10.6–11.4)].*[820] **Weight:** 1000g *[35.3oz].*[580] **Intermembral index:** NA. **Adult brain weight:** NA.

Habitat Deciduous dry forest.[580]

Diet Leaves, fruit, seeds, flowers.[580]

Life History NA.

Locomotion Vertical clinging and leaping.[580]

Social Structure Solitary foraging. Territories are defended. **Group size:** Sleep 1–3.[580] **Home range:** 1ha.[580] **Night range:** NA.

Behavior Nocturnal and arboreal.[438] The territorial display includes branch shaking and loud calls.[580] **Vocalizations:** The maternal call to an infant is described as *tchen.*[662] The distant communication call sounds like *oooai* followed by *oui oui oui.*[662]

Madagascar

White-footed Sportive Lemur *Lepilemur leucopus*

L. NASH (NATURAL HABITAT)

Taxonomy Disputed. Chromosomal evidence has split what was once considered to be 1 species, *L. mustelinus,* into 7 species.[315]

Distinguishing Characteristics White-footed sportive lemurs have a light gray body, with brownish fur around the head and shoulders. The tail is light brown above and white underneath.[580]

Physical Characteristics Head and body length: 250mm *[9.8in].*[580] **Tail length:** NA. **Weight:** ♀ 580g *[20.5oz],* ♂ 544g *[19.2oz].*[436] **Intermembral index:** NA. **Adult brain weight:** NA. Females weigh 12% more than males.[436]

Habitat Spiny desert, gallery forest, riverine areas.[580]

Diet Primarily leaves of 2 species of *Alluaudia* and flowers.[370]

Life History NA. Birth season: Sep–Nov. Offspring: 1.[580] Copulation occurs only during mating season. In other seasons, the testes regress and the vulva closes.[371]

Locomotion Vertical clinging and leaping.[580]

Social Structure Variable. Solitary or 1 male–1 female pairs.[580] Males have territories that overlap those of females.[580] Each gender defends its own territory.[580] All adults examined had scars, which indicate territorial fighting.[371] **Group size:** NA. **Home range:** 0.2–0.46ha.[371] **Night range:** 320–500m *[11.3–17.6ft].*[580]

Behavior Nocturnal and arboreal. Foraging consists of rapid leaps, then 10 minutes of feeding and 10–20 minutes of rest. These lemurs often sit at their territory borders and watch for intruders.[436] **Scent marking:** Urination on surveillance posts.[371]

Vocalizations: Vocalizations are used to defend territory.[371] The male distance call is *hein hein hein,* and the response is a *hee.*[662] The maternal call sounds like a kiss.[662]

Madagascar

Small-toothed Sportive Lemur *Lepilemur microdon*

Taxonomy Disputed. This species may still be a subspecies of *L. mustelinus.*[315]

Distinguishing Characteristics Small-toothed sportive lemurs have chestnut-colored forelimbs and shoulders and a dark stripe down the back.[580] Their molar teeth are much smaller than those of *L. mustelinus.*[580]

Physical Characteristics Head and body length: ♂ 260mm *[10.2in].*[673] **Tail length:** ♀ 270mm *[10.6in].*[673] **Weight:** ♂ 970g *[34.2oz].*[673] **Intermembral index:** NA. **Adult brain weight:** NA.

Habitat Secondary forest with dense saplings and bamboo.[673]

Diet NA.

Life History NA.

Locomotion Vertical clinging and leaping.[580]

Social Structure Solitary foraging. **Group size:** Sleep 1–2.[673] **Home range:** >0.5ha.[673] **Night range:** NA.

Behavior Nocturnal and arboreal. Small-toothed sportive lemurs are active during winter and have been observed to rest in the sun next to their tree cavity during the day and return to the hole before dusk.[673] Humans hunt for them in tree holes, for food.[673] **Sleeping site:** Multiple sites in cavities 2–8 meters above the ground in the largest trees in the forest. A female and her juvenile offspring share sleeping sites.[963]

Madagascar

L. PORTER (NATURAL HABITAT)

These lemurs may rest in the sunshine for part of the day.

Weasel Sportive Lemur *Lepilemur mustelinus*

Taxonomy Disputed. Chromosomal evidence has split what was once considered to be 1 species, *L. mustelinus,* into 7 species.[315]

Distinguishing Characteristics The back and head of weasel sportive lemurs are chestnut brown with a black median stripe. The face is gray or brown, with lighter throat and cheeks. The tail is dark toward the tip.[580] In this species the cecum is larger than the colon; in other primates the cecum is much smaller than and often insignificant relative to the colon.[514]

Physical Characteristics Head and body length: NA. **Tail length:** NA. **Weight:** ♀ 617g (583.3–650.7) *[21.8oz (20.6–23.0)],* ♂ 593.8g (584.6–603) *[21.0oz (20.6–21.3)].*[435] **Intermembral index:** NA. **Adult brain weight:** 9.5g *[0.3oz].*[346]

Habitat Humid forest.[580]

Diet Leaves, fruit, flowers. A diet high in alkaloids can be tolerated.[580]

Life History Weaning: 75[346]–120d.[662] **Sexual maturity:** 18[662]–21mo.[346] **Estrus cycle:** NA. **Gestation:** 130[662]–150d.[438] **Age 1st**

S. NASH

birth: NA. **Birth interval:** NA. **Life span:** NA.

Locomotion Vertical clinging and leaping.[580]

Social Structure Solitary.[692] **Group size:** NA. **Home range:** 1.5ha.[335] **Night range:** NA.

Behavior Nocturnal and arboreal. **Sleeping site:** Dry season—tree holes 6–12 meters *[39ft]* above ground. Wet season—leaf nests and liana.[580]

Madagascar

Red-tailed Sportive Lemur *Lepilemur ruficaudatus*

O. BLOXAM (NATURAL HABITAT)

Taxonomy Disputed. Chromosomal evidence has split what was once considered to be 1 species, *L. mustelinus,* into 7 species.[368]

Distinguishing Characteristics The back of red-tailed sportive lemurs is gray, with light brown. The tail is reddish, and the face and throat are pale gray to pale brown.[580]

Physical Characteristics Head and body length: 280mm *[11.0in].*[580] **Tail length:** 250–260mm *[9.8–10.2in].*[580] **Weight:** ♀ 845g (607–915) *[29.8oz (21.4–32.3)],* ♂ 764g (600–900) *[26.9oz (21.2–31.7)].*[437] **Intermembral index:** NA. **Adult brain weight:** NA.

Habitat Deciduous dry forest.[580]

Diet Leaves, fruit.[370] In captivity, these lemurs consume 71 kilocalories per week per 100 grams *[3.5oz]* of body weight, or half the intake (140kcal) of Coquerel's dwarf lemur *(Microcebus*

coquereli). Red-tailed sportive lemurs get more protein, however, because they eat young leaves, which have a lot of protein.[368] The observed caecotrophy may increase the digestion of long-chain ß-linked carbohydrates.[368]

Life History Weaning: NA. **Sexual maturity:** NA. **Estrus cycle:** NA. **Gestation:** 120–150d.[664] **Age 1st birth:** NA. **Birth interval:** NA. **Life span:** NA. **Mating season:** May–Jul. **Offspring:** 1.[580] The vagina is closed except during estrus. During the breeding season, body weight decreases slightly[368] and the testicles enlarge.[664] The photoperiod triggers the physiological changes for breeding.[580] Infants are carried by mouth and parked while the mother forages.[580]

Locomotion Vertical clinging and leaping.[580]

Red-tailed Sportive Lemur *continued*

Social Structure NA. **Group size**:
NA. **Home range**: NA.
Night range: NA.

Behavior Nocturnal and arboreal.
Red-tailed sportive lemurs do not
hibernate.[580] In intensively logged
forests, sportive lemurs are extir-
pated because of the dietary lack of energy to move
between more distant trees.[261] **Association:** If the
woolly lemur *(Avahi laniger)* is present within the
range of these sportive lemurs, the competition forces
the latter to eat lower-quality leaves.[261] **Vocalizations:**
The distant communication call is "Boako Boako."[662]

Madagascar

Northern Sportive Lemur *Lepilemur septentrionalis*

Taxonomy Disputed. Chromosomal
evidence has split what was once
considered to be 1 species, *L.
mustelinus,* into 7 species.[315]

Distinguishing Characteristics
The head and body of northern
sportive lemurs are light gray to
brown, with a dark median stripe.
The tail tip is a darker brown.[580]

**Physical Characteristics Head and
body length:** 278mm *[10.9in].*[820]
Tail length: 247mm *[9.7in].*[820]
Weight: ♀ 700g *[24.7oz].*[580]
Intermembral index: NA. **Adult brain
weight:** NA.

Habitat Deciduous dry and humid
forest.[580]

Diet Leaves.[580]

Life History NA.

Locomotion Vertical clinging and
leaping.[580]

Social Structure Solitary.[580] Adults
forage alone and rarely associate
during nocturnal activities.[580]
Group size: NA. **Home range:**
1ha.[580] **Night range:** NA.

Behavior Nocturnal and arboreal.
Northern sportive lemurs are
found in high densities in limestone canyons of the Ankarana
range in northern Madagascar.[892]
Sleeping site: Leaf tangles and
tree holes.

Madagascar

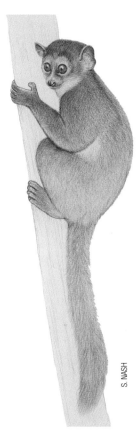
S. NASH

Ring-tailed Lemur *Lemur catta*

Taxonomy Disputed. Since 1988 this genus, which used to include all the species now in *Eulemur,* has included only 1 monotypic species.[315]

Distinguishing Characteristics Ringtails get their name from their conspicuous black-and-white-banded tail. They have a gray to brown back, white underparts, a white face with dark brown triangular eye patches, and white ears.[580] Males have a fingernail-like spur near each wrist.[580]

Physical Characteristics Head and body length: 425mm *[16.7in]*.[580] **Tail length:** 600mm *[23.6in]*.[580] **Weight:** ♀ 2678g *[94.5oz]*, ♂ 2705g *[95.4oz]*.[436] **Intermembral index:** 70.[234] **Adult brain weight:** 25.6g *[0.9oz]*.[346]

Habitat Scrub, spiny desert, dry and gallery forest.[898]

Diet Fruit, 70%; leaves, 25%; flowers, 5%; bark, sap, herbs.[580]

Life History Weaning: 105d.[731] **Juvenile:** 16mo.[655] **Sexual maturity:** 21–30mo.[692] **Estrus cycle:** 39d.[655] **Gestation:** 134–138d.[580] **Age 1st birth:** 36mo.[580] **Birth interval:** 12–24mo.[692] **Life span:** 27y. Mating season: Mid-Apr.[580] Birth season: Aug–Sep.[580] Offspring: 1, occasionally 2.[438] Females are receptive for only 24 hours per year.[726] Infants are carried ventrally for the first 2 weeks, then on the back. Infant mortality is 40%.[438]

Madagascar

NATURAL HABITAT

Locomotion Quadrupedal; terrestrial running.[580]

Social Structure Multimale-multifemale social structure with 1 alpha female.[726] Ringtails defend resources[421] but are not strictly territorial.[726] **Emigration:** Males emigrate periodically.[726] **Group size:** 17 (5–30).[692] **Home range:** 6–23.1ha.[580] **Day range:** 1377m *[4826ft]*.[421]

Behavior Diurnal and arboreal. Ringtails are more terrestrial than other lemurs.[580] On sunny mornings they often sit in treetops, "sunbathing." Females are extremely dominant and win all disputes with males.[421] Dominance relationships are not always transitive, as in anthropoid monkeys. If *A* dominates *B* and *B* dominates *C, A* is not necessarily dominant to *C*.[655] In captivity,

female dominance begins during puberty, at about 16 months.[655] Females have dominance relationships but not dominance hierarchies.[655] An individual may be targeted by 2 or 3 adversaries who attack her and reverse her dominance, or if she is subordinate, she may be expelled from the group permanently.[655] Targeted aggression, in which 1 female persistently harasses and attacks a target individual, may last for days or weeks. Even sisters, grandmothers, or granddaughters are targeted, but usually more distantly related members are driven from the group.[655] No female ever dominates her mother.[655] But the status of close kin is not affected by close kinship—if a daughter's rank changes, her mother's is

Ring-tailed Lemur *continued*

unaffected.[655] An adolescent female that does not become dominant to an adult female will be evicted.[655] This species is said to be the only primate in which the infants "grapple" for dominance.[655] Ringtails do not have a stable hierarchy.[655] **Mating:** Males fight viciously during the breeding season. Females seek out males that are new troop members.[726] **Scent marking:** Males have "stink fights" in which they face each other and wave their tails, which have been rubbed on their strong-smelling wrist glands.[731] These lemurs scent-mark with the brachial gland on the inner arm. The spur on the forearm is rubbed on the gland and then used to scar branches and leave scent.[731] **Vocalizations:** 22.[347] The cohesion call is a *miaouw* given before troop moves. The territorial loud call, given by males, is a long plaintive howl that carries 1km *[3281ft]*.[370] The alarm call is *oua-oua,* uttered by the whole group in unison. There is one call for aerial predators and another for terrestrial ones.[580] Rapid staccato grunts are made by 2 aggressive individuals. The distress call is a high-pitched, piercing call made during a fight.[370]

NATURAL HABITAT

Ringtails are quadrupedal and spend more time on the ground than other lemurs.

Crowned Lemur *Eulemur coronatus*

Female crowned lemur.

Taxonomy Monotypic species.[580]

Distinguishing Characteristics
Crowned lemurs are sexually dichromatic. Males have a dark gray brown back and a dark tail; females are light gray. Both sexes have an orange V shape on the crown. Males have more orange at the top of the head, and they have black in the side of the V.[580]

Physical Characteristics Head and body length: 340mm *[13.4in]*.[580] **Tail length:** 450mm *[17.7in]*.[580] **Weight:** ♀ 1687g *[59.5oz]*, ♂ 1712g *[60.4oz]*.[436] **Intermembral index:** 69.[234] **Adult brain weight:** NA.

Habitat Semideciduous dry tropical forest, scrub, savanna,[892] edges of humid forest.[898]

Diet Leaves, fruit, flowers. Pollen[370] and occasionally insects are also reportedly eaten.[580]

Life History Weaning: NA. **Sexual maturity:** 20mo.[433] **Estrus cycle:** 34d.[433] **Gestation:** 125d.[433] **Age 1st birth:** NA. **Birth interval:** NA. **Life span:** NA. Mating season: May–Jun. Birth season: Mid-Sep–Oct.[580] Offspring: 1–2.[433] Copulation occurs on 1 day of each cycle.[433]

Locomotion Quadrupedal, vertical clinging and leaping.[892]

Social Structure Multimale-multifemale. Crowned lemurs live in uncohesive groups of 2 adult males, 2 females, and offspring, with no discernible hierarchy.[892] While foraging, they may split into smaller groups.[580] **Group size:** 5–6, up to 15.[580] **Home range:** NA. **Day range:** NA.

Behavior Cathemeral and arboreal. Crowned lemurs have been found at all levels of forest. They prefer thick cover[580] and come to the ground only to eat fallen fruit.[892] They are active at night as well as during the day (cathemeral).[892] Females are dominant over males, but leadership during travel rotates.[892] **Association:** Sympatric with 9 other species but associates with brown lemurs *(E. fulvus sanfordi)*.[892] **Scent marking:** Males have 3 types: anogenital, head, and hand marking.[434] They head-mark by rubbing their forehead on objects, a behavior that increases 5 days before females come into estrus.[434] **Vocalizations:** Loud shrieks are heard at night.[892] Crowned lemurs grunt-shriek (alarm-call) at the fossa *(Cryptoprocta ferox),* a ground predator that causes them to flee higher into the trees.[892]

Madagascar

This male is in the typical posture of a vertical clinger and leaper.

ENDANGERED

USESA

Brown Lemur *Eulemur fulvus*

ENDANGERED   USESA

Female and infant collared brown lemur
(E. f. collaris).

Male collared brown lemur *(E. f. collaris)*.

Madagascar

Brown lemurs can leap at least 5 meters
(16 feet).

smelling secretions from their scrotum. Brown lemur scent has been experimentally shown to identify the gender and individual identity of the marker but not its subspecies.[341] **Vocalizations:** Brown lemurs wag their tail and grunt at terrestrial predators.[811] They give loud alarm barks when they see a large hawk *(Gymnogenys radiata)*.[811]

less than 2% on the ground.[811] They salivate on millipedes, then roll them between the hands for 5–6 minutes before eating them.[646] **Association:** *E. f. sanfordi* associates with crowned lemurs *(E. coronatus)*.[892] *E. f. fulvus* sometimes travels with mongoose lemurs *(E. mongoz)*.[580] **Scent marking:** Brown lemurs scent-mark during sexual behavior, alarm response to human observers, territorial defense, and travel.[341] Males have strong-

Taxonomy Disputed. 6 subspecies, each with a distinct head pattern.[580]

Distinguishing Characteristics *E. f. fulvus* is not sexually dichromatic, but in the other 5 subspecies, coat color varies from gray to reddish and the head pattern varies with gender.[580]

Physical Characteristics Head and body length: 400–500mm *[15.7–19.7in]*.[580] **Tail length:** 500–550mm *[19.7–21.7in]*.[580] **Weight:** ♀ 2147–2495g *[75.7–88.0oz]*, ♂ 2110–2459g *[74.4–86.7oz]*.[436] **Intermembral index:** 72.[234] **Adult brain weight:** 29.2g *[1.0oz]*.[346]

Habitat Primary closed canopy forest and deciduous forest above 400m *[1312ft]*. Brown lemurs occur in lower numbers in deciduous forest.[898]

Diet Fruit, mature leaves (39 species), flowers (18 species), bark, sap, dirt (red clay, black soil),[646] and insects, especially centipedes and millipedes.[580] Ten tree species account for 60% of the diet. *E. f. fulvus* can tolerate higher levels of toxic plant compounds than other prosimians can.[646]

Life History Weaning: 135d.[346] **Sexual maturity:** ♀ 10mo, ♂ 23.3mo.[346] **Estrus cycle:** NA. **Gestation:** 120d.[438] **Age 1st birth:** 27.8mo.[346] **Birth interval:** 12mo.[692] **Life span:** 30.8y.[346] **Mating season:** Jun. **Birth season:** Sep–Oct.[580] **Offspring:** 1–2.[438]

Locomotion Quadrupedal.[580]

Social Structure Multimale-multifemale cohesive groups without a "noticeable hierarchy."[580] Agonistic intergroup interactions have been reported to be not strictly territorial.[580] **Group size:** 13–18.[580] **Home range:** 0.75–7ha,[898] 20ha,[580] up to 100ha.[645] **Day range:** 125–150m *[410–492ft]*,[811] 457–1471m *[1499–4826ft]*.[580]

Behavior Cathemeral and arboreal. Brown lemurs spend 95% of their time in upper forest layers and

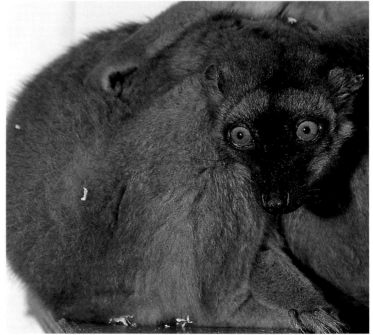

Female brown lemur *(E. f. fulvus)*.

Brown Lemur *continued*

Male white-fronted brown lemur *(E. f. albifrons)*.

This male red-fronted brown lemur *(E. f. rufus)* has a collar and a tag so that researchers can identify him.

Female red-fronted brown lemur *(E. f. rufus)*.

Black Lemur *Eulemur macaco*

Taxonomy 2 subspecies.[898]

Distinguishing Characteristics
Black lemurs are named after the males, which are all black. Females have a brown back, white underparts, a gray head, and long white ear tufts. *E. m. macaco* has brown eyes; *E. m. flavifrons* has bright blue eyes and no ear tufts.[580]

Physical Characteristics Head and body length: 410mm *[16.1in]*.[580]
Tail length: 550mm *[21.7in]*.[580]
Weight: ♀ 2487g *[87.7oz]*, ♂ 2403g *[84.8oz]*.[436] **Intermembral index:** 71.[234] **Adult brain weight:** 25.6g *[0.9oz]*.[346]

Habitat Primary and secondary humid forest; timber, coffee, and cashew plantations.[580]

Diet Fruit (including palm and pandanus fruit[126]), seeds, leaves, animal matter, flowers, nectar,[464] mushrooms, and millipedes.[126]

Life History Infant: 4.5mo.[346]
Weaning: NA. **Sexual maturity:** 24mo.[346] **Estrus cycle:** 33d.[346]

Gestation: 125d.[126] **Age 1st birth:** 24mo.[126] **Birth interval:** 12mo.[692]
Life span: 27.1y.[346] Birth season: Sep–Nov.[580] Offspring: 1–2. Females carry infants for 180 days.[438]

Locomotion Quadrupedal;[580]some suspensory.[126] Black lemurs feed in the terminal branches of trees by suspending themselves upside down from the hind feet.[126]

Social Structure Multimale-multifemale groups. Groups reportedly have more males than females; different groups may join together at night.[580] **Group size:** 7–10, up to 15.[580] **Home range:** 5.3ha.[126] **Day range:** NA.

Behavior Cathemeral and arboreal. Black lemurs are less active during the day than at night, when the *Parkia* tree's night-blooming flowers are available for nectar feeding.[946] Agonistic behavior over a piece of large fruit involves "spats and high pitched squeaks."
Mating: Males are aggressive during mating season and may leave to follow neighboring groups.[126]
One male was observed to mate 6 times in 30 min-

Female Sclater's black lemur *(E. m. flavifrons)*.

Female black lemur *(E. m. macaco)*.

Black Lemur *continued*

Male Sclater's black lemurs *(E. m. flavifrons).*

utes.[126] **Vocalizations:** Black lemurs purr contentedly while being groomed.[126] They have a cohesion call; recognition grunts, which sound somewhat like a duck's quacking; and territorial calls given by the whole troop at dusk.[464] When an aerial predator is sighted, black lemurs make a rasping loud call that ends with a descending scream-whistle; the group responds by "diving" to lower branches.[126] When startled, they give an "alarm hack" vocalization.[126] They have been observed mobbing snakes and dogs.[126]

Madagascar

Mongoose Lemur *Eulemur mongoz*

Male.

D. HARING

Taxonomy Monotypic species.

Distinguishing Characteristics Mongoose lemurs are sexually dichromatic. Males have a gray brown back and rufous hair around the face. Females have a gray to brown back, a white beard, and a dark gray face.[580]

Physical Characteristics Head and body length: 350mm *[13.8in].*[580] **Tail length:** 480mm *[18.9in].*[580] **Weight:** ♀ 1658g *[58.5oz],* ♂ 1682g *[59.3oz].*[436] **Intermembral index:** 72.[234] **Adult brain weight:** 21.8g *[0.8oz].*[346]

Habitat Primary and secondary dry forest, scrub.[898]

Diet Fruit, leaves, flowers, nectar.[580] The nectar from the flowers of kapok trees is a preferred food source.[909]

Life History Weaning: 5mo.[909] **Sexual maturity:** NA. **Estrus cycle:** 37d.[346] **Gestation:** 128d.[180] **Age 1st birth:** NA. **Birth interval:** NA.

Life span: NA. **Birth season:** Mid-Oct.[580]

Locomotion Quadrupedal.

Social Structure Variable. 1 male−1 female family groups with offspring[464] and larger groups.[580] Home ranges overlap.[580] **Group size:** 3−4.[580] **Home range:** 0.5−1ha.[898] **Day range:** 610m *[4826ft].*[903]

Behavior Cathemeral and arboreal. Mongoose lemurs have a variable activity period; some times of the year they are nocturnal.[464] They are important pollinators.[580] **Association:** This species sometimes travels with brown lemurs *(E. fulvus fulvus).*[580] **Scent marking:** Encounters between different groups of *E. mongoz* involve vocalization and excited scent marking. **Vocalizations:** These lemurs have a distinct contact call and response from infant to mother.[662] The alarm call is interspersed with grunts somewhat like those of a pig;[662] the whole group may alarm-call in unison.[662] During intergroup encounters, they vocalize and scent-mark.[580]

Madagascar

Female and infant.

D. HARING

Red-bellied Lemur *Eulemur rubriventer*

Taxonomy Monotypic species.[580]

Distinguishing Characteristics Red-bellied lemurs have a dark chestnut brown body and a black tail. Males have white patches below the eyes. Females have whitish underparts and reduced white patches.

Physical Characteristics Head and body length: 400mm *[15.7in].*[580] **Tail length:** 500mm *[19.7in].*[580] **Weight:** ♀ 2139g *[75.4oz]*, ♂ 2267g *[80.0oz].*[436] **Intermembral index:** 68.[234] **Adult brain weight:** 27.2g *[19.7oz].*[346]

Habitat Primary rain forest at medium and higher altitudes.[580]

Diet Fruit, leaves, animal matter, flowers.[646] A total of 67 plant species are used as food,[580] 10 of which make up 50% of the diet.[646]

Life History Weaning: 4–5mo. **Sexual maturity:** NA. **Estrus cycle:** NA. **Gestation:** 123d.[438] **Age 1st birth:** NA. **Birth interval:** NA. **Life span:** NA. Birth season: Sep–Oct.[580] Both genders carry infants for equal amounts of time for the first 35 days; then only the male carries the infant until it is 100 days old.[580] Infant mortality is 50%.[580]

Locomotion Quadrupedal; some vertical clinging and leaping.

Social Structure 1 male–1 female family groups.[419] **Group size:** 3 (2–5).[898] **Home range:** 19ha.[645] **Day range:** NA.

Behavior Cathemeral and arboreal. Red-bellied lemurs feed on fewer plant species than brown lemurs *(E. fulvus rufus)* do, and they lick nectar rather than eat whole flowers. In order to detoxify large millipedes, red-bellied lemurs roll them between their hands and salivate on them.[646]

Madagascar

R. A. MITTERMEIER (NATURAL HABITAT)

The male is distinguished from the female by the white patches under his eyes.

Ruffed Lemur *Varecia variegata*

Taxonomy 2 very distinctive subspecies.[580]

Distinguishing Characteristics Ruffed lemurs get their name from the long, thick hair on their ears.[580] The black-and-white form *(V. v. variegata)* has variable amounts of black and white, depending on where it is found; in the south it is whiter, and in the north, blacker. The red form *(V. v. rubra)* is mostly red, with a black crown and a white nape.[580]

Physical Characteristics Head and body length: 500mm *[19.7in].*[585] **Tail length:** 600mm *[23.6in].*[585] **Weight:** ♀ 3512g *[123.9oz]*, ♂ 3471g *[122.4oz].*[436] **Intermembral index:** 72.[234] **Adult brain weight:** 34.2g *[1.2oz].*[346]

Habitat Primary rain forest[898] up to 1200m *[3937ft].*[580] Ruffed lemurs are found only in undisturbed primary forest. They are one of the first species to disappear after selective logging, probably because they eat large fruits from large trees, which are the first to be cut.[948]

Diet Fruit, seeds, leaves, nectar.[580] Ruffed lemurs pass seeds in 2–3 hours.[167]

Life History Infant: 30d.[346] **Weaning:** 90d.[346]

The black-and-white ruffed lemur *(V. v. variegata)* has a very loud alarm bark, which is given by all the members of the group.

Ruffed Lemur *continued*

The red ruffed lemur *(V. v. rubra)* lives only on the Masoala Peninsula in northeastern Madagascar.

Juvenile: 4.5–20mo. **Sexual maturity:** 20mo. **Estrus cycle:** 30d.[346] **Gestation:** 90–102d.[580] **Age 1st birth:** NA. **Birth interval:** 12mo.[692] **Life span:** 19y. Mating season: May–Jul.[580] Birth season: Sep–Oct.[580] Offspring: Usually 2.[580] Ruffed lemurs are the largest lemur to have 3 pairs of nipples and multiple infants. Neonates are parked in nests for the first week and, rather than clinging, are carried in the mother's mouth.[585]

Madagascar

Locomotion Quadrupedal and suspensory. Ruffed lemurs often hang upside down by their feet to feed.[510]

Social Structure Variable. 1 male–1 female, multimale-multifemale, and communities.[912] Ruffed lemurs are territorial; intertroop aggression has been reported.[580] **Group size:** 5–16,[580] up to 32.[912] **Home range:** ♂ 17.1ha, ♀ 22.1ha.[912] **Day range:** 400–2300m *[1312–7546ft].*[912]

Behavior Diurnal and arboreal. Ruffed lemurs pursue sitting birds of prey[510] and confront carnivores on the ground. This response may be to distract predators from a nest with young.[510] Females form the core of the group and defend territory. During the cooler winter, ruffed lemurs travel less, sun themselves, and feed more.[585]

Vocalizations: 6 relating to predation, including an all-clear regrouping call.[510] The raucous loud call is a roaring bark given by all group members when mobbing an intruder.[580]

The ruffed lemur often suspends itself by its hind limbs to feed.

Golden Bamboo Lemur *Hapalemur aureus*

Taxonomy Disputed. Monotypic species discovered in 1986.[580] Some taxonomists place this genus in the same family with *Lepilemur.*[820]

Distinguishing Characteristics Golden bamboo lemurs have a reddish to brown back, a dark face surrounded by golden cheek hair,[580] and a pink nose. They have brachial glands on the inner forearm but no ear tufts.[548]

Physical Characteristics Head and body length: ♀ 395mm *[15.6in]*,[548] ♂ 340mm *[13.4in]*.[580] **Tail length:** ♀ 410mm *[16.1in]*,[580] ♂ 370mm *[14.6in]*.[548] **Weight:** ♀ 1500g *[52.9oz]*, ♂ 1660g *[58.5oz]*.[293] **Intermembral index:** NA. **Adult brain weight:** NA.

Habitat Evergreen and bamboo forest.[580]

Diet Mature bamboo, 85%;[915] new shoots of bamboo.[580] Golden bamboo lemurs eat the growing shoots of one type of bamboo (*Cephalostachyam viguieri*), which contains "15 mg of cyanide per 100g fresh weight bamboo."[293] Individuals eat 500 grams *[17.6oz]* of bamboo per day, which should contain 12 times the lethal dose of cyanide. How these lemurs avoid the effects of cyanide poisoning is unknown.[293]

Life History Weaning: 4–5mo. **Sexual maturity:** NA. **Estrus cycle:** NA. **Gestation:** NA. **Age 1st birth:** NA. **Birth interval:** 12mo.[915] **Life span:** NA. **Birth season:** Nov–Dec.[912]

Locomotion Vertical clinging and leaping.[580]

Social Structure 1 male–1 female family groups.[915] **Group size:** 2–4.[912] **Home range:** 80ha.[580] **Day range:** 400m *[1312ft]*.[580]

Behavior Diurnal or cathemeral; arboreal.[580] **Association:** Sympatric with lesser and greater bamboo lemurs (*H. griseus, H. simus*) in Ranomafana National Park. All 3 species eat bamboo but select different plant parts because of different chemical compounds in the parts.[293] **Vocalizations:** The contact call is a grunt. The loud call, a raucous ascending cry, is usually given at daybreak by the male or the male and female.[548]

Madagascar

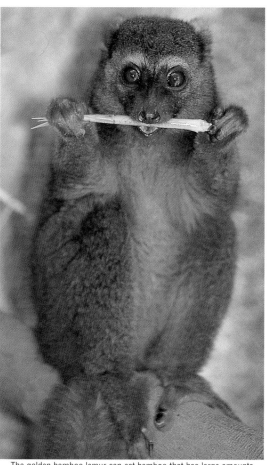

The golden bamboo lemur can eat bamboo that has large amounts of naturally occurring cyanide.

Lesser Bamboo Lemur *Hapalemur griseus*

Taxonomy Disputed. 3 subspecies. *H. g. alaotrensis* may be a separate species.[315]

Distinguishing Characteristics Lesser bamboo lemurs have a gray body, a variable amount of red on the head,[580] and no ear tufts.[436]

Physical Characteristics Head and body length: 280mm *[11.0in]*.[580] **Tail length:** 304–406mm *[12.0–16.0in]*. **Weight:** ♀ 892g *[31.5oz]*, ♂ 939g *[33.1oz]*.[436] **Intermembral index:** 64.4.[425] **Adult brain weight:** 14.7g *[0.5oz]*.[346]

Habitat Dense bamboo in tropical rain forest.

Diet Bamboo shoots, stems, and leaves, 90%. Soil is reportedly eaten every day.[905] The Lake Alaotra bamboo lemur, or bandro (*H. g. alaotrensis*), eats 10 species of plants, including the shoots and leaves of *Phragmites communis*.[216]

Life History Weaning: 4mo.[910] **Sexual maturity:** 24mo.[346] **Estrus cycle:** NA. **Gestation:** 137d[180] (135[910]–140[580]). **Age 1st birth:** 28.6mo.[346] **Birth interval:** NA. **Life span:** NA. **Birth season:** Sep.[912] **Offspring:** 1–2. Lesser bamboo lemurs carry their young by mouth and park them for short periods.[580] In captive studies, the male carries the infant up to 20% of the time.[580]

Locomotion Vertical clinging and leaping.[580]

Social Structure 1 male–1 female family groups.[912] **Group size:** 3–6.[464] **Home range:** 6–10ha, up to 15ha.[580] **Day range:** 425m *[1394ft]*.[905]

Behavior Diurnal or crepuscular; arboreal.[580] Lesser bamboo lemurs spend 48% of the day feeding. Bandros (*H. g. alaotrensis*) live only in the papyrus and

Lesser Bamboo Lemur *continued*

reed beds by Lake Alaotra, in territorial groups.[216] They are critically endangered[577] because of habitat destruction and capture for food and for pets. As pets they live an average of only 2 months. **Association:** Sympatric with golden and greater bamboo lemurs *(H. aureus, H. simus)* in Ranomafana National Park. **Vocalizations:** A mating call is given by both males and females.[662] **Sleeping site:** Emergent trees. The group sleeps in close contact.

Madagascar

T. MUTSCHLER (NATURAL HABITAT)

The bandro *(H. g. alaotrensis)* is critically endangered.

Greater Bamboo Lemur *Hapalemur simus*

ENDANGERED

USESA

This greater bamboo lemur is eating the stem of a giant bamboo.

Adult brain weight: NA.

Habitat Dense bamboo in primary rain forest.[898]

Diet Stems and soft pith of giant bamboo, 98%; flowers, fruits, leaves.[580] Bamboo stems and soft pith have no toxins but are high in fiber and low in protein.[915]

Life History NA. Birth season: Nov.[912]

Locomotion Vertical clinging and leaping.[580]

Social Structure Variable. 1 male–multifemale, multimale-multifemale. **Group size:** 4–12,[580] up to 30.[766] **Home range:** 100ha.[915] **Day range:** NA.

Behavior Cathemeral and arboreal. Greater bamboo lemurs are reported to be active mostly at night.[580] Their population numbers less than 1000.[315]

Association: Sympatric with golden and lesser bamboo lemurs *(H. aureus, H. griseus)* in Ranomafana National Park. **Vocalizations:** The loud call of these lemurs is a whistle.[548] They have a raucous ascending call that begins with a motorlike purr.[912]

Madagascar

NATURAL HABITAT

NATURAL HABITAT

Taxonomy Monotypic species.[580]

Distinguishing Characteristics

Greater bamboo lemurs have a reddish gray body,[412] a broad blunt face, and white ear tufts. They have brachial glands near the elbow[580] and males have glands on the side of the neck.[412] The spleen is Y-shaped, which is more like that of monotremes and marsupials than of other prosimians.[412]

Physical Characteristics Head and body length: 450mm *[17.7in]*.[412]

Tail length: 439mm *[17.3in]*.[912]

Weight: ♀ 2150g *[75.8oz]*, ♂ 2400g *[84.6oz]*.[580]

Intermembral index: 65.[234]

46

Woolly Lemur *Avahi laniger*

Taxonomy Disputed. 2 subspecies, which have been recognized by some as 2 species *(A. laniger, A. occidentalis)*.[580]

Distinguishing Characteristics Woolly lemurs have a gray brown to reddish body, with white on the back of the thighs.[580] The tail is reddish. The face is brownish, with variable amounts of white on the throat and cheeks and around the eyes. The ears are small and hidden.[898]

Physical Characteristics Head and body length: ♀ 292mm *[11.5in]*, ♂ 265–270mm *[10.4–10.6in]*.[436] **Tail length:** 330–370mm *[13.0–14.6in]*.[436] **Weight:** ♀ 1316g *[46.4oz]*, ♂ 1033g *[36.4oz]*.[580] **Intermembral index:** 57.5.[425] **Adult brain weight:** 10g *[0.4oz]*.[464]

Habitat Seasonal dry forest, rain forest.[898]

Diet Leaves (19 species), flowers, fruit, bark. One species *(Haronga madagascarensis)* accounts for 42% of leaves eaten.[436]

Life History Weaning: 150d.[346] **Sexual maturity:** 12mo.[436] **Estrus cycle:** NA. **Gestation:** 120–150d. **Age 1st birth:** NA. **Birth interval:** NA. **Life span:** NA. Birth season: Aug–Sep.[436] No care of infants by males has been reported.[909] Infants are carried on the female's belly for 1 week and on the back for 3–4 months.[909]

Locomotion Vertical clinging and leaping; some suspensory.[580]

Social Structure Monogamous, 1 male–1 female pairs. Both members of the pair defend a common territory.[436] **Group size:** 2–5.[580] **Home range:** 1.4ha.[436] **Night range:** 300–621m *[984–2038ft]*.[464]

Behavior Nocturnal and arboreal. The male and female stay in

Madagascar

close contact, resting or grooming, 40% of the night.[334] Activity budget: Resting, 59.5%; feeding, 22%; traveling, 13.5%; grooming, 5%.[334]

Vocalizations: The distant communication call is a prolonged and

Woolly lemurs are monogamous and huddle close together like these two.

very high-pitched whistle that is often answered by an individual from another territory. The "alarm call of this species sounds like its Malagasy name 'Aha Hy.' "[662]

Sleeping site: Pairs sleep huddled together at a height of 3 meters *[10ft]* and often use the same tree for days.[334] They leave their sleeping tree at about 5:45 p.m. and return between 5:05 and 5:40 a.m.[334]

Diademed Sifaka *Propithecus diadema*

Diademed sifaka *(P. d. diadema)*.

Taxonomy 4 subspecies.[580]

Distinguishing Characteristics The color of diademed sifakas varies with subspecies, from the all-black *P. d. perrieri* to the almost all-white *P. d. candidus*. *P. d. edwardsi* is black and white, and *P. d. diadema* is white, gray, and golden.[580]

Physical Characteristics Head and body length: 480–520mm *[18.9–20.5in]*.[580] **Tail length:** 465mm *[18.3in]*.[580] **Weight:** 5633[436]–7250g[580] *[198.7–255.7oz]*. **Intermembral index:** 64.[234] **Adult brain weight:** 37g *[1.3oz]*.[346]

Habitat Humid rain forest to dry seasonal forest, depending on the subspecies.[898]

Diet Young leaves, 26%; seeds, 25%; fruit, 20%; flower stamens and petioles, 3%. Vine leaves make up 15% of the diet. Fruits are eaten when ripe.[554] Up to 27 species of plants are eaten per day.[580]

Life History Weaning: 180d.[554] **Sexual maturity:** ♀ 48mo,[905] ♂ 60mo.[913] **Estrus cycle:** NA. **Gestation:** 179d.[913] **Age 1st birth:** 48mo.[554] **Birth interval:** 24mo.[913] **Life span:** 20y.[913] Mating season: Nov–Jan.[554] Birth season: May–Jul. Estrus is triggered by photoperiodicity.[913] Females are receptive for only 1 day. Infant mortality before 1 year old is 56%.[554]

Locomotion Vertical clinging and leaping; some suspensory. Sifakas often suspend themselves by their hind feet while foraging.[580]

Social Structure Multimale-multifemale[419] groups with 1–3 females and 1-2 males. **Emigration:** Males emigrate at age 5 to a neighboring group.[436] Females may emigrate or remain in their natal troop.[436] Exclusive territories are maintained for years, but little territorial aggression has been observed.[554] **Group size:** 3–9.[913] **Home range:** 20–30ha.[580] **Day range:** 1000m *[3281ft]*.[580]

Behavior Diurnal and arboreal. Diademed sifaka males rarely carry, play with, or groom infants.[308] Three

47

Diademed Sifaka *continued*

infanticides by 2 different males have been observed.[913] Resident males do not defend the infants in their group.[913] The fossa *(Cryptoprocta ferox)* is the main predator upon Milne-Edwards' sifakas *(P. d. edwardsi).*[913] **Mating:** Diademed sifakas mate several times during their short estrus period.[913] **Vocalizations:** Diademed sifakas have a *roo-han* contact call, and their alarm call is a grunting *on-ee.*[662] If surprised, they make a *vouiff,* which is thought to be an alarm and cohesion call.[662] **Sleeping site:** The breeding male may sleep with the females and offspring or with the other male.[913]

Madagascar

Milne-Edwards' sifaka *(P. d. edwardsi).*

Golden-crowned Sifaka *Propithecus tattersalli*

D. HARING

The golden-crowned sifaka was discovered in 1988.

Taxonomy Monotypic species.[580] First discovered and described in 1988.[759]

Distinguishing Characteristics Golden-crowned sifakas have a mostly white body, with a golden crown between furry, tufted ears.[759]

Physical Characteristics Head and body length: ♀ 455mm *[17.9in],*[759] ♂ 500mm *[19.7in].*[436] **Tail length:** ♀ 449mm *[17.7in],*[759] ♂ 400mm *[15.7in].*[436] **Weight:** 3300g (2100–3800) *[116.4oz (74.1–134.0)].*[759] **Intermembral index:** NA. **Adult brain weight:** NA.

Habitat Rain forest and deciduous dry forest.[580]

Diet Fruit, 37%; young leaves, 22%; mature leaves, 17%; fruit pulp, 9%.[554] As with many primate species, young leaves are preferred over mature leaves, because the former are higher in protein.[554]

Life History Weaning: 5[580]–6mo.[554] **Sexual maturity:** NA. **Estrus cycle:** NA. **Gestation:** 176d.[554] **Age 1st birth:** NA. **Birth interval:** NA. **Life span:** NA. **Mating season:** Late Jan. **Birth season:** Late Jun–

Jul.[580] Infants are weaned in December.[580]

Locomotion Vertical clinging and leaping.

Social Structure Variable. 1 male–1 female, multimale-multifemale.[580] These sifakas usually live in small groups with 2 adults of each sex, but only 1 female breeds.[419] **Emigration:** During mating season, males may change groups.[580] **Group size:** 5 (3–10).[580] **Home range:** 9–12ha.[580] **Day range:** 400–1200m *[1312–3937ft].*[580]

Behavior Diurnal and arboreal. During the dry season, golden-crowned sifakas are forced to feed on less nutritious mature foliage and conserve energy by ranging less far and interacting less with other group members.[553] Infants are weaned during the wet season, when nutritious foods are most abundant.[554] **Vocalizations:** A loud *sifak* alarm call is given when ground predators such as fossas *(Cryptoprocta ferox),* boa constrictors, or humans are sighted.[580] This species also gives "churrs and whinnies" when disturbed.[759] **Sleeping site:** High in emergent trees.[580]

Madagascar

Verreaux's Sifaka *Propithecus verreauxi*

Sifakas leap from one vertical support to another.

Taxonomy 4 subspecies.

Distinguishing Characteristics
Verreaux's sifakas generally have a white body, with a brown or black crown. Some subspecies have brown or black forearms and thighs.[580]

Physical Characteristics Head and body length: 425–450mm *[16.7–17.7in]*.[580] **Tail length:** 560–600mm *[22.0–23.6in]*.[580] **Weight:** ♀ 3480g *[122.7oz]*,[180] ♂ 3637g *[128.3oz]*.[436] **Intermembral index:** 59.[234] **Adult brain weight:** 27.5g *[1.0oz]*.[346]

Habitat Galley and evergreen forest to dry deciduous forest to spiny desert.[898]

Diet Leaves, fruit, flowers, bark.[370] Verreaux's sifakas consume up to 102 species, which represent 49–91% of the plant species available in their habitat.[937]

Life History Weaning: 180d.[346] **Sexual maturity:** 30–42mo.[692] **Estrus cycle:** 10–21d.[689] **Gestation:** 130[689]–155d.[438] **Age 1st birth:** NA. **Birth interval:** 12–24mo.[692] **Life span:** NA. Mating season: Jan–Mar.[848] Birth season: Aug–Sep.[689] Females ovulate only once and are receptive for only 12–36 hours.[689] Infants are carried for 195 days.[346] They cling to the abdomen of the female for 6–12 weeks and then ride on her back.[848]

Locomotion Vertical clinging and leaping.[580]

Social Structure Variable. 1 male–1 female family groups or multimale-multifemale groups.[689] These sifakas are territorial, but only 24–51% of their territory is used exclusively.[937] They often defend food sources rather than strict boundaries.[937] **Group size:** 2–12.[580] **Home range:** 2–8.5ha.[692] **Day range:** 550–1100m *[1805–3609ft]*.[937]

Behavior Diurnal and arboreal. Females are dominant over males, but no submissive behavior has been reported. Males actively maintain proximity to females.[472] In the wet season, sifakas spend more time feeding and range farther.[937] Males fight each other for dominance only during mating season.[689]

Mating: The female presents to the male by rolling up her tail. The male mounts 1–6 times (24–53 thrusts).[848] Females mate only with the male that is dominant during mating season.[689]

Vocalizations: The name "sifaka" comes from the very noisy guttural barking call that all troop members give at once during territorial confrontations.[580]

Madagascar

ENDANGERED
USESA

The male (right) is scent-marking the branch with his throat gland.

Sifakas hop bipedally on the ground.

Indri *Indri indri*

The indri uses its long legs for leaping.

Taxonomy Monotypic genus and species. The Malagasy common name is "Babakoto," meaning "little father" or "man of the forest."[580]

Distinguishing Characteristics Indris are the largest living lemur. Their body has varying amounts of black and white, and they have a very short tail.[580]

Physical Characteristics Head and body length: 600mm *[23.6in]*.[580] **Tail length:** 50mm *[2.0in]*.[580] **Weight:** ♀ 7135g *[251.7oz]*,[939] ♂ 5825g *[205.4oz]*.[939] **Intermembral index:** 64.3.[425] **Adult brain weight:** 34.5g *[1.2oz]*.[346]

Habitat Humid montane forest[898] up to 1300m *[4265ft]*. Indris use all levels of forest from 2 to 40 meters *[7–131ft]*. In October–December they stay at lower levels to avoid horseflies.[830]

Diet Young leaves, 50–75%; fruit, 25%; unripe seeds, 10%; buds;[370] occasionally soil.[670] Indris eat 62 species of plants, with 50% of their feeding on 5 species.[670]

Life History Weaning: 180[670]–365d.[346] **Sexual maturity:** 48–84mo.[692] **Estrus cycle:** NA. **Gestation:** 120–150d.[580] **Age 1st birth:** 84–108mo.[580] **Birth interval:** 24–36mo.[670] **Life span:** NA. Birth

season: May.[580] Offspring: 1. Infants are carried for 225 days,[670] ventrally for the first 3–4 weeks and then on the back.[830]

Locomotion Vertical clinging and leaping.

Social Structure Monogamous, 1 male–1 female family groups. Females are dominant.[580] Males actively defend territories,[670] which for the 4 years of one study were maintained unchanged.[670] **Group size:** 2–6.[580] **Home range:** 18ha.[580] **Day range:** 300–700m *[984–2296ft]*.[580]

Behavior Diurnal and arboreal. Females displace males and feed in the higher levels of trees.[670] **Mating:** Indris mate in a face-to-face (ventroventral) position while hanging under a branch.[830] **Vocalizations:** The

Indris are the largest living lemurs and have the shortest tails.

best-known vocalization is the loud call, which carries over 1km *[3281ft]*,[580] lasts several minutes, and is heard only between 8 and 11:30 am.[830] Described "as a very resonant plaintive howl,"[662] it is a territorial call performed by the whole group[662] and transmits information about the group's location and composition.[662] Indris produce a "honk" as a contact call if one indri is separated from the group and an alarm call if a terrestrial predator is sighted.[662]

Madagascar

Aye-aye *Daubentonia madagascariensis*

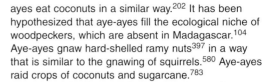

Note the second aye-aye.

Taxonomy Aye-ayes are the only primate with a monotypic family, genus, and species. They are thought to be distantly related to the family Indridae.[580]

Distinguishing Characteristics The aye-aye's body has coarse black hair with white tips and shorter soft off-white hair. The face is pale, with dark eye rings, black ears,[436] and a pink nose. The tail is long and bushy.[580] Aye-ayes are the largest nocturnal primate.[580]

Physical Characteristics Head and body length: 400mm *[15.7in]*.[580] **Tail length:** 400mm *[15.7in]*.[580] **Weight:** ♀ 2573g *[90.7oz]*, ♂ 2800g *[98.8oz]*.[436] **Intermembral index:** 71.[234] **Adult brain weight:** 45.2g *[1.6oz]*.[346] **Baculum length:** 28mm *[1.1in]*.[436] Aye-ayes have the largest brain-to-body-weight ratio of any prosimian.[104]

Habitat Primary forest, secondary forest, spiny desert, rain forest, dry forest, tree plantations.[898]

Diet Animal matter, seeds, fruit, nectar,[202] fungus.[785] Aye-ayes find insect larvae, especially cerambycid beetles,[202] in trees by tapping the wood with their middle finger and listening for hollow cavities. They gnaw the wood with their rodentlike incisors. The larvae are killed and the juice extracted by the elongated middle finger, which is then licked. Aye-ayes eat coconuts in a similar way.[202] It has been hypothesized that aye-ayes fill the ecological niche of woodpeckers, which are absent in Madagascar.[104] Aye-ayes gnaw hard-shelled ramy nuts[397] in a way that is similar to the gnawing of squirrels.[580] Aye-ayes raid crops of coconuts and sugarcane.[783]

Life History Weaning: NA. **Sexual maturity:** NA. **Estrus cycle:** NA. **Gestation:** 172d.[912] **Age 1st birth:** 36–48mo.[783] **Birth interval:** 24–36mo.[783] **Life span:** NA. Mating season: Year-round, but females may not cycle all year.[785] Offspring: 1. Females have estrous swellings and a 3-day period of receptivity. The infant is dependent for 2 years.[785]

Locomotion Quadrupedal.[784]

Social Structure Solitary foraging; complex social interactions.[783] The male's home range is 3–6 times larger than the female's and overlaps the female's by 40–75%.[785] Female ranges rarely overlap.[785]

Note the elongated finger.

Group size: 1–2.[785] **Home range:** ♀ 31.7–39.5ha, ♂ 126–214ha.[785] **Night range:** 1500[580]–4391.5m[785] *[4922–14,409ft]*.

Behavior Nocturnal and arboreal. Aye-ayes have not been observed to groom each other.[662] They have a wider distribution than was published in the 1970s, but they occur at low densities. Local people, who believe that aye-ayes are a bad omen, still kill them on sight.[783] **Mating:** The female calls repeatedly during her 3-day receptive period. She is often surrounded by up to 6 males. Aye-ayes mate for 1 hour while hanging upside down from a branch.[785] **Vocalizations:** The contact call is a prolonged resonant *creeee*.[662] When illuminated in the forest at night, aye-ayes make "repeated grunting sounds" of alarm. They make a "powerful hissing noise" if disturbed in their nest during the day.[662] **Sleeping site:** During the day, nests hidden in vine tangles high in trees. In one study, up to 100 nests were used by 8 adults, and some nests were used by different individuals on succeeding days.[580]

Madagascar

Tarsiers

Tarsiers share some characteristics of both the prosimians and the anthropoids, while maintaining characteristics unique to themselves. Taxonomists have classified them as intermediate between both groups and have assigned them to their own infraorder, which has just one living genus—*Tarsius.* Their relatives in the fossil record are found going back to the Eocene epoch, from 54 to 36 million years ago.[234]

Tarsiers are small, weighing only 113–142 grams (4–5 ounces). Like many prosimians, they are nocturnal and have grooming claws and a bicornuate uterus. Like anthropoids, they do not have a tapetum (a reflective layer in their eyes), and their eye sockets have postorbital closure rather than the postorbital bar of prosimians. In tarsiers the internal structures of the nose and ears and the blood supply to the brain and to a developing fetus are more like those of monkeys than of lorises. The monthly sexual swellings of female tarsiers are also similar to those in anthropoids.[234]

Unique among primates, tarsiers have only 2, rather than 4, incisors in their lower jaw. Their dental formula is

$$\frac{2.1.3.3.}{1.1.3.3.} \times 2 = 34.$$

Tarsiers are named for their special elongated tarsal bones, which form their ankles and enable them to leap 3 meters (almost 10 feet) from tree to tree.[234] They have a long, partly hairless tail that arcs over their back when they hop on the ground.

The eyes are the tarsiers' most notable features. Each eye is bigger than the entire brain. Tarsiers can rotate their head almost 180 degrees in each direction, like owls. All tarsiers hunt at night, exclusively for animal prey. Their diet includes primarily insects such as cockroaches and crickets and sometimes reptiles, birds, and bats. One tarsier species is found in the Philippines. Four species are currently recognized in Indonesia and another *(Tarsius sangirensis)* has been proposed, based on recently gathered field data on vocalizations, measurements, and genetics.[955]

Western Tarsier *Tarsius bancanus*

Taxonomy 2 subspecies. The male genital morphology suggests that the subspecies may be recognized as a species after further study.[938]

Distinguishing Characteristics The color of western tarsiers varies with the subspecies. They are ivory yellow in Sumatra and golden orange or rusty brown in Borneo.[654] The last 25% of the tail has a tuft of long hair (64mm *[2.5in]*).[654]

Physical Characteristics Head and body length: 128.8mm *[5.1in].*[654] **Tail length:** 217mm *[8.5in].*[654] **Weight:** ♀ 111g (100–119) *[3.9oz (3.5–4.2)]*, ♂ 120g (108–131) *[4.2oz (3.8–4.6)].*[606] **Intermembral index:** 52.[234] **Adult brain weight:** NA.

Habitat Secondary growth, forest edges.[618]

Diet Animal prey, 100%. Prey includes beetles, 35%; ants, 21%; locusts, 16%; cicadas, 10%; cockroaches, 8%; vertebrates (birds, bats, and snakes), 11%;[41] mantids, moths. Western tarsiers have been observed to eat the following birds: warblers, spider hunters, kingfishers, pittas.[616] One individual is reported to have eaten a neurotoxic snake.[616]

Life History Weaning: 60–90d.[606] **Subadult:** 20mo.[170] **Sexual maturity:** 13mo.[41] **Estrus cycle:** 20.8–27.2d.[914] **Gestation:** 178d.[695] **Age 1st birth:** 24mo.[170] **Birth interval:** NA. **Life span:** 8–12y.[41] Birth seasons: 2 in captivity, Apr–May and Nov–Dec.[170] Offspring: 1.[170] Females show an estrous swelling for 6–9 days.[914] The fetal growth rate is one of the slowest for all mammals.[695] Infants weigh 24–25 grams *[0.8–0.9oz]*,[170] or 20% of the mother's weight.[695] They eat their first solid food within 1 week of birth and catch their first insect at 4 weeks.[616]

Locomotion Vertical clinging and leaping, >65%;[938] climbing, 27.6%; walking, 3.2%; hopping, 2.3%.[144] Western tarsiers can leap noiselessly more than 1000 times per night.[938]

Social Structure Variable. Solitary[144] and 1 male–1 female

pairs.[41] A female may forage with a juvenile but is otherwise reported to be solitary.[144] Males travel farther than females and forage alone.[144] One male was found in the same home range after 7 years.[616] Males and females share territorial marking but are rarely in contact with each other.[41] **Group size:** 1–3.[144] **Home range:** 0.9–1.6ha,[616] 4.5–11.2ha.[144] **Night range:** 1800m *[5905ft].*[144]

Behavior Nocturnal and arboreal. Western tarsiers forage low in trees (0–2.8m *[0–9.2ft]*), using small-diameter vertical tree trunks (20–40mm *[0.8–1.6in]*).[144] If scared, they retreat higher up in the trees (to 4m *[13ft]* or above).[616] They can jump 2–5.6 meters *[7–18ft]* from tree to tree, or 45 times their head and body length.[616] They can rotate their heads 180° like an owl. Tarsiers ambush their prey by moving noiselessly[616] and jumping to the ground to catch it. Most prey is caught on the ground, but 4% is flying insects caught by "cantilevering" (springing out and grabbing insects with the hands while holding on to a support with the feet).[144] Tarsiers wake up at sunset and rest for 10–20 minutes before starting to leap. They rest often between leaping and feeding.[616] Activity budget: Foraging, 60%; travel, 27%; rest, 11%; grooming, 1%.[941] **Mating:** The male courtship call is given only when a female is in estrus.[436] The female solicits copulation once or twice a night during the 1-to-3-day peak of estrus.[436] Before mating, the male chases the female, and both "display by suspending themselves by their hands and spreading their legs."[848] Copulation lasts 60–90 seconds.[436] **Scent marking:** Marks

D. HARING

Tarsiers are faunivorous. They eat only animal prey. Note the grooming claw on the third toe.

are made with urine, which has a strong smell that humans can detect.[616] Both males and females scent-mark with epigastric glands.[616] Females rub their genitals on tree trunks, especially when in estrus.[616] **Vocalizations:** 7 types, some of which are beyond the range of human hearing and have been distinguished via sound recordings.[616] Western tarsiers call most often in the early morning.[144] **Sleeping site:** Usually a branch with foliage, found before sunrise.[41] The male sleeps alone; the female, with her offspring.[41]

Indonesia, Sumatra, Borneo

Dian's Tarsier *Tarsius dianae*

Taxonomy Disputed. Dian's tarsiers, first discovered in 1991,[618] were named for Dian Fossey and Diana, the goddess of hunting.

Distinguishing Characteristics Dian's tarsiers have a grayish buff body, a black spot on both sides of the nose, and a naked field at the base of the ears. The fingernails are keeled and pointed, and the eyes are more symmetrical than in other tarsiers.[618]

Physical Characteristics Head and body length: ♀ 115–121mm *[4.5–4.8in]*, ♂ 113–121mm *[4.4–4.8in]*.[941] **Tail length:** ♀ 228–231mm *[9.0–9.1in]*, ♂ 210–235mm *[8.3–9.3in]*.[941] **Weight:** ♀ 105–110g *[3.7–3.9oz]*, ♂ 75–104g *[2.6–3.7oz]*.[941] **Intermembral index:** NA. **Adult brain weight:** NA. Chromosomal diploid number: 46.[618]

Habitat Primary forest.[618]

Diet Animal prey (moths, crickets, lizards), 100%.[941]

Life History NA.

Locomotion Leaping, 58%; climbing, 22%; walking, 11%; hopping, 6%; canting, 4%.[941]

Social Structure NA. **Group size:** NA. **Home range:** 0.5–0.8ha.[941] **Night range:** NA.

Behavior Nocturnal and arboreal. Dian's tarsier spends most of its time foraging between 1.6 and 3 meters *[5–10ft]*.[941] Activity budget: Foraging, 44%; travel, 28%; rest, 21%; grooming, 7%.[941] **Scent marking:** Marks are made with urine.[941] **Vocalizations:** Males and females duet at night and at dawn. The male call is distinct from the female's.[618] **Sleeping site:** Dense foliage and tree cavities.[941]

B. WALTON BBC NATURAL HISTORY UNIT

Dian's tarsier was first described in 1991.

Sulawesi
(Indonesia)

Pygmy Tarsier *Tarsius pumilus*

Taxonomy Disputed. Elevated from a subspecies in 1985, based on museum specimens.[593]

Distinguishing Characteristics Pygmy tarsiers have a gray or brown body and a red face.[130] Behind the small ears is a buff-colored spot of fur.[617] The fingers have small terminal pads, and the nails are laterally compressed into points.

Physical Characteristics Head and body length: 95–98mm *[3.7–3.9in]*.[593] **Tail length:** 203–205mm *[8.0–8.1in]*.[593] **Weight:** NA. **Intermembral index:** NA. **Adult brain weight:** NA. Pygmy tarsiers are 75% the size of other tarsiers.[593]

Habitat Montane and cloud forest up to 2200 meters *[7218ft]* above sea level. The forest has trees covered with liverworts and mosses.[593]

Diet Animal prey.[593]

Life History NA.

Locomotion NA.

Social Structure NA. **Group size:** NA. **Home range:** NA. **Night range:** NA.

Behavior Assumed to be nocturnal and arboreal like other tarsiers. No studies of this species have been undertaken. Microwear of the lower incisors indicates that pygmy tarsiers may groom their long hair with their mouth.[593]

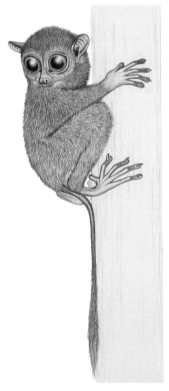

S. NASH

Sulawesi
(Indonesia)

Spectral Tarsier *Tarsius spectrum*

This Sangihe Island tarsier *(T. s. sangirensis)* was proposed as a valid species in 1996.[955]

Spectral tarsier *(T. s. spectrum).*

fer to hunt below 9 meters *[30ft]*[320] and often hop to the ground to catch prey.[938] They travel at heights of 1–3 meters *[3–10ft]* and up to 20 meters *[66ft]*.[938] A group forages together and is often in visual and tactile contact.[41] Subadult siblings stay in closer contact with the infant than do other members, including the mother.[320] Infants are reportedly clumsy and likely to fall 2–5 meters *[7–16ft]*.[320] **Scent marking:** Marks are made at a height of 1 meter *[3ft]* or above.[938] **Vocalizations:** 12. Each gender has a specific dawn call.[320] A territorial vocal chorus is given by the entire group usually just before dawn.[513] **Sleeping site:** Trees or tangles;[619] 1 main site and 2 alternates.[321] Males, females, and young sleep together.[619]

Sulawesi
(Indonesia)

Taxonomy Disputed. 4 subspecies.[313] *T. s. sangirensis* may be a separate species.[315]

Distinguishing Characteristics Spectral tarsiers have a white spot behind each ear and long hair on the last third to half of the tail.[130] The tail tuft is 124mm *[4.9in]* long.[938] Spectral tarsiers are believed to be the least specialized of all tarsiers.[616]

Physical Characteristics Head and body length: 117mm (112–151) *[4.6in (4.4–5.9)]*.[942] **Tail length:** 241mm *[9.5in]*.[942] **Weight:** ♀ 108g (94–114) *[3.8oz (3.3–4.0)]*, ♂ 126g (122–132) *[4.4oz (4.3–4.7)]*.[322] **Intermembral index:** NA. **Adult brain weight:** 3.8g *[0.1oz]*.[346]

Habitat Primary and secondary tropical forest, thorn scrub, coastal mangrove glades, montane forest[41] up to 1300m *[4265ft]*.[619]

Diet Insects, 100%. Vertebrate prey has not been reported.[41]

Life History Weaning: 72–85d.[322] **Sexual maturity:** 14mo.[346] **Estrus cycle:** 24d.[346] **Gestation:** 181–194d.[322] **Age 1st birth:** NA. **Birth interval:** NA. **Life span:** 12y.[708] Birth peaks: Apr–May, Nov–Dec. Offspring: 1.[606] Infants are carried in the mother's mouth and parked for 43% of the night.[320]

Locomotion Vertical clinging and leaping, 56.1%; quadrupedal, 36.3%; climbing, 7.6%.[322]

Social Structure Variable. 1 male–1 female (80%) and multimale-multifemale groups (20%) with polygynous matings and more than 1 breeding female. Territories overlap but have an exclusive core area of 0.25–0.6 hectare near the sleeping tree.[619] **Group size:** 2–6.[323] **Home range:** 2–4ha,[321] 1–10ha.[619] **Night range:** NA.

Behavior Nocturnal and arboreal. Spectral tarsiers pre-

Young spectral tarsier *(T. s. dentatus).*

Philippine Tarsier *Tarsius syrichta*

Taxonomy 3 subspecies.[313]

Distinguishing Characteristics
Philippine tarsiers have gray fur and a nearly naked tail.[346] The middle finger is elongated.[130]

Physical Characteristics Head and body length: 124mm (118–140) *[4.9in (4.6–5.5)]*.[942] **Tail length:** 232mm *[9.1in]*.[942] **Weight:** ♀ 117g *[4.6oz]*, ♂ 134g *[5.3oz]*.[436] **Intermembral index:** 58.[234] **Adult brain weight:** 4g *[0.2oz]*.[346] Males are larger than females.[436] Tarsiers have the highest infant-weight-to-maternal-weight ratio of any primate that gives birth to a single infant.[909]

Habitat Secondary lowland and coastal forest.[616]

Diet Animal prey (mostly insects; lizards). Philippine tarsiers in captivity will eat live shrimp and fish in a bowl of water.[616]

Life History Weaning: 60d.[909] **Sexual maturity:** NA. **Estrus cycle:** 23.5d.[848] **Gestation:** 180d.[436] **Age 1st birth:** NA. **Birth interval:** NA. **Life span:** 13.5y. **Mating:** Year-round.[848] Offspring: 1. After copulation in captivity, vaginal plugs have been observed.[436] The infant is born with open eyes and is carried by the female in her mouth when disturbed in captivity.[616] The female parks her infant while foraging.

Locomotion Vertical clinging and leaping.[616]

Social Structure Groups are believed to be larger than 1 male and 1 female.[616] **Group size:** >2.[616] **Home range:** NA. **Night range:** NA.

Behavior Nocturnal and arboreal. Philippine tarsiers use hollows close to the ground for hiding.[616] In captivity, individuals may huddle together or intertwine their tails.[616] No male parental care has been observed.[909] These tarsiers have intestinal parasites and external parasites, such as *Trichomonas*.[616] **Scent marking:** Males have an epigastric gland used for scent marking.[616] **Vocalizations:** The loud call is a loud, piercing single note.[616] The contented call is a "soft sweet bird-like trill."[616] Several individuals make a "chirping locust-like communication."[616] Females have a vocalization that signals their approaching sexual receptivity.[848]

Tarsiers hunt at low levels of the forest and often leap to the ground to catch insects like crickets.

Philippines, Samar Island

Higher Primates

The suborder Anthropoidea includes all of the simian primates. All the species in this suborder are descended from the same distant ancestor, which means they form a natural phylogeny or clade. Unlike prosimians, the anthropoids do not have a grooming claw, a tooth comb, or a tapetum layer in their eyes. Anthropoids have a solid lower jaw and rely more on sight than on smell. The latter trait is reflected in cranial features such as closed eye sockets (postorbital closure), a short snout without an olfactory recess, and fused frontal bones. All higher primates also have stereoscopic color vision. Other features that separate anthropoids from prosimians have to do with the arteries that supply blood to the brain and with the anatomy of the ear. Anthropoids are usually larger in body size than prosimians, have hind limbs and forelimbs of equal length, and have shorter trunks.[234]

Anthropoidea is divided into two infraorders: Platyrrhini and Catarrhini. These can be remembered easily by where each occurs geographically. Platyrrhines are from South and Central America and are often referred to as the Neotropical or New World monkeys. The monkeys and apes of the suborder Catarrhini are found in Africa and Asia and are often referred to as Old World monkeys. The name "Platyrrhini" means "broad flat nose."[599] Platyrrhini species in general have rounded nostrils that face toward their ears. The Catarrhini have narrow nostrils that face downward.[234]

The New World monkeys have evolved over the last 30 million years into two distinct forms that have been classified as two families—Callitrichidae and Cebidae—and placed in one superfamily, Ceboidea.

Black-and-gold howler *Alouatta caraya*

Buffy-headed marmoset *Callithrix flaviceps*

Neotropical Primates

New World Monkeys

Neotropical monkeys weigh 100 grams–10 kilograms (3 ounces–22 pounds). They have 3 premolars rather than 2, giving them a dental formula of

$$\frac{2.1.3.3.}{2.1.3.3.} \times 2 = 36.$$

There are small anatomical differences between them and the catarrhines with regard to the skull and teeth,[234] the details of which are too lengthy to describe here. As a rule of thumb, if a primate has a prehensile tail that it can use as a fifth hand or if a primate is small, diurnal, and squirrel-like, it is a platyrrhine.

Marmosets and Tamarins   *Family:* Callitrichidae

The Callitrichidae are all small (under 1 kilogram, or under 2 pounds) and do not have a prehensile tail. Their reduced size has necessitated the loss of one set of molars, giving them a dental formula of

$$\frac{2.1.3.2.}{2.1.3.2.} \times 2 = 32.$$

Their ancestral primate nails have evolved to look like claws on all but their large toe. This adaptation enables them to climb the trunks of trees like squirrels. Most species produce twins. Parental care is shared by both sexes, and the males carry the infants most of the time.

There are four genera of callitrichids. The marmosets of the genus *Callimico* are thought to be the closest to the callitrichid common ancestor because they have retained the third molar and have single births. Pygmy marmosets *(Callithrix pygmaea)* are the smallest Neotropical primates, weighing in at only 3.5 ounces (100 grams). The marmosets of the genus *Callithrix* can be distinguished from the tamarins *(Saguinus)* only by the lower front teeth. *Callithrix* members have enlarged incisors that are the same size as the canines and are used for gouging holes in trees from which exudates (gums, resins, saps) aretaken. Marmosets live in larger groups (average of 9) than tamarins (average

5.5),[226] and both rely on dense secondary growth and edge habitats for insect foraging and for shelter.[722] Depending on the season, fruit and nectar are also a major part of the diet of callitrichids.

The social structure of callitrichids varies from one male–one female to multimale-multifemale groups. Usually only the dominant female breeds and the young are raised communally by the whole group. The other females in the group are unable to come into estrus and conceive as long as they are within the range of a dominant female's pheromones. These pheromones suppress the hormonal system of the subordinate females to the point that they cannot ovulate.[200]

The lion tamarins *(Leontopithecus)* are a little larger than the other callitrichids and do not eat exudates. They are adapted to use their long fingers for searching out insects from under bark, in tree holes, and in vegetation.[721] They sleep in cavities found in large trees. Lion tamarins are some of the most endangered species in the world because their natural habitat is primary lowland forest in southeastern Brazil, the same habitat that is preferred by humans for agricultural development.

Callitrichidae

Goeldi's Monkey *Callimico goeldii*

Taxonomy Disputed. Monotypic.[723] Formerly in its own family, Callimiconidae.[315]

Distinguishing Characteristics Goeldi's monkeys are entirely black, with long hair that sticks out to make them look "disheveled."[198] This is the only small Neotropical species with 36 teeth; all other marmosets and tamarins have 32 teeth.

Physical Characteristics Head and body length: 222mm (216–232) *[0.2in (8.5–9.1)].*[600] **Tail length:** 255–324mm *[10.0–12.8in].*[198] **Weight:** 400–535g *[14.1–18.9oz].*[247] **Intermembral index:** 69.[234] **Adult brain weight:** 10.8g *[0.4oz].*[346]

Habitat Tropical mixed-level rain forest with undergrowth and bamboo; not in floodplain forests.[672]

Diet Fruit, animal prey, exudates.[672] In the dry season Goeldi's monkeys depend entirely on 2 keystone plants.[672]

Life History Weaning: 65d.[346] **Sexual maturity:** ♀ 8.5mo, ♂ 16.5mo.[346] **Estrus cycle:** 22d (21–24).[349] **Gestation:** 154d (150–165).[349] **Age 1st birth:** 16mo.[346] **Birth interval:** 6–12mo.[349] **Life span:** 17.9y.[708] Birth season: Sep–Nov. Offspring: 1.[226] Females have a postpartum estrus (1–10d).[349] The female carries the infant for the first 23 days; then the male shares the carrying. Infants eat their first solid food at day 38 and travel independently by day 46.[265]

Locomotion Quadrupedal; vertical clinging and leaping to 4m *[13ft].*[672]

Social Structure Variable. Monogamous pairs; multimale-multifemale groups with 1 breeding pair and often more than 1 breeding female.[672] Group cohesion is strong; all members are usually within 15 meters *[49ft]* of one another and often in single file. They rest separately rather than in close contact.[672] **Group size:** 2–8.[349] **Home range:** 30–60ha.[296] **Day range:** 2000m *[6562ft].*[296]

Behavior Diurnal and arboreal. Goeldi's monkeys prefer to travel and forage below 5 meters *[16ft]* (88% of sightings), but they do feed in the tops of tall fruit trees, as high as 30 meters *[98ft].*[672] If disturbed, they stay below 5 meters *[16ft]* to flee and hide. When surprised, they reportedly park their infants and escape.[587] They jump to the ground to catch insects, especially grasshoppers.[672] During their midday rest they groom each other's fur (allogrooming).[349] Tongue flicking is an aggressive threat display.[349] Allomothering has been observed.[531] Activity budget: Travel and foraging, 60%; rest, 16%. **Association:** Goeldi's monkeys share their home ranges with red-bellied tamarins *(Saguinus labiatus)*[672] and associate 50% of the time with saddle-back tamarins *(S. fuscicollis).*[672] They are often found in the same area as Spix's black-mantled tamarins *(S. nigricollis).* **Mating:** The male monitors the female's urine to ascertain her reproductive status.[351] The male flicks his tongue while mounting the female; she turns around to place her hand on his head. **Scent marking:** These monkeys mark trails with urine and scent-mark with sternal secretions that leave a light brown smear.[351] **Vocalizations:** 7,[672] including a shrill long-distance call;[672] a loud, distinct feeding call given at canopy fruit trees; a "Chok" mobbing call; and an infant play vocalization.[531] The *tschoq* call is believed to be unique to this species.[351] When excited, these monkeys exhibit piloerection and give *chick chick* calls.[351] Tamarins *(Saguinus* spp.) often answer the calls of Goeldi's monkeys.[672]

Sleeping site: Tangles below 15 meters *[49ft]* high.[349]

Brazil, Bolivia, Peru, Colombia

Bare-ear Marmoset *Callithrix argentata*

Taxonomy Disputed. 5 subspecies *(C. a. argentata, C. a. emiliae, C. a. leucippe, C. a. melanura,* and *C. a. saterei)* are considered full species by some taxonomists.[723] A new subspecies, *C. a. marcai,* was described in 1995.[949]

Distinguishing Characteristics The body color of bare-ear marmosets varies with the subspecies, from white to dark brown. The hairless ears and face are pink or mottled. The tail is black, except in *C. a. leucippe,* which has a silvery tail.[600] The cecum is specialized for exudate digestion.

Physical Characteristics Head and body length: ♀ 206mm (199–212) *[8.1in (7.8–8.3)],* ♂ 215mm (200–235) *[8.5in (7.9–9.3)].*[600] **Tail length:** ♀ 305mm (302–308) *[12.0in (11.9–12.1)],* ♂ 322mm (295–342) *[12.7in (11.6–13.5)].*[600] **Weight:** ♀ 320g *[11.3oz],* ♂ 357g *[12.6oz].*[246] **Intermembral index:** 76.[234] **Adult**

C. a. melanura.

Bare-ear Marmoset *continued*

C. a. saterei.

brain weight: NA.

Habitat Tropical rain forest, deciduous dry forest and seasonally flooded white-river forest (varzea) up to 900m *[2952ft]*.[898]

Diet Fruit, animal prey, gums.[198] More fruit is eaten than exudates.

Life History Weaning: NA. **Sexual maturity:** NA. **Estrus cycle:** NA. **Gestation:** NA. **Age 1st birth:** 20.4mo.[708] **Birth interval:** NA. **Life span:** NA. Offspring: Normally 2.[708]

Locomotion Quadrupedal.

Social Structure NA. **Group size:**

NA. **Home range:** NA. **Day range:** NA.

Behavior Diurnal and arboreal. The threat display of bare-ear marmosets includes lowered eyebrows, lip smacking, and rapid tongue flicking.[351] In savanna habitats, groups will cross grassland from one tree clump to another.[198] **Association:** In parts of its range, this species associates with golden-handed tamarins *(Saguinus midas)*.[351] *C. a. emiliae* associates with saddleback tamarins *(Saguinus fuscicollis),* especially when gum feeding is prevalent.[500] **Mating:** Both sexes rhythmically lip-smack before mating.[226] **Scent marking:** Glands in the circumgenital[351] and sternal regions[786] are used to scent-mark. **Vocalizations:** When highly excited, these marmosets make a *tsik* call; when slightly alarmed, they produce a trill.[351]

C. a. argentata (above) has a black tail. *C. a. leucippe* looks similar but has a whitish tail.

C. a. emiliae.

They have a special play vocalization described as *ee-ee*.[786] **Sleeping site:** Tree hollows, dense vegetation, and vine tangles.[351]

Brazil, Bolivia

Buffy Tufted–eared Marmoset *Callithrix aurita*

Taxonomy Disputed. Elevated from a subspecies of *C. jacchus* in 1988.[315]

Distinguishing Characteristics Buffy tufted–eared marmosets have black body fur with rufous speckling, a white blaze on the forehead, a rufous crown, and white to brown ear tufts.[600]

Physical Characteristics Head and body length: NA. **Tail length:** NA. **Weight:** 400–450g *[14.1–15.9oz]*.[264] **Intermembral index:** NA. **Adult brain weight:** NA.

Habitat Upland evergreen and semideciduous forest above 400–500m *[1312–1641ft]*.[722]

Diet Fruit, animal prey, [722]

exudates.[950]

Life History NA.

Locomotion Quadrupedal.

Social Structure NA. **Group size:** NA. **Home range:** 11–16ha.[722] **Day range:** NA.

Behavior Diurnal and arboreal. Unlike other marmosets, this species has lower incisors that are poorly adapted for gouging trees to produce sap.[264] Instead it uses its lower front teeth to remove tree bark and eat termites and wood-boring insects.[264] This species prefers to forage near the ground (0–5m *[0–16ft]*) in forests with dense bamboo.[351]

Brazil

Buffy-headed Marmoset *Callithrix flaviceps*

NATURAL HABITAT

Brazil

Taxonomy Disputed. Elevated from a subspecies of *C. jacchus* in 1988.[315]

Distinguishing Characteristics Buffy-headed marmosets get their name from their yellowish buff-colored head and short yellow ear tufts.[600]

Physical Characteristics Head and body length: 231mm (222–248) *[9.1in (8.7–9.8)]*.[351] **Tail length:** 322mm (298–350) *[12.7in (11.7–13.8)]*.[351] **Weight:** 406g *[14.3oz]*.[247] **Intermembral index:** NA. **Adult brain weight:** NA.

Habitat Highland evergreen and semideciduous forest above 400m *[1312ft]*.[763]

Diet Gums, 65.8%; animal prey (66% orthopterans), 19.8%; fruit, 14.4%; seeds. These marmosets use 30 species of plants, but 74.6% of feeding time is spent on 2 species.[219]

Life History Weaning: NA. **Sexual maturity:** NA. **Estrus cycle:** NA. **Gestation:** NA. **Age 1st birth:** NA. **Birth interval:** 6mo.[226] **Life span:** NA. Newborn twins are equivalent to 14–24% of the mother's body weight. The cooperative breeding seen in many callitrichids has evolved in response to this high-energy demand on the females.[223] The postpartum estrus may influence the male's desire to be near the female and carry her infants in the first week. When 2 months old, infants are able to move independently and often are not carried. At 3 months they are carried only when "there is some threat."[223] At 9 months juveniles begin to capture insects.[221]

Locomotion Quadrupedal branch running; vertical clinging and leaping.

Social Structure Multimale-multifemale. **Emigration:** Males and females emigrate to form new groups. **Group size:** 9.8 (5–15).[226] **Home range:** 35.5ha.[226] **Day range:** 1200m *[3937ft]*.[226]

Behavior Diurnal and arboreal. Buffy-headed marmosets have the largest home range of any marmoset species with a known home range. They forage 50% of the time below 3 meters *[10ft]* in dense vegetation.[219] All members of the group provide infant care, including carrying and food sharing. This helps not only the infants but also the caregivers by inclusive fitness, infant care experience, and intrasexual competition among males to show females how good the caregivers' skills are.[223] Adults share food with immatures 85% of the time.[223] In this species' 3-tiered hierarchy, breeding females rank the highest, then the males and subordinates. There is very little male–male aggression[222] or intragroup aggression.[763] If a dominant female is overthrown, she remains in the group and cares for infants.[763] Activity budget: Prey foraging, 26.8%; rest, 24.8%; social activities, 11.6%; plant foraging, 10.6%. **Vocalizations:** 4 vocalizations related to predator avoidance.[227] Adults make specific vocalizations to call an infant in order to transfer animal food.[221]

Sleeping site: A group in a study used 49 sleeping trees.[227] Body temperature may fall 4°C during sleep at night.[227]

Geoffroy's Tufted-eared Marmoset *Callithrix geoffroyi*

Brazil

Taxonomy Disputed. Elevated from a subspecies of *C. jacchus* in 1988.[315]

Distinguishing Characteristics The forehead, cheeks, temples, and throat of Geoffroy's tufted-eared marmosets are white; the ears have black tufts. The body is blackish brown, with dark brown underparts, and the tail is ringed.[509]

Physical Characteristics Head and body length: 198mm *[7.8in]*.[351] **Tail length:** 290mm *[11.4in]*.[351] **Weight:** ♀ 190g *[6.7oz]*;[246] ♂ 230–350g *[8.1–12.3oz]*.[247] **Intermembral index:** NA. **Adult brain weight:** NA.

Habitat Secondary lowland, evergreen, and semideciduous forest and forest edge up to 500m *[1641ft]*.[722] Human-disturbed forest is preferred over mature forest.[722]

Diet Fruit, animal matter, gums.[351]

The infants are carried by all menbers of the troop.

Geoffroy's Tufted-eared Marmoset *continued*

Life History NA.
Locomotion Quadrupedal branch running; some vertical clinging and leaping.[351]
Social Structure NA. **Group size**: 8–10.[226] **Home range**: NA. **Day range**: NA.
Behavior Diurnal and arboreal. Geoffroy's tufted-ear marmosets reportedly follow army ant swarms in order to catch the insects flushed from hiding by the ants.[224] **Association**: These marmosets occasionally feed with masked titi monkeys *(Callicebus personatus)*.[450] **Mating**: The male coils his tail as a sexual display during copulation.[587]

Marmosets groom with their hands.

Tassel-eared Marmoset *Callithrix humeralifer*

Taxonomy Disputed. All 3 sub-species *(C. h. chrysoleuca, C. h. humeralifer, C. h. intermedia)* are considered full species by some taxonomists.[723]

Distinguishing Characteristics The coloration of tassel-eared marmosets varies with the sub-species. They all have white ear tufts in a fan shape.[600] The tail is banded distinctly or faintly.[198]

Physical Characteristics Head and body length: 215mm (198–239) *[8.5in (7.8–9.4)]*.[351] **Tail length**: 355mm (320–398) *[14.0in (12.6–15.7)]*.[351] **Weight**: ♀ 310g *[10.9oz]*, ♂ 280g *[9.9oz]*.[246] **Intermembral index**: NA. **Adult brain weight**: NA.

Habitat Secondary forest with dense vines.[198]

Diet Fruit, 82.5%; exudates,

C. h. humeralifer.

17%;[722] animal prey, insects. These marmosets use 52 species of fruit trees.[722] When fruit is not abundant, they rely on exudates for 25–59% of feeding time.[722]

Life History Weaning: NA. **Sexual maturity**: NA. **Estrus cycle**: NA. **Gestation**: NA. **Age 1st birth**: NA. **Birth interval**: NA. **Life span**: 12–13y.[723]

Locomotion Quadrupedal.

Social Structure Multimale-multifemale.[296] **Group size**: 11.5 (8–15).[226] **Home range**: 13[198]–28.3ha.[226] **Day range**: 740–1500m *[2428–4922ft]*.[722]

Behavior Diurnal and arboreal. Tassel-eared marmosets have been observed following army ants in order to catch the insects they disturb.[722] **Scent marking**: These marmosets reportedly

Tassel-eared Marmoset *continued*

C. h. chrysoleuca.

S.NASH

C. h. intermedia.

scent-mark by "rubbing tree branches with the inside of arms."[200] Suprapubic, sternal, and circumgenital marking has also been observed.[950] **Vocalizations:** The tongue is rapidly vibrated to produce "cricket-like" calls.[351] **Sleeping site:** Vine-covered trees.[198]

Common Marmoset *Callithrix jacchus*

Social play begins at four weeks of age.

Taxonomy Disputed. Used to include *C. aurita, C. flaviceps, C. geoffroyi, C. kuhlii,* and *C. penicillata.*[315]

Distinguishing Characteristics Common marmosets have large white ear tufts. The tail has alternating dark wide bands and pale narrow bands.[351]

Physical Characteristics Head and body length: ♀ 185mm (173–198) *[0.2in (6.8–7.8)]*, ♂ 188mm (158–207) *[7.4in (6.2–8.1)]*.[351] **Tail length:** ♀ 274mm (243–303) *[10.8in (9.6–11.9)]*, ♂ 280mm (247–312) *[11.0in (9.7–12.3)]*.[351] **Weight:** ♀ 236g *[8.3oz]*, ♂ 256g *[9.0oz]*.[246] **Intermembral index:** 75.[234] **Adult brain weight:** 7.9g *[0.3oz]*.[598] Canine length: ♂ 2.06mm *[0.1in]*, ♀ 2.11 *[0.1in]*.[490] The cecum is specialized for digesting exudates.[226]

Habitat Scrub, swamp, tree plantations.[351] The core area of each home range has a higher density of trees with exudate production. A home range needs 50 gum trees to support this species.[728]

Diet Fruit, gums, animal prey. Common marmosets eat more gum (15% of total diet) than other marmosets.[226]

Life History Infant: 2–3mo.[920] **Weaning:** 2mo.[893] **Juvenile:** 5–10mo.[920] **Subadult:** 10–15mo.[920] **Sexual maturity:** ♀ 12mo,[893] ♂ 16.7mo.[346] **Estrus cycle:** 13–15d.[598] **Gestation:** 148d.[528] **Age 1st birth:** 20–24mo.[398] **Birth interval:** 5.1mo. **Life span:** 11.7y.[708] Mating season: Yes.[351] Birth season: Yes.[351] Offspring: 1 (12.5%), 2 (62.5%), 3 (21.4%), 4 (3.6%). A postpartum estrus occurs at 9–10 days after a birth.[920] Infants first leave their carrier at 2 weeks. The first social play with others is at 4 weeks; by week 7, twins have been observed fighting. In week 8, infants first scent-mark.

Locomotion Quadrupedal branch running; some clinging and leaping.[351]

Social Structure Variable. Multimale–1 female, 1 male–multifemale, multimale-multifemale.[528] **Group size:** 8.9 (3–15),[226] up to 20.[466] **Home range:** 0.5–6.5ha.[351] **Day range:** 500–1000m *[1641–3281ft]*.[226]

Behavior Diurnal and arboreal. Common marmosets are most active in early morning and late evening; during midday they take naps and groom extensively. Larger groups, which share allomaternal care and territorial defense, may be more successful than smaller groups, given that more juveniles were found in groups that had more than 2 males.[466] Exudate-producing trees may have hundreds of round or slit-like holes 10–15mm wide by 20–200mm long *[0.4–0.6 x 0.8–7.9in]* gouged by these marmosets.[198] This species has been imported to several South American regions where it has adapted to local conditions, such as parks in Rio de Janeiro. **Mating:** The proceptive behavior of the female is to stare at a male and flick her tongue in and out. During mating, the female looks back over her shoulder and opens her mouth. **Vocalizations:** 9.[347] The contact calls are a *phee* and a twitter. A *tsik* is a mobbing call. A squeal denotes submission to a dominant.[893] Infants have a play vocalization.[531] **Sleeping site:** Vine tangles or a tree branch. The group sleeps together.[198]

Brazil

Wied's Tufted-eared Marmoset *Callithrix kuhlii*

Taxonomy Disputed. Elevated from a subspecies of *C. jacchus* in 1988.[315]

Distinguishing Characteristics Wied's tufted-eared marmosets have an agouti body, with transverse bands of alternating black and gray. The crown and ears are black; the forehead, cheeks, and throat are white.[351]

Physical Characteristics Head and body length: NA. **Tail length:** NA. **Weight:** 350–400g *[12.3–14.1oz].*[264] **Intermembral index:** NA. **Adult brain weight:** NA.

Habitat Secondary, lowland, evergreen, and semideciduous forest and forest edge.[722]

Diet Fruit, 67.4%; gums, 32.6%.[722]

Life History NA.

Locomotion Quadrupedal branch running; some vertical clinging and leaping.

Social Structure NA. **Group size:** NA. **Home range:** 12ha.[722] **Day range:** 974m *[3195ft].*[818]

Behavior Diurnal and arboreal. Wied's marmosets forage at heights of 6–13 meters *[20–43ft].*[351] They are reported to catch insects that have been disturbed by army ants.[722] **Association:** This species occasionally associates with golden-headed lion tamarins *(Leontopithecus chrysomelas)*[351] and may associate with black lion tamarins *(L. chrysopygus).*[721] **Mating:** A species-specific silent open-mouth display initiates mating.[750] **Scent marking:** The sternal gland is not used to scent-mark. In captivity, these marmosets reportedly mark with genitalia and urinate more at feeding time.[248]

Brazil

Maues Marmoset *Callithrix mauesi*

Maues marmoset was first discovered in 1992.

Taxonomy Disputed. First described in 1992,[579] this species is related to *C. humeralifer.*[317]

Distinguishing Characteristics Maues marmosets have a dark mantle and erect ear tufts. The back is banded.[579]

Physical Characteristics Head and body length: ♀ 211–226mm *[8.3–8.9in],* ♂ 207mm (198–226) *[8.1in (7.8–8.9)].*[579] **Tail length:** ♀ 339–376mm *[13.3–14.8in],* ♂ 350mm (341–376) *[13.8in (13.4–14.8)].*[579] **Weight:** NA. **Intermembral index:** NA. **Adult brain weight:** NA.

Habitat Primary rain forest.[579]

Diet NA.

Life History NA.

Locomotion Quadrupedal.

Social Structure NA. **Group size:** NA. **Home range:** NA. **Day range:** NA.

Behavior Diurnal and arboreal. Discovered in 1992, Maues marmosets were named after the Maues River, on whose banks they live.[579] No long-term field study about their behavior had been published as of 1995.

Brazil

65

Black-headed Marmoset *Callithrix nigriceps*

Taxonomy Disputed. First described in 1992,[229] this species is related to *C. argentata.*

Distinguishing Characteristics Black-headed marmosets get their name from their black crown. They have a hairless black face with mottling, yellow hips and thighs, a brown to black rump, and an all-black tail. The underparts are yellow to orange. Males have a white hairless scrotum.[228]

Physical Characteristics Head and body length: 200mm (193–220) *[7.9in (7.6–8.7)].*[228] **Tail length:** 320mm (314–327) *[12.6in (12.4–12.9)].*[228] **Weight:** 370g (330–400) *[13.0oz (11.6–14.1)].*[228] **Intermembral index:** NA. **Adult brain weight:** NA. This species is more robust than *C. argentata,*[228] and its gut morphology is more specialized.[229]

Habitat Lowland rain forest and edge.[228]

Diet Gum, fruit, seeds, insects (based on gut analysis).[229]

Life History NA.

Locomotion Quadrupedal.

Social Structure NA. **Group size:** NA. **Home range:** NA. **Day range:** NA.

Behavior Diurnal and arboreal. No field studies have yet been published on this recently described species. Its range is subject to human disturbance from cattle ranching, logging, and colonization that follows the trans-Amazon highway. It has no protected area.[225]

Brazil

S. FERRARI
Large insects are a favorite food of marmosets and tamarins.

Black Tufted–eared Marmoset *Callithrix penicillata*

Notice the ear tufts.

Taxonomy Disputed. Elevated from a subspecies of *C. jacchus* in 1988.[315]

Distinguishing Characteristics Black tufted–eared marmosets have black ear tufts and a white forehead with light facial hair. The back and tail are banded.[351]

Physical Characteristics Head and body length: 202–225mm *[8.0–8.9in].*[351] **Tail length:** 287–325mm *[11.3–12.8in].*[351] **Weight:** ♀ 182g *[6.4oz]*, ♂ 225g *[7.9oz].*[246] **Intermembral index:** 75.[234] **Adult brain weight:** NA.

Habitat Secondary forest and often near cacao plantations.[351]

Diet Gums, 70%; fruit, 30%; animal prey (insects).[722] In captivity these marmosets reportedly catch and eat sparrows that fly into their cage.[351]

Life History NA.

Locomotion Quadrupedal; some vertical clinging and leaping.[351]

Social Structure NA. **Group size:** 6.6 (3–9).[226] **Home range:** 1.25–10ha.[226] **Day range:** 1000m *[3281ft].*[226]

Behavior Diurnal and arboreal.

Because black tufted–eared marmosets eat more gum than fruit, they have a smaller home range than other callitrichids. Gums are a more reliable food source than fruit in the seasonal habitat in which this species lives. In captivity, when threatening a human, these marmosets put their ear tufts forward, then "fluff up their fur to make themselves look larger and make a rapid and continuous vocalization."[351] **Association:** These marmosets occasionally associate with golden-headed lion tamarins *(Leontopithecus chrysomelas).*[351] **Scent marking:** Marks are made most often near the sap holes where this species feeds.[234] **Vocalizations:** At least 4 are recognized by humans, including an alarm call, a threat call, and a loud, piercing contact call.[351] **Sleeping site:** The group sleeps huddled together.[351]

Brazil

Pygmy Marmoset *Callithrix pygmaea*

At 3 ounces, pygmy marmosets are the smallest Neotropical primate.

Taxonomy Disputed. Until 1989, this species was in its own genus, *Cebuella*.[315]

Distinguishing Characteristics
Pygmy marmosets are the smallest South American primate. They have a tawny agouti body[226] and a tawny gold gray head.[198]

Physical Characteristics Head and body length: 136mm (117–152) *[5.4in (4.6–6.0)]*.[351] **Tail length:** 202mm (172–229) *[8.0in (6.8–9.0)]*.[351] **Weight:** ♀ 126g[246] (112–140)[247] *[4.4oz (4.0–4.9)]*, ♂ 130g[246] (99–160)[247] *[4.6oz (3.5–5.6)]*. **Intermembral index:** 83.2.[425] **Adult brain weight:** 4.2g *[0.1oz]*.[346]

Habitat Floodplain forest near rivers;[765] edges of agricultural fields; secondary growth;[765] bamboo thickets.[898]

Diet Gum, 67%; fruit, nectar, animal prey. Exudates from 38 tree species and 19 vine species and fruit from 5 species are eaten.[764] Nectar is eaten during the dry season.[826]

Life History Infant: 1–5mo.[346] **Weaning:** 3mo.[346] **Juvenile:** 5–12mo.[764] **Subadult:** 12–16mo.[764] **Sexual maturity:** 24mo.[346] **Estrus cycle:** NA. **Gestation:** 119–142d.[351] **Age 1st birth:** 22.8mo.[708] **Birth interval:** 5–7mo.[764] **Life span:** 11.7y.[708] Births: Year-round. Birth peaks: Oct–Jan,[764] May–Jun.[296]

Offspring: Usually 2, occasionally 3.[351] The male carries the infants.[826] Females have a postpartum estrus at 3 weeks.[764] Infant mortality is 33%, mostly in the first 2 months.

Locomotion Quadrupedal; vertical clinging and leaping. Pygmy marmosets can leap up to 5 meters *[16ft]*.[351]

Social Structure 1 male–1 female monogamous family groups with the offspring of up to 4 litters.[765] If there are 2 males, 1 male is dominant and restricts the other's access to the female.[764] Of the population, 17% are solitary or in pairs.[765] **Group size:** 6.4 (1–15).[765] **Home range:** 0.1–0.5ha.[764] **Day range:** NA.

Behavior Diurnal and arboreal.[198] Pygmy marmosets gouge 10–20mm *[0.4–0.8in]* holes in the bark of trees and revisit them each day to eat sap (exudate).[198] They gouge new holes daily to produce a steady supply of gums. Researchers have counted up to 1700 new holes in 6 months.[826] These marmosets regularly move home ranges, depending on exudate availability.[765] They forage for insects in the crowns of small and medium-sized trees and in vine tangles below 20 meters *[66ft]*.[198] Though they do not forage on the ground, they will go to the ground to catch grasshoppers.[765] Hostile displays include piloerection and ritualized expressions. Pygmy marmosets display a threat by turning and lifting their tail to show their genitalia.[763] Infants are occasionally left unattended on tree branches.[1003] **Association:** In the dry season, tamarins (*Saguinus* spp.) may visit the gum trees of pygmy marmosets to steal gum.[198] Pygmy marmosets are found in the same area as Spix's black-mantled tamarins *(S. nigricollis)*. **Mating:** In captivity, mating lasts 4–10 seconds; then the male grooms the female for up to an hour.[351] **Vocalizations:** 15,[347] including "a call produced prior to defecation by all members of the group."[669] The long-distance contact call is a trill, the submissive call is a *tsik*, and the fear call is a screech.[669] The alarm call is a raspy chirp, and often the whole group gives a chorus call.[198] Pygmy marmosets chatter when enraged.[669]

Brazil, Ecuador, Peru

Pygmy marmosets gouge holes in trees for sap.

Bare-faced Tamarin *Saguinus bicolor*

The subspecies *S. b. bicolor* is critically endangered.

Taxonomy 3 subspecies.[723]

Distinguishing Characteristics Bare-faced tamarins get their name from their black hairless face and ears. Fur color varies with the subspecies.[198]

Physical Characteristics Head and body length: 208–283mm *[8.2–11.1in]*.[198] **Tail length:** 335–420mm *[13.2–16.5in]*.[198] **Weight:** ♀ 430g *[15.2oz]*, ♂ 430g *[15.2oz]*.[246] **Intermembral index:** NA. **Adult brain weight:** NA.

Habitat Secondary forest, swamp, edge,[898] white sand forest.[190]

Diet Fruit, 96.1%;[198] gum, animal prey, flowers.[190] The fruits of 21 plants are eaten. Seedpod gums are eaten in the dry season.[190]

Life History Weaning: NA. **Sexual maturity:** NA. **Estrus cycle:** NA. **Gestation:** NA. **Age 1st birth:** NA. **Birth interval:** 6.5mo.[351] **Life span:** 8y.[351] **Offspring:** Usually 2.[351]

Locomotion Quadrupedal; some clinging and leaping.[351]

Social Structure Multimale–multifemale family groups.[198] **Group size:** 2–8. **Home range:** 12ha.[190] **Day range:** NA.

Behavior Diurnal and arboreal. Bare-faced tamarins use a stealthy approach to hunt and capture large insects on leaves and branches on all levels of the canopy.[351] They feed at all heights of the forest, from the ground to over 20 meters *[66ft]*, but they prefer 10–12 meters *[33–39ft]*. *S. b. bicolor* has the smallest range of any Amazon primate, corresponding with the ever-expanding city of Manaus. **Vocalizations:** Whistles and chirps.[198]

S. b. martinsi.

Brazil

S. b. bicolor.

Saddleback Tamarin *Saguinus fuscicollis*

The color pattern of saddleback tamarins is variable.

Taxonomy 12 subspecies.[723] *S. tripartitus* was formerly included as a subspecies.[315]

Distinguishing Characteristics Saddleback tamarins have large bare ears and a hairy face. Coat color varies with the subspecies.[827] This species is the smallest and most widely distributed tamarin.[201]

Physical Characteristics Head and body length: ♀ 220mm (204–260) *[8.7in (8.0–10.2)]*, ♂ 213mm (175–270) *[8.4in (6.9–10.6)]*.[600] **Tail length:** ♀ 324mm (246–350) *[12.8in (9.7–13.8)]*, ♂ 318mm (250–365) *[12.5in (9.8–14.4)]*.[600]

Weight: ♀ 403g *[14.2oz]*, ♂ 387g *[13.6oz]*.[246] **Intermembral index:** 75.[351] **Adult brain weight:** 9.3g *[0.3oz]*.[346]

Habitat Primary, secondary, and lowland forest and seasonally flooded white-river forest (varzea).[898]

Diet Wet season—fruit, 96%; sap, 3%; petioles, 1%. Dry season—nectar, 75%; fruit, 16%; sap, 9%.[826] Animal prey: Orthoptera, 61%; Hemiptera, 7%; frogs, 3%; Lepidoptera larvae, 3%; Coleoptera larvae, 3%; Coleoptera adults, 3%; galls, 3%; millipedes, 3%; lizards, 1%.[826]

Saddleback Tamarin *continued*

Life History Weaning: 3mo.[346]
Sexual maturity: NA. **Estrus cycle:**
15.5d.[351] **Gestation:** 145–152d.[351]
Age 1st birth: 18mo.[398] **Birth interval:** 6–12mo.[1] **Life span:** NA.
Mating season: Apr–Oct. Birth
season: Sep–Mar.[829] Offpring:
1 (22%), 2 (74%), 3 (2%).
Locomotion Quadrupedal; some
vertical clinging and leaping
between tree trunks.[263]
Social Structure Variable.
Multimale–1 female, multimale–
multifemale.[829] All-male groups
and solitary males have been
observed.[296] Home ranges overlap 21–79%.[265] **Group size:** Range
of means 5–6.8.[226] **Home range:**
30–100ha.[198] **Day range:**
1000–1800m *[3281–5906ft].*[265]
Behavior Diurnal and arboreal.
Saddleback tamarins prefer
dense viny growth and like to be
less than 15 meters *[49ft]* high in
trees. They leap to catch terrestrial prey. They forage for animal
prey about 1.5 hours per day and
catch about 10 animals per day
on tree trunks. All group members help rear infants.[829]
Association: Saddlebacks are
sympatric with and subordinate to
emperor tamarins *(S. imperator)*
most of the time;[826] both species
defend their common territory
with vocalizations. The direct
aggression that does occur
between neighboring groups is
always between conspecifics.[826]
Sympatric species have larger
populations than allopatric callitrichids.[500] Saddlebacks associate with *Callithrix argentata
emiliae,* especially when gum
feeding is prevalent;[500] with
Goeldi's monkeys *(Callimico
goeldii)* 50% of the time;[672] often
with dusky titis *(Callicebus
moloch);* and occasionally with
collared titis *(C. torquatus).*[456]
They also form groups with Spix's
black-mantled tamarins
(Saguinus nigricolis), mustached
tamarins *(S. mystax),* and redbellied tamarins *(S. labiatus).* In
Bolivia, *S. f. weddelli* associates
with *S. labiatus,* and they defend
a common territory.[88] **Mating:** The
dominant male of the group may
suppress the subordinate males'
sexual behavior and stop the sub-

ordinate females from ovulating.
The dominant female suppresses
the ovulation of subordinate
females.[1] The twin offspring need
the care of more than the breeding pair to survive; if subordinates
bred, there would be competition
for caregivers.[1] Laboratory studies have shown that subordinate
males have significantly lower
plasma testosterone concentrations. Subordinate females have
very low plasma progesterone
concentrations and conjugated
estrogens,[1] caused by suppression of pituitary LH (luteinizing
hormone) secretions.[1] If a subordinate female is removed from a
group, ovulation begins after 9
days.[1] If she is returned to the
group, ovulation stops.[1] **Scent
marking:** The scent glands do not
develop fully without gonadal hormones.[1] In captive experiments,
saddleback tamarins, using only
scent marks, can discriminate the
scent marker's gender, social status, and subspecies.[200] They
react to marks for up to 48
hours.[200] The "major volatile constituents [of scent marks] are
squalene and 15 esters of
N-butyric acid," as well as several
organic acids.[200] **Vocalizations:**
13.[347] The most notable are a
soft trill contact call and a longdistance loud whistle.[198] This
species responds to the alarm
calls of emperor tamarins
(S. imperator) and vice versa.[826]
Sleeping site: Holes in trees and
tangles.[500]

Brazil, Bolivia, Peru,
Ecuador, Colombia

Tamarins often leap from one vertical trunk to another.

Red-crested Tamarin *Saguinus geoffroyi*

Taxonomy Disputed. Elevated from a subspecies of *S. oedipus* in 1988.[315]

Distinguishing Characteristics
Red-crested tamarins have a variegated yellow and black body, an almost bare face, and a triangular crown of short white fur on a reddish head. The tail is red with a black tip.[198]

Physical Characteristics Head and body length: ♀ 247mm *[9.7in]*, ♂ 252mm *[9.9in]*.[351] **Tail length:** 345mm (315–396) *[13.6in (12.4–15.6)]*.[351] **Weight:** ♀ 544g *[19.2oz]*, ♂ 546g *[19.3oz]*.[246] **Intermembral index:** NA. **Adult brain weight:** 10.5g *[0.4oz]*.[346] **Canine length:** ♂ 2.8mm *[0.1in]*, ♀ 2.9mm *[0.1in]*.[490]

Habitat Primary, secondary, moist tropical, and dry forest.[761] Red-crested tamarins prefer secondary growth with large trees and dense understory and are often seen near shifting cultivated areas.[761]

Diet Fruit, 60%; animal prey, 30%; flowers, gums, buds.[351] In a 1970s study, large grasshoppers made up 65% or more of the stomach contents.[265] Females eat exudates during gestation and lactation for calcium.[761]

Life History Weaning: 2–3mo. **Sexual maturity:** 24mo. **Estrus cycle:** NA. **Gestation:** 140–180d. **Age 1st birth:** NA. **Birth interval:** NA. **Life span:** 13y.[351] Mating season: Jan–Feb.[351] Birth season:

Apr–Jun.[1003]

Locomotion Quadrupedal and leaping.[263] These tamarins suspend themselves from branches with just their hind limbs.[198]

Social Structure Variable. Multimale–multifemale. Home ranges overlap 13–83%.[761] **Group size:** Range

of means 3.4–6.9.[226] **Home range:** 9.4–43ha.[226] **Day range:** 2100m *[6890ft]*.[226]

Behavior Diurnal and arboreal. Red-crested tamarins travel through all parts of the canopy on supports of all sizes,[263] but they prefer tangles.[351] They forage for insects in the low shrub layer (1–5m *[3–16ft]*).[346] If they were not hunted for the pet trade, they would coexist close to humans—they are found in a park in Panama City.[761] Red-crested tamarins allogroom with their teeth and hands.[351] They can swim if forced into water.[351] **Scent marking:** These tamarins scent-mark the tree branches that are used as regular trails where their range overlaps with another group's.[200] **Vocalizations:** 10.[351] A long whistle is a long-distance intragroup call. Trills are given in hostile situations and at potential predators. Long rasps are heard when there is a violent agonistic encounter.[351] The long call is of a much lower frequency (1.0–1.5kHz) than that of other callitrichids (5.0–10.0kHz).[763] **Sleeping site:** Large emergent trees.[198]

Costa Rica, Colombia

Emperor Tamarin *Saguinus imperator*

Taxonomy 2 subspecies.[723]

Distinguishing Characteristics
Emperor tamarins get their name from the regal appearance of their long, white mustache. They have a black head, a grayish brown body, a red orange tail, and whitish underparts.[226]

Physical Characteristics Head and body length: 230–255mm *[9.1–10.0in]*.[351] **Tail length:** 390–415mm *[15.4–16.3in]*.[351] **Weight:** 450g *[15.9oz]*.[247] **Intermembral index:** 75.[234] **Adult brain weight:** NA.

Habitat Primarily lowland, ever-

green, and broadleaf forest up to 300m *[984ft]*.[898] This species is seldom found in flooded forest.[898]

Diet Wet season—fruit, 97%; nectar, 1%; sap, 1%; fungi, 1%. Dry season—nectar, 52%; fruit, 41%; flowers, 4%; sap, 2%.[826] Animal prey: Orthoptera, 57%; Lepidoptera larvae, 16%; Lepidoptera pupae, 10%; Coleoptera adults, 7%; ants, 4%; Coleoptera larvae, 2%; snails, 2%;[826] frogs, 2%.[898] Emperor tamarins eat gums during the late dry season and early wet season (Sep–Dec).

This emperor tamarin is scent-marking a branch with glands on the side of his face.

Emperor Tamarin *continued*

attacking it. These tamarins patrol their territories. Confrontations occur about once a week and may last up to 6 hours.[826] **Association:** Emperor tamarins share territory with and are dominant to saddleback tamarins *(S. fuscicollis)* most of the time.[826] They occasionally associate with dusky titis *(Callicebus moloch)*.[826] **Vocalizations:** Vocalizations include quavering whistles, chirps, and long, descending whistles.[198] Emperor tamarins announce their presence with loud vocalizations, particularly near territorial boundaries. They respond to the alarm calls of saddleback tamarins *(S. fuscicollis)* and vice versa.[826] **Sleeping site:** Large vine-covered and isolated trees.[198] The group members sleep close together.[898]

Brazil, Peru

Life History Weaning: NA. **Sexual maturity:** NA. **Estrus cycle:** NA. **Gestation:** 140–145d.[351] **Age 1st birth:** NA. **Birth interval:** NA. **Life span:** 17y.[625] Offspring: 2.

Locomotion Quadrupedal; some vertical clinging and leaping.[351]

Social Structure Multimale–1 female.[826] **Group size:** 4.[903] **Home range:** 30–40ha.[898] **Day range:**

1400m *[4593ft]*.[226]

Behavior Diurnal and arboreal. Emperor tamarins spend 3.3 hours per day foraging for animal prey.[898] They forage for insects on leaves, vines, and branches in lower and middle levels (15m *[49ft]*) of the forest.[265] They hunt mobile insect prey by scanning leaves around it and then quickly

Mottled-face Tamarin *Saguinus inustus*

Taxonomy Monotypic.[803]

Distinguishing Characteristics Mottled-face tamarins have a black body and naked black ears. The muzzle has a white patch of skin on each side, and the genitalia are white.[198]

Physical Characteristics Head and body length: 233mm (208–254) *[0.2in (8.2–10.0)]*.[351] **Tail length:** 366mm (330–410) *[14.4in (13.0–16.1)]*.[351] **Weight:** NA. **Intermembral index:** NA. **Adult brain weight:** NA.

Habitat Rain forest.[898]

Diet NA.

Life History NA.

Locomotion Quadrupedal branch running; some vertical clinging

and leaping.[351]

Social Structure NA. **Group size:** Small.[898] **Home range:** NA. **Day range:** NA.

Behavior Presumed to be diurnal and arboreal. No field studies had been published as of 1995.

Brazil, Colombia

71

Red-bellied Tamarin *Saguinus labiatus*

LOWER RISK   CITES II

Taxonomy 2 subspecies.[723]

Distinguishing Characteristics Red-bellied tamarins have white hair around the lips and nose. The back and tail are black, with silvery gray highlights. There is a white triangle on the nape of the neck, and the underparts are bright reddish orange.[198]

Physical Characteristics Head and body length: 261mm (234–300) *[10.3in (9.2–11.8)]*.[351] **Tail length:** 387mm (345–410) *[15.2in (13.6–16.1)]*.[351] **Weight:** ♀ 455g (380–495) *[16.0oz (13.4–17.5)]*, ♂ 460g (397–505) *[16.2oz (14.0–17.8)]*.[88] **Intermembral index:** 76.[234] **Adult brain weight:** NA.

Habitat Primary and secondary forest, swamps, flooded forest.[351]

Diet Fruit, insects, exudates, nectar.[88] The fleshy, sweet fruits of 19 species are eaten in tree crowns.[351]

Life History Weaning: NA. **Sexual maturity:** NA. **Estrus cycle:** NA. **Gestation:** 140–150d. **Age 1st birth:** NA. **Birth interval:** NA. **Life span:** NA. Birth peak: Oct–Dec.

Locomotion Quadrupedal; some vertical clinging and leaping.

Social Structure Variable. Multimale-multifemale, multimale–1 female.[88] Most groups have only 1 reproductive female. Solitary males have been observed.[296] Group home ranges overlap 40%.[265] **Group size:** 4.2 (2–13). **Home range:** 33.5ha.[88] **Day range:** 1487m *[4878ft]*.[88]

Behavior Diurnal and arboreal. Red-bellied tamarins travel and forage at heights of 3–32 meters *[10–105ft]*. They spend 90% of their time above 10 meters *[33ft]*. They travel with saddleback tamarins *(S. fuscicollis)*, which forage lower in the forest.[88] Antipredator behavior includes having sentinels and mobbing ground predators.[101] In a captive study, these tamarins showed a ritualized behavior that "emphasize[s] areas of contrasting pelage [fur]: the leg stand is accompanied by piloerection that displays the orange coloration of the ventrum [underparts]."[125] These tamarins have cohesive social groups without any clear dominance hierarchy. Reconciliation behaviors are not observed after agonistic encounters.[729] In captivity, males groom females more often than vice versa.[531] Researchers report that it took 1 week for a male immigrating into a new group to be fully integrated and allowed to sleep with the group.[101] Allomothering has been observed.[531] **Association:** Sympatric with Goeldi's monkeys *(Callimico goeldii)*.[672] Red-bellied tamarins form groups with saddleback tamarins *(S. fuscicollis)*;[500] in Bolivia they associate with *S. fuscicollis weddelli*, and both species defend a common territory.[88] **Scent marking:** Females scent-mark more than males.[125]

Vocalizations: Chirps and birdlike whistles.[198] Infants have a specific play vocalization.[531] **Sleeping site:** Triple forks of trees about 12–18 meters *[39–59ft]* above the ground.[88]

Brazil, Peru

Silvery-brown Bare-faced Tamarin *Saguinus leucopus*

Taxonomy Monotypic. In the *S. oedipus* group.[723]

Distinguishing Characteristics
Silvery-brown bare-faced tamarins have a buff brown body, whitish arms and legs, reddish orange underparts, and a blackish tail. The face is a silvery color.[351]

Physical Characteristics Head and body length: ♂ 244mm (224–263) *[9.6in (8.8–10.4)]*, ♀ 241mm (232–250) *[9.5in (9.1–9.8)]*.[351] **Tail length:** ♂ 389mm (360–410) *[15.3in (14.2–16.1)]*, ♀ 375mm (347–405) *[14.8in (13.7–15.9)]*.[351] **Weight:** 440g *[15.5oz]*.[246] **Intermembral index:** 74.[234] **Adult brain weight:** NA.

Habitat Primary and secondary forest near streams up to 1500m *[4921ft]*.[898] Low, thick secondary growth and edge habitats are preferred.[351]

Diet Primarily fruit.[198]

Life History NA. Infants have been seen in June.[351]

Locomotion Quadrupedal; some climbing and leaping.

Social Structure NA. **Group size:** 2–15. **Home range:** NA. **Day range:** NA.[898]

Behavior Diurnal and arboreal.[198] Silvery-brown bare-faced tamarins are more active and agile than squirrels.[351] They use all heights of the forest and are often the only primate left in habitats disturbed by humans.[198] These tamarins will come to the aid of wounded troop members.[351] **Scent marking:** Marks are made with the circumgenital, suprapubic, and sternal areas.[200] The scent-marking chemistry is closer to that of cotton-top tamarins *(S. oedipus)* than to that of saddleback tamarins *(S. fuscicollis)*.[200] **Vocalizations:** The most common call is *tee tee,* which is said to be "shrill and somewhat melancholic."[351]

Colombia

J. K. HAMPTON AND S. HAMPTON

Golden-handed Tamarin *Saguinus midas*

Brazil, Guyana, French Guiana, Suriname

Taxonomy 2 subspecies.[723]

Distinguishing Characteristics
Color varies with the subspecies. *S. m. midas* has yellowish red hands and feet; a black head, face, and underparts; and a black back variegated with yellow. *S. m. niger* has black hands and feet. Infants have pale fur on their face that turns to black with age.[198]

Physical Characteristics Head and body length: 240mm (217–278) *[9.4in (8.5–10.9)]*.[351] **Tail length:** 392mm (330–440) *[15.4in (13.0–17.3)]*.[351] **Weight:** ♀ 432g *[15.2oz]*, ♂ 586g *[20.7oz]*.[246] **Intermembral index:** 74.[234] **Adult brain weight:** 10.4g *[0.4oz]*.[346]

Habitat Primary and secondary forest, edges, swamps; not in flood forest. Open high canopy is preferred.[351]

Diet Fruit, seeds, insects, other animal prey.[351]

Life History Weaning: 2.3mo.[346] **Sexual maturity:** 20mo.[346] **Estrus cycle:** 16d.[346] **Gestation:** 140–168d. **Age 1st birth:** 24mo.[708] **Birth interval:** 8mo.[346] **Life span:** 13.2y.[708] **Birth seasons:** Spring, summer. **Offspring:** 2 (75%), 1 (25%). Males do most of the infant carrying, starting a few days after the infant's birth.[12]

Locomotion Quadrupedal.

Social Structure Variable. 1 male–multifemale, multimale-multifemale. **Group size:** 5 (2–12).[226] **Home range:** 9ha.[898] **Day range:** NA.

Behavior Diurnal and arboreal. Golden-handed tamarins forage at 5–25 meters *[16–82ft]*.[198] They prefer large branches and can leap up to 8 meters *[26ft]*.[351] Individuals solicit grooming by "lying down with throat exposed to groomer."[351] The threat display includes wrinkling the muzzle, head shaking, baring the teeth, and calling.[351] The breeding female dominates the group. The males do not threaten her.[351]

Association: This species associates with bare-ear marmosets *(Callithrix argentata)*.[351] **Mating:** These tamarins have been observed to "mate up to a few hours of giving birth and two days after."[351] Mating is preceded by "mock fighting and tonguing." **Scent marking:** Scent marks are made before and after mating and during threat displays.[351] **Vocalizations:** 8.[351] The contact call is *pi-pi-pi.* The warning call for potential predators is *di-ah.*[351]

Mustached Tamarin *Saguinus mystax*

Taxonomy 3 subspecies.[723]

Distinguishing Characteristics Mustached tamarins have a black head and a white mustache. The back and hind legs are brownish, and the tail is black. Males have white genitals.[198]

Physical Characteristics Head and body length: 258mm (248–272) *[10.2in (9.8–10.7)]*.[351] **Tail length:** 386mm (372–423) *[15.2in (14.6–16.7)]*.[351] **Weight:** ♀ 508–640g *[17.9–22.6oz]*, ♂ 491–643g *[17.3–22.7oz]*.[247] **Intermembral index:** 76.[234] **Adult brain weight:** NA.

Habitat Primary and secondary forest, swamps, and inundated areas up to 610m *[2001ft]*.[898]

Diet Fruit, insects, and exudates that contain high levels of calcium and crude protein.[265]

Life History Weaning: NA. **Sexual maturity:** ♀15–17mo, ♂ 17–18mo.[296] **Estrus cycle:** NA. **Gestation:** 140–150d.[296] **Age 1st birth:** NA. **Birth interval:** 11–20mo.[764] **Life span:** 12-20y.[265] **Birth seasons:** Nov–Feb.

Locomotion Quadrupedal.[263]

Social Structure Multimale–multifemale.[265] Groups of 3 males and up to 4 females have been reported.[265] **Group size:** 5.3 (1–16).[226] **Home range:** 25–35ha,[764] 47.5ha.[226] **Day range:** 1700–2000m *[5577–6562ft]*.[265]

Behavior Diurnal and arboreal. Moustached tamarins spend the majority of the daytime foraging for mobile insect prey. They eat exudates from September to December, when females are pregnant.[265] "Aggression is minimal in fruit trees which have small to medium crowns and produce a small amount of fruit each day in a scattered and patchy distribution."[265]

Association: Mustached tamarins associate with saddleback tamarins *(S. fuscicollis)* but use higher levels of forest, foraging for insects on leaves at an average height of 15 meters *[49ft]*.[265] **Scent marking:** Marks are made by urinating into the hands. Individuals reportedly may rub their cheeks in the urine of their sexual partners.[200] **Vocalizations:** The tongue is used to produce a trill call.[351] Whistles and chirps are also produced.[198]

Brazil, Peru

Spix's Black-mantled Tamarin *Saguinus nigricollis*

Taxonomy 2 subspecies.[723]

Distinguishing Characteristics The blackish brown mantle of Spix's black-mantled tamarins reaches to the midback and occasionally beyond. These tamarins have hairless ears and grayish white hair around the muzzle. The thighs, rump, lower back, and underparts vary from reddish brown to olivaceous.[509]

Physical Characteristics Head and body length: 220–226mm *[8.7–8.9in]*.[598] **Tail length:** 356–361mm *[14.0–14.2in]*.[598] **Weight:** ♀ 480g *[16.9oz]*, ♂ 470g *[16.6oz]*.[246] **Intermembral index:** NA. **Adult brain weight:** 8.9g *[0.3oz]*.[346]

Habitat Primary and secondary high tropical forest and edges up to 914m *[2998ft]*.[898]

Diet Fruit, seeds, animal prey, flowers, gums, resins.[403] These tamarins feed on 41 plant species.[155]

Life History Infant: 0–2mo.[403] **Weaning:** 2.8 mo.[346] **Juvenile:** 2–14mo. **Subadult:** 14–18mo.[403] **Sexual maturity:** NA. **Estrus cycle:** NA. **Gestation:** NA. **Age 1st birth:** NA. **Birth interval:** 8.4mo. **Life span:** 13.9y.[708] Births: Year-round. Offspring: 1 (21%), 2 (78%), 3 (1%).[351]

Locomotion Quadrupedal; some vertical clinging and leaping.[351]

Social Structure Variable. Multimale–1 female, multimale-multifemale. In large groups a dominance hierarchy has been reported.[721] Group size: 6.3 (4–12).[226] **Home range:** 30–50ha.[351] **Day range:** 1000m *[3281ft]*.[265]

Behavior Diurnal and arboreal.[198] Black-mantled tamarins are reported to be the only tamarin to form large noisy groups. These groups last only for short periods and number up to 40 individuals.[198] Groups may merge and forage together for up to 1.5 days.[403] They spend 34.8% of the day foraging for insects and 17% feeding on plant foods. They forage on tree trunks for hidden prey[265] and on the ground within

R. A. MITTERMEIER

Tamarins stretch out to dissipate heat on hot days.

2–3 meters *[7–10ft]* of trees.[403] Flying insects are caught with the mouth; larger insects, with the hands.[403] These tamarins are fond of large grasshoppers (60–80mm *[2.4–3.1in]*), which they eat headfirst, taking an average of 5 minutes (1.4–13.5min) to finish.[403] Most agonistic behavior is seen during insect hunting.[403] Adults will share insects with infants and juveniles.[403] These tamarins use their hands and mouth to groom each other.[403] **Association:** Spix's black-mantled tamarins associate with *Pithecia monachus*[403] and are often found in the same area as squirrel monkeys *(Saimiri)*, pygmy marmosets *(Callithrix pygmaea)*, sakis *(Pithecia)*, saddleback tamarins

(Saguinus fuscicollis), and Goeldi's monkeys *(Callimico goeldii)*. White-throated toucans *(Ramphastos tucanus)* follow Spix's black-mantled tamarins when they are foraging.[403] **Scent marking:** These tamarins use the chest and genital region to scent-mark branches and each other's backs.[403] Scent marking is not believed to be territorial in this species.[403] **Vocalizations:** Adults vocalize at ground predators while the other group members flee.[403]

Sleeping site: Vine tangles.[403] This species rests 2 or 3 times per day for 60–90 minutes.[403]

Brazil, Peru, Ecuador

Cotton-top Tamarin *Saguinus oedipus*

This tamarin is getting a very thorough grooming.

Taxonomy Disputed. Formerly included *S. geoffroyi* as a subspecies.

Distinguishing Characteristics Cotton-top tamarins have a long, whitish, fanlike crest on the top of their black head. The back is brown, and the tail is red toward the base. The underparts, limbs, and feet are a cream to white color.[198]

Physical Characteristics Head and body length: 232mm (206–243) *[9.1in (8.1–9.6)]*.[351] **Tail length:** 372mm (333–402) *[14.6in (13.1–15.8)]*.[351] **Weight:** ♀ 430g *[15.2oz]*, ♂ 411g *[14.5oz]*.[246] **Intermembral index:** 74.[234] **Adult brain weight:** 9g *[0.3oz]*.[346]

Habitat Secondary wet and dry forest and low vine tangles from sea level to 1500m *[4921ft]*.[198]

Diet Fruit, seeds, gums, and animal matter, including insects, mice, and birds.[351]

Life History Infant: 1–7mo.[920] **Juvenile:** 7–14mo.[920] **Subadult:** 14–21mo.[920] **Sexual maturity:** 18mo.[346] **Estrus cycle:** 15d.[351] **Gestation:** 140d.[351] **Age 1st birth:** 33mo. **Birth interval:** 8mo.[351] **Life span:** 13.5y.[708] Mating season: Jan–Feb. Birth season: Apr–Jun. Offspring: 1 (34%), 2 (64%), 3 (2%). The female carries her infants for 1–2 weeks before letting the male take over the task entirely. In a captive study, females with no experience rearing younger siblings had 100% mortality of their first infants.[903] Infants develop as follows: day 3, eyes open; day 20, can walk;[351] week 5, eat solid food (captivity); week 12, independent but adults share food; weeks 20–30, stop sleeping on parent's back.[351]

Locomotion Quadrupedal; some clinging and leaping.[351]

Cotton-top Tamarin *continued*

Social Structure Multimale–multifemale.[351] **Group size:** 7.4[819] (3–13).[226] **Home range:** 7.8–10ha,[226] 26–43ha.[296] **Day range:** 1700m *[5577ft]*.[226]

Behavior Diurnal and arboreal.[198] Cotton-top tamarins are active from 1 hour after sunrise. They do not rest at midday but go to sleep before sunset. When they catch a bird, they bite the head first, then remove the beak.[351] They stand bipedally to display aggression and dominance.[351] Raising the hair on the crown is a mild threat.[351] Allomothering has been observed. In the 1960s–1970s, more than 30,000 cotton-top tamarins were exported from Colombia for pets and biomedical research on colon cancer, colitis, and Epstein-Barr virus.[576] **Mating:** There is no obvious presenting behavior before mating. The male clasps the female's flanks while mounting. In a captive study, a male attempted to mate when the female was in labor.[351] **Scent marking:** Females scent-mark and urine-mark more often than males.[351] This species does not sternal-mark.[200] **Vocalizations:** 9.[351] A short high-pitched *te* is a contact call. These tamarins have a submission squeal, an alarm trill, a high-pitched whistle for aerial predators, a "Tsik" when mobbing a potential predator,[351] and an infant play vocalization.[531] The long call is a much lower frequency (1–1.5kHz) than that of other callitrichids, except red-crested tamarins *(S. geoffroyi)*.[763] **Sleeping site:** High tree forks and vine tangles.[198]

Colombia, Panama

Golden-mantled Saddleback Tamarin *Saguinus tripartitus*

Taxonomy Disputed. Monotypic.[723] Elevated from a subspecies of *S. fuscicollis* in 1988.[315]

Distinguishing Characteristics Golden-mantled saddleback tamarins have a black head, golden shoulders, and a variegated gray, white, or orange back. The underparts are orange, and the tail is black above with orange below.[198]

Physical Characteristics Head and body length: 218–240mm *[0.2–9.4in]*.[351] **Tail length:** 316–341mm *[12.4–13.4in]*.[351] **Weight:** NA. **Intermembral index:** NA. **Adult brain weight:** NA.

Habitat Lowland evergreen forest.[198]

Diet Fruit, insects.[351]

Life History Weaning: NA. **Sexual maturity:** NA. **Estrus cycle:** NA. **Gestation:** NA. **Age 1st birth:** NA. **Birth interval:** NA. **Life span:** 6y.[351]

Locomotion Quadrupedal; some clinging and leaping.

Social Structure NA.

Behavior Presumed to be diurnal and arboreal. No field studies had been published as of 1995.

P. OXFORD - NATURAL SCIENCE PHOTOS

Ecuador

Black-faced Lion Tamarin *Leontopithecus caissara*

Taxonomy Disputed. Elevated from a subspecies of *L. chrysopygus*.[315] Discovered and described in 1990.

Distinguishing Characteristics Black-faced lion tamarins have a golden body and a black face.[509]

Physical Characteristics Head and body length: NA. **Tail length:** NA. **Weight:** NA. **Intermembral index:** NA. **Adult brain weight:** NA.

Habitat Primary lowland coastal forest (restinga),[721] with many epiphytic bromeliads[660] and palms.[721]

Diet NA.

Life History NA.

Locomotion Quadrupedal.

Social Structure NA.

Behavior Diurnal and arboreal. Researchers survey for black-faced lion tamarins by playing tape recordings of their calls, to which the tamarins respond by calling and approaching the suspected intruders to their territory.[660]

Brazil

Z. KOCH (NATURAL HABITAT)

The black-faced lion tamarin was first discovered in 1990.

Golden-headed Lion Tamarin *Leontopithecus chrysomelas*

Taxonomy Disputed. Elevated from a subspecies of *L. rosalia* in 1984.[315]

Distinguishing Characteristics Golden-headed lion tamarins have a black body, with golden fringe around the face. They have golden-colored fur on the arms, legs, and part of the tail.[509]

Physical Characteristics Head and body length: ♂ 257mm (240–290) *[10.1in (9.4–11.4)]*.[351] **Tail length:** 376mm (360–400) *[14.8in (14.2–15.7)]*.[351] **Weight:** ♀ 480–590g *[16.9–20.8oz]*, ♂ 540–700g *[19.0–24.7oz]*.[247] **Intermembral index:** NA. **Adult brain weight:** NA.

Habitat Lowland forest, swamp,[898] semideciduous[721] and tall ever-green forest from sea level to 112m *[367ft]*,[898] occasionally, shaded cacao plantations.[721]

Diet Fruit,[721] gums, nectar, and animal prey, including large insects.[818] Nectar is important in

August–November.[721] Exudates from the pods of a leguminous tree (*Parkia* sp.) are eaten.[264]

Life History NA.

Locomotion Quadrupedal climbing.[351]

Social Structure Variable. 1 male–1 female, multimale-multifemale.[721] **Group size:** 6.7[721] (5–8).[226] **Home range:** 36ha.[226] **Day range:** 1410–2175m *[4626.2–7136.2ft]*.[721]

Behavior Diurnal and arboreal. Golden-headed lion tamarins forage at a height of 12–20 meters *[39–66ft]*. They search for insects in epiphytic bromeliads, in the leaf litter of vine tangles, on bark, and in tree holes and crevices.[315] They do not come to the ground.[721] **Association:** These tamarins occasionally associate with Wied's tufted-eared marmosets (*Callithrix kuhlii*) and black tufted–eared marmosets (*C. penicillata*).[351] *C. kuhlii* forages at

Golden-headed Lion Tamarin *continued*

lower levels (6–13m *[20–43ft]*).[721] **Sleeping site:** Tree holes found only in the primary forest within the species' range.[721] The lack of primary forest with suitable big trees with tree holes may be why this species' distribution is so small.

Brazil

Black Lion Tamarin *Leontopithecus chrysopygus*

Taxonomy Disputed. Elevated from a subspecies of *L. rosalia* in 1984.[315]

Distinguishing Characteristics Black lion tamarins are black, with a gold rump and gold at the base of the tail.[198]

Physical Characteristics Head and body length: 294mm (255–330) *[11.6in (10.0–13.0)]*.[351] **Tail length:** 376mm (330–400) *[14.8in (13.0–15.7)]*.[351] **Weight:** 540–690g *[19.0–24.3oz]*.[264] **Intermembral index:** NA. **Adult brain weight:** NA.

Habitat Lowland semideciduous riparian forest from sea level to 100m *[328ft]*.[721] This forest has no epiphytes. The dry season is April–September.[721]

Diet Fruit (16 species, including cactus fruit), gums (4 species), animal prey. Wet season—fruit, 90%; dry season—gums, 55%.[721] These tamarins lick gums from palm tree trunks.[264]

Life History NA.

Locomotion Quadrupedal; climbing.[351]

Social Structure Variable. Multimale, multifemale, 1 male, or 1 female. The group has 2–3 adults.[721] **Group size:** 3.6[721] (2–7).[898] **Home range:** 66–200ha.[721] **Day range:** 2289m (2061–2611) *[7510ft (6762–8567)]*.[721]

Behavior Diurnal and arboreal. Black lion tamarins come to the ground to forage for prey in the leaf litter and to catch insects.[721] Their specialized hands have long fingers for probing crevices to find mobile prey. Their inland habitat lacks bromeliads and has a distinct dry season; it is ecologically different from the habitat of the other 3 species of lion tamarins, which are found near the coast. Consequently, this species has smaller groups with larger home ranges.[721]

Association: This species may associate with Wied's tufted-eared marmosets *(Callithrix kuhlii)*.[721] **Sleeping site:** Tree holes.[721]

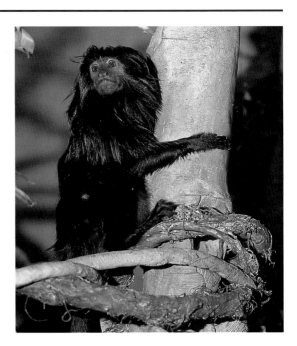

Brazil

Golden Lion Tamarin *Leontopithecus rosalia*

The golden lion tamarin is one of the few primates to have been reintroduced to its natural habitat from captivity.

give allomothering care to an infant until it is 1–3 weeks old.[2] In the early 1980s it was realized that more golden lion tamarins were living in captivity than in the wild, and one of the first primate reintroduction programs was started. As of 1990, 75 individuals had been reintroduced, with a survival rate of 36%. The number of infants born was 34, and the total number of individuals living in the wild from the reintroduction was 53. The program points out the difficulties of reintroducing primates used to cage life to the rigors of their natural habitat.[742] **Mating:** These tamarins have multiple-mount matings. **Scent marking:** Males and females scent-mark in equal amounts in captivity. Sternal marking is more common than circumgenital marking. Vocalizations from neighboring groups lead to more scent marking and return calls from both genders.[511] Captive tamarins scent-mark more after heavy rain.[351] **Vocalizations:** 10. The most common are a shrill whistle, chirring,[351] descending trills, whines, and clucks.[198] **Sleeping site:** Tree holes.[198]

Brazil

Taxonomy Disputed. Included all *Leontopithecus* species as subspecies until 1984.[313]

Distinguishing Characteristics Golden lion tamarins are reddish, golden, orange, or buffy all over and have a dark, hairless face.[463]

Physical Characteristics Head and body length: 261mm (200–336) *[10.3in (7.9–13.2)].*[463] **Tail length:** 370mm (315–400) *[14.6in (12.4–15.7)].*[463] **Weight:** ♀ 361–794g *[12.7–28.0oz]*, ♂ 437–710g *[15.4–25.0oz].*[247] **Intermembral index:** 86.7.[425] **Adult brain weight:** 12.9g *[0.5oz].*[346]

Habitat Primary and secondary lowland forest from sea level to 300m *[984ft].*[898] The dry season is May–July.[721]

Diet Fruit (38 species), 78%; nectar, flowers (3 species), exudates (3 species), and animal prey, including insects and reptiles.[94] A poor captive diet (lacking carotenoids) results in "dull faded" fur.[351]

Life History Infant: 0–4mo.[920] **Weaning:** 3mo.[346] **Juvenile:** 4–9mo.[920] **Subadult:** 9–12mo.[920] **Sexual maturity:** 24mo. **Estrus cycle:** NA. **Gestation:** 129d (125–132d).[171] **Age 1st birth:** 29mo.[708] **Birth interval:** 6–12mo. **Life span:** 14.2y.[708] Mating season: May–Jul. Birth season: Sep–Mar. Offspring: 1–2.[171] Male body weight increases in May

before breeding and decreases in June (12%).[171] Litters are produced during the rainy season, when food is more abundant.[171] In high-fruit years, 2 litters are possible.[171] The female's milk is richer in protein and ash than the milk of other simian groups.[351] Postpartum estrus occurs 3–10 days after a birth.[171]

Locomotion Quadrupedal.

Social Structure Variable. Multimale, multifemale, or both. Males are dominant to females.[721] **Emigration:** Both genders disperse from their natal group.[721] **Group size:** 5.8 (2–16).[898] **Home range:** 20–40ha.[721] **Day range:** 1339–1533m *[4393–5030ft].*[721]

Behavior Diurnal and arboreal.[198] Golden lion tamarins are active at 5–10 meters *[16–33ft]* in primary forest with bromeliads. The ovulation of subordinate females is suppressed behaviorally rather than physiologically by the dominant female.[2] Approximately 10% of the groups in the wild have 2 breeding females.[1003] In captivity, the dominant male shows intense aggression toward unfamiliar animals.[2] Captive experiments suggest that adult food sharing with infants enables the infants to feed on hard-to-find foods to ensure adequate nutrition.[676] The nonbreeding group members do not

Northern woolly spider monkey *Brachyteles hypoxanthus*

Cebids

Family: **Cebidae**

The family Cebidae is a diverse group of Neotropical monkeys that are difficult to generalize about as a whole. They differ from the Callitrichidae in having three premolar teeth, a larger body, and greater weight (over 1000 grams, or over 2 pounds).

Night Monkeys   *Subfamily:* Aotinae

The social structure of the Aotinae subfamily is based on monogamy, which occurs in only 3% of mammals and 14% of primate species.[940] The males carry the infant almost exclusively for the first two months.[940] Night monkeys *(Aotus)* get their common name from their nighttime activity pattern and are also called owl monkeys because of their owl-like hooting vocalizations. *Aotus* species are the only higher primates to exploit this nocturnal niche. Though they do not have the tapetum lucidum of prosimians to enable them to see better in the dark, they have larger eyes relative to body size than any other anthropoids.[234] Taxonomists recognize 10 species of night monkeys by chromosomal analysis,[315] but only the 2 species that are visually distinct are included in this text.

Titi Monkeys   *Subfamily:* Callicebinae

Titi monkeys *(Callicebus)* are diurnal, frugivorous, and monogamous primates that have a thick fluffy fur, small canine teeth, a short face, and a long tail. The large number of titi monkeys (13 species) is due in part to their isolation in small refuges caused by the climatic drying that occurred during the Pleistocene epoch (in the last 1.5 million years).[720]

Capuchins and Squirrel Monkeys   *Subfamily:* Cebinae

The two genera of the subfamily Cebinae, *Cebus* and *Saimiri,* are the least specialized of the Cebidae.[409] They are recognizable to many people because, until recently, they were sold as pets or trained to entertain people. The common name for the capuchin monkeys *(Cebus)* comes from the similarity of one *Cebus* species to the costume worn by the Capuchin order of monks from Spain. The other common name for the genus, "organ grinder monkeys," comes from their being trained to dance and collect money for an organ grinder. Their manipulative ability, semiprehensile tail, and intelligence are adaptations for foraging for hidden insects and dispersed fruits and hard nuts. The squirrel monkeys *(Saimiri)* have a prehensile tail only in infancy. During the breeding season, squirrel monkey males put on weight to appear larger and fight viciously among themselves for high rank. The females prefer to mate with large high-ranking males.

Sakis and Uacaris   *Subfamily:* Pitheciinae

The subfamily Pitheciinae is made up of three genera: *Pithecia, Chiropotes,* and *Cacajao.* All weigh 1000–4100 grams (2.2–9 pounds). They use their specialized procumbent incisors and stout canines for processing seeds, fruits, and hard nuts.[452] They do not have a prehensile tail. The saki monkeys *(Pithecia)* are great leapers. The bearded sakis *(Chiropotes),* which prefer the upper canopy, live in large groups, and the males and females look alike. The uacaris *(Cacajao)* have short tails.

Howler Monkeys   *Subfamily:* Alouattinae

All of the howler monkeys of the subfamily Alouattinae share the same genus, *Alouatta.* Howler monkeys get their name from their loud territorial vocalizations, which are produced in part by an enlarged hyoid bone that is hollow and resonates. All howlers have a prehensile tail and grasp branches between their second and third fingers because their thumbs are diminutive. They are sexually dimorphic—the males are invariably larger than the females. Howlers are the only New World monkeys to have a diet that includes mature leaves, which are widely available and easy to find but are of low quality in terms of energy gain. This diet has led them to minimize the amount of energy they expend. Unlike the atelines, they rest a great deal, and they have smaller brains (50.4 grams, or 1.8 ounces)[680] and small home ranges.[790]

Cebids *continued*

Spider Monkeys and
Woolly Monkeys   *Subfamily:* Atelinae

Members of the subfamily Atelinae have the largest body size of Neotropical primates: up to 12 kilograms (26.4 pounds). The subfamily is made up of three genera: *Ateles, Lagothrix,* and *Brachyteles.* Near the end of the underside of the long prehensile tail of these monkeys is a hairless patch of skin which has a unique pattern of grooves like a fingerprint and enhances the tail's ability to

A hairless patch of skin enhances the grasp of the tail.

grasp branches. Females of this family have an elongated clitoris used for distributing scent. The diet of the atelines consists of high-quality, energy-rich fruit, which is hard to find and widely scattered. These monkeys have adapted to this diet by evolving larger brains (107 grams, or 3.8 ounces)[680] than howler monkeys (Alouattinae), and by traveling long distances in big home ranges to maximize their fruit intake.[790]

The spider monkeys *(Ateles)* have long, shaggy hair, slender limbs, and a thumb that either has been reduced to a stump or is entirely absent. They use their hooklike hands to suspend themselves beneath branches while feeding and to travel by swinging from tree to tree.

The woolly monkeys *(Lagothrix)* have short dense fur and a short thumb that is not opposable. Their face resembles that of a human *(Homo sapiens),* because they have a rounded skull, shortened nasal bones, and robust brow-ridges. Infant woolly monkeys are often kept as pets. Like most primate pets, they become aggressive and dangerous as they mature, often attacking their caretakers as well as strangers.

The muriqui *(Brachyteles)* is often called the woolly spider monkey. Its rounded head and dense fur are similar to those of woolly monkeys, and its slender arms are similar to those of spider monkeys. Its premolars and molars are larger than those of the other two Atelinae genera, but it has shorter canines.

All of the species in this subfamily are threatened by human hunting. Because of their large body size, they provide more meat per bullet. The woolly spider monkey is critically endangered because of hunting and because 95% of its Atlantic coastal forest habitat has been destroyed since European settlers arrived in Brazil 500 years ago.

The feet of the white-bellied spider monkey.

Spider monkeys' thumbs are reduced or entirely absent.

82

Southern Red-necked Night Monkey *Aotus nigriceps*

Taxonomy Disputed. This genus used to contain 1 species with 10 subspecies, but chromosomal evidence led to the elevation of all 10 to species level in 1983.[353] Not all 10 species are included in this book, because only 2 forms are obviously recognizable: gray-necked and red-necked. The 5 species of red-necked night monkeys are *A. azarai, A. infulatus, A. miconax, A. nancymaae,* and *A. nigriceps.*[353]

Distinguishing Characteristics The neck, throat, and underparts of southern red-necked night monkeys are orange. These monkeys have black and white facial markings above their large eyes; a grizzled gray to brown back; and a black-tipped nonprehensile tail.[198]

Physical Characteristics Head and body length:
♀ 341mm (300–350) *[13.4in (11.8–13.8)],*
♂ 346mm (296–420) *[13.6in (11.7–16.5)].*[600] **Tail length:** ♀ 373mm (290–440) *[14.7in (11.4–17.3)],*
♂ 354mm (250–430) *[13.9in (9.8–16.9)].*[600] **Weight:**
♀ 780–1100g *[27.5–38.8oz],* ♂ 825–1050g *[29.1–37.0oz].*[908] **Intermembral index:** 74.[234] **Adult brain weight:** 18.2g *[0.6oz].*[346]

Habitat Primary and secondary lowland rain forest to montane forest from sea level to 3200m *[10,499ft].*[907] These night monkeys spend 72% of their time above 10m *[33ft].*

Diet Fruit, 65–75%; leaves (including buds, flowers, and sap), 5–30%; animal prey, 5–20%.[911]

Life History Weaning: 8mo.[909] **Juvenile:** 8–24mo.[907] **Subadult:** 24–36mo.[907] **Sexual maturity:** NA. **Estrus cycle:** NA. **Gestation:** 120[911]–133d.[909] **Age 1st birth:** 30mo.[907] **Birth interval:** 5.53–13.97mo.[176] **Life span:** 20y.[28] **Mating season:** Aug (Paraguay).[911] **Birth season:** Jan–Feb.[907] **Offspring:** 1.[908] The male carries the infant from birth to 5 months up to 81% of the time.[907] Father–offspring conflict about carrying was observed between months 4 and 5. Male parental care is necessary because females are pregnant or lactating year-round.[911]

Locomotion Quadrupedal walking and leaping up to 5m *[16ft].*[600]

Social Structure Monogamous 1 male–1 female family groups.[907] Males and females share in territorial defense. Intergroup encounters include chases, whoops, and wrestles that may last up to 10 minutes.[911] **Emigration:** Males and females emigrate at 3 years of age.[909] **Group size:** 3 (2–5).[907] **Home range:** 3.1ha.[907] **Night range:** 252–829m *[827–2720ft].*[907]

Behavior Nocturnal and arboreal. Night monkeys are most active 1 hour after dusk and 1 hour before dawn and are more than twice as active during a full moon.[907] They travel at midlevels of the forest, resting for long periods between feedings on soft fruit. The male plays with the infant and juveniles, but allogrooming is rare.[908] Night monkeys *(A. azarai)* in Paraguay are more active in daylight because great horned owls prey on them at night.[911] Activity budget: Feeding, 53%; travel, 21%; rest, 22%; agonistic behavior, 4%.[907] **Association:** *A. miconax* is found in the same montane habitat as Andean titis *(Callicebus oenanthe).* **Mating:** Subadult males give a hoot vocalization for 1–2 hours at their territorial border during the full moon to attract a mate.[911] **Scent marking:** Night monkeys urine-wash and scent-rub with their caudal gland, located at the base of the tail. They scent-mark more for intergroup communication than for sexual advertisement.[911]

Vocalizations: 6.[347] Calls are owl-like hoots that are simple and low frequency in order to carry farther in the forest.[911] Frequent contact calls are given for group cohesion. "Sneeze grunts alert the group to danger."[907] Both male and female vocalize but do not duet.[908] **Sleeping site:** During the day, holes in trees and vine tangles.[907] A single sleeping tree was used 57% of the time during the year of a study.[907]

Brazil, Peru, Bolivia, Paraguay

C. JANSON (NATURAL HABITAT)

Northern Gray-necked Owl Monkey *Aotus trivirgatus*

Some authorities consider this owl monkey
(Aotus lemurinus griseimembra) to be its own species.

Taxonomy Disputed. This genus used to contain 1
species with 10 subspecies, but chromosomal evidence led to the elevation of all 10 to species level in
1983.[353; 315] Not all 10 species are included in this
book, because only 2 forms are obviously recognizable: gray-necked and red-necked. The 5 species of
gray-necked owl monkeys are *A. brumbacki, A.
hershkovitzi, A. lemurinus, A. trivirgatus,* and *A.
vociferans.*[245]

Distinguishing Characteristics Gray-necked owl monkeys have a gray neck, pale yellow underparts, black
and white markings above the eyes, and a gray nonprehensile tail. The back is agouti to brown.[198]

Physical Characteristics Head and body length:
♀ 341mm (300–350) *[13.4in (11.8–13.8)]*, ♂ 346mm
(296–420) *[13.6in (11.7–16.5)]*.[600] **Tail length:**
♀ 373mm (290–440) *[14.7in (11.4–17.3)]*, ♂ 354mm
(250–430) *[13.9in (9.8–16.9)]*.[600] **Weight:** ♀ 920g
[32.4oz], ♂ 950g *[33.5oz]*.[234] **Intermembral index:**
74.[234] **Adult brain weight:** 18.2g *[0.6oz]*.[346]

Habitat Primary and secondary rain forest to dry forest.
These monkeys are often found near human
settlements.[198]

Diet Fruit, flowers, leaves, insects.[198]

Life History Weaning: NA. **Sexual maturity:** NA. **Estrus**

cycle: NA. **Gestation:** 120d.[911] **Age
1st birth:** NA. **Birth interval:**
5.53–13.97mo.[176] **Life span:**
20y.[708] Offspring: 1.

Locomotion Quadrupedal.

Social Structure Monogamous 1
male–1 female family groups.[198]
Group size: 2–5.[198] **Home range:**
10ha.[198] **Night range:** NA.

Behavior Nocturnal and arboreal.
Gray-necked owl monkeys are
active in the upper half of the forest but prefer habitats with vines.
They are truly monogamous,
which is rare among mam-
mals.[911] These species are susceptible to malaria and have been
used in laboratories to study
malaria and herpes virus.[176] The
nocturnal monkeys of *Aotus* have
the largest eyes of any South
American primate, with more rods
and fewer cones. Other adaptations in the eye, lens, and brain
enhance night vision.[911] **Scent
marking:** These monkeys use

urine and glandular secretions
more for intergroup communication than for sexual advertisement.[911] **Vocalizations:** 6.[708] The
alarm call is a "soft metallic click"
with tonal grunts. On moonlit
nights these monkeys make an
owl-like hoot of 2–4 syllables—
hence their common name.[198]
Sleeping site: Vine tangles and
tree hollows.[198]

Brazil

Northern gray-necked owl monkey *(Aotus trivirgatus).*

R. A. MITTERMEIER

Brown Titi Monkey *Callicebus brunneus*

Taxonomy Disputed. May be a subspecies of *C. cupreus*.[315]

Distinguishing Characteristics
Brown titi monkeys have a reddish brown back, a blackish head and limbs, reddish brown sideburns, and a blackish tail with a pale tip. The underparts are not sharply defined.[357]

Physical Characteristics Head and body length: ♀ 312mm *[12.3in]*, ♂ 317mm (300–345) *[12.5in (11.8–13.6)]*.[357] **Tail length:** ♀ 410mm (380–440) *[16.1in (15.0–17.3)]*, ♂ 397mm (371–420) *[15.6in (14.6–16.5)]*.[357] **Weight:** ♀ 850g *[30.0oz]*, ♂ 845g *[29.8oz]*.[357] **Intermembral index:** 75.[365] **Adult brain weight:** NA.

Habitat Riverine and flooded forest.[450]

Diet Fruit, 47%; leaves, 28%; insects, 15%: flowers, 2%.[452] Figs and *Brosimum rubescens* are important fruit sources.[450]

Life History NA. Offspring: 1.[357]

Locomotion Quadrupedal.

Social Structure Monogamous 1 male–1 female with offspring.[450] **Group size:** 2–4.[450] **Home range:** 3–4.8ha.[357] **Day range:** 593–786m *[1946–2579ft]*.[357]

Behavior Diurnal and arboreal. Brown titis spend 48% of their time below 10 meters *[33ft]* in the understory but do feed in the exposed canopy.[450] They rest for 2 hours at midday and feed on leaves during the last 2 hours of the day.[450] Six percent of a population tested positive for malaria.[153] **Association:** Partially sympatric with chestnut-bellied titis (*C. caligatus*), red titis (*C. cupreus*), and Bolivian gray titis (*C. donacophilus*).[357] **Sleeping site:** Trees with vine tangles. The same tree is often used on different days.[450]

Brazil, Peru

L.C. MARIGO

Titi monkeys are monogamous and often huddle together with their tails entwined.

Chestnut-bellied Titi Monkey *Callicebus caligatus*

S. NASH

Taxonomy Disputed. Elevated from a subspecies of *C. cupreus*.[315]

Distinguishing Characteristics
The dusky back of chestnut-bellied titi monkeys contrasts with their reddish underparts and tail, which has a mixture of buff and gray hairs. The forehead is blackish.[357]

Physical Characteristics Head and body length: ♀ 326mm (300–390) *[12.8in (11.8–15.4)]*, ♂ 342mm (310–410) *[13.5in (12.2–16.1)]*.[357] **Tail length:** ♀ 411mm (385–460) *[16.2in (15.2–18.1)]*, ♂ 422mm (380–476) *[16.6in (15.0–18.7)]*.[357] **Weight:** NA. **Intermembral index:** 75.[364] **Adult brain weight:** NA.

Habitat Riverbanks.[364]

Diet NA.

Life History NA.

Locomotion NA.

Social Structure NA.

Behavior Presumed to be diurnal and arboreal. No field studies had been published as of 1995. **Association:** Partially sympatric with brown titis (*C. brunneus*), red titis (*C. cupreus*),[357] *C. dubius*, and collared titis (*C. torquatus*).[357]

Brazil, Peru

Ashy Titi Monkey *Callicebus cinerascens*

S. NASH

Brazil

Taxonomy Disputed. Elevated from a subspecies of *C. moloch* in 1990.[313]

Distinguishing Characteristics The body, head, and tail of ashy titis are "gray to blackish agouti with a contrasting tawny agouti at mid dorsum."[357]

Physical Characteristics Head and body length: ♀ 342mm (320–380) *[13.5in (12.6–15.0)]*, ♂ 330mm *[13.0in]*.[357] **Tail length:** ♀ 450mm (390–480) *[17.7in (15.4–18.9)]*, ♂ 480mm *[18.9in]*.[357] **Weight:** NA. **Intermembral index:** 75.[364] **Adult brain weight:** NA.

Habitat Riverine areas.[364]

Diet NA.

Life History NA.

Locomotion NA.

Social Structure NA.

Behavior Presumed to be diurnal and arboreal. No field studies had been published as of 1995.

Association: Partially sympatric with dusky titis *(C. moloch)*.[357]

Red Titi Monkey *Callicebus cupreus*

Taxonomy Disputed. 3 subspecies.[723]

Distinguishing Characteristics The back of red titis is buffy brown agouti, and the underparts are reddish orange. The tail is agouti, with a buffy tip.[357]

Physical Characteristics Head and body length: ♀ 328mm (304–347) *[12.9in (12.0–13.7)]*, ♂ 336mm (312–360) *[13.2in (12.3–14.2)]*.[357] **Tail length:** ♀ 414mm (334–444) *[16.3in (13.1–17.5)]*, ♂ 428mm (405–450) *[16.9in (15.9–17.7)]*.[357] **Weight:** ♂ 1163g *[41.0oz]*, ♀ 1178g *[41.5oz]*.[357] **Intermembral index:** 75.[364] **Adult brain weight:** NA.

Habitat Tropical rain forest.

Diet NA.

Life History NA.

Locomotion Quadrupedal.

Social Structure NA.

Behavior Diurnal and arboreal. Red titi monkeys prefer to forage in the lower levels of the canopy.[452]

Association: Partially sympatric with brown titis *(C. brunneus)*,[357] chestnut-bellied titis *(C. caligatus)*, and collared titis *(C. torquatus)*.[357]

Brazil, Peru

R. A. MITTERMEIER

Bolivian Gray Titi Monkey *Callicebus donacophilus*

Taxonomy Disputed. 2 subspecies.[723] Elevated to a full species in 1990.[315]

Distinguishing Characteristics
The back of Bolivian gray titis is grayish agouti to buffy orange agouti, the chest and belly are orange, and the tail is a mixture of blackish buff, with orange at the base. These titis have conspicuous whitish ear tufts.[357]

Physical Characteristics Head and body length: ♀ 340mm (305–420) *[13.4in (12.0–16.5)]*, ♂ 311mm (278–330) *[12.2in (10.9–13.0)]*.[357]
Tail length: ♀ 440mm (410–460) *[17.3in (16.1–18.1)]*, ♂ 411mm (372–445) *[16.2in (14.6–17.5)]*.[357]
Weight: 800g *[28.2oz]*.[247]
Intermembral index: 75.[364] **Adult brain weight:** NA.
Habitat Moist forest.[364]
Diet NA.
Life History NA.
Locomotion Quadrupedal.
Social Structure NA.
Behavior Presumed to be diurnal and arboreal. Although Bolivian gray titis are occasionally seen in zoos, no field studies had been published as of 1995. **Association:** Partially sympatric with brown titis *(C. brunneus)*.[357]

Bolivia

Titi Monkey *Callicebus dubius*

S. NASH

Taxonomy Disputed. Tentatively declared a full species in 1990 but could be a hybrid of the red titi *(C. cupreus)* and the chestnut-bellied titi *(C. caligatus)*.[357]

Distinguishing Characteristics
This titi has a reddish body, a buffy whitish forehead, and black forearms, crown, and sideburns.[357]

Physical Characteristics Head and body length: ♀ 385mm (370–400) *[15.2in (14.6–15.7)]*, ♂ 370mm *[14.6in]*.[357] **Tail length:** ♀ 415mm (390–440) *[16.3in (15.4–17.3)]*, ♂ 470mm *[18.5in]*.[357] **Weight:** NA. **Intermembral index:** NA. **Adult brain weight:** NA.
Habitat Primary lowland rain forest.
Diet NA.
Life History NA.
Locomotion Quadrupedal.
Social Structure NA.
Behavior Presumed to be diurnal and arboreal. No field studies of this species had been published

as of 1995. **Association:** Partially sympatric with chestnut-bellied titis *(C. caligatus)*.[357]

Brazil

Hoffmann's Titi Monkey *Callicebus hoffmannsi*

LOWER RISK

CITES II

Brazil

R. A. MITTERMEIER

(283–360) [12.7in (11.1–14.2)].[357]
Tail length: ♀ 465mm (420–513)
[18.3in (16.5–20.2)], ♂ 453mm
(400–525) [17.8in (15.7–20.7)].[357]
Weight: 920g [32.4oz].[247]
Intermembral index: 75.[364] **Adult
brain weight:** NA.
Habitat Swamps and river
edges.[364]
Diet NA.
Life History NA.

Locomotion NA.
Social Structure NA.
Behavior Presumed to be diurnal
and arboreal. No field studies had
been published as of 1995.
Association: Not sympatric with
other *Callicebus* species.[357]

Taxonomy Disputed. 2 subspecies.[723] Elevated from a
subspecies of *C. moloch* in 1990.[357]
Distinguishing Characteristics The head and body of
Hoffmann's titis are grayish to blackish agouti. The
throat, sideburns, and underparts are yellowish red,
and the tail is blackish agouti with a buffy tip.[357]
Physical Characteristics Head and body length: ♀
316mm (271–351) [12.4in (10.7–13.8)], ♂ 322mm

Titi Monkey *Callicebus modestus*

LOWER RISK

CITES II

S. NASH

Bolivia

Taxonomy Disputed. Elevated from
a subspecies of *C. cupreus* in
1990.[357] Known only from 2
museum specimens.[357]
Distinguishing Characteristics
This titi has a grizzled brown
body, a reddish brown forehead,
and white ear tufts. The tail is
blackish agouti.[357]
**Physical Characteristics Head and
body length:** ♂ 315mm [12.4in].[357]
Tail length: ♂ 400mm [15.7in].[357]
Weight: NA. **Intermembral index:**
75.[364] **Adult brain weight:** NA.
Habitat Lowland and riverine rain
forest.[357]
Diet NA.
Life History NA.
Locomotion NA.
Social Structure NA.
Behavior Presumed to be diurnal
and arboreal. No field studies had
been published as of 1995.
Association: Not sympatric with
other *Callicebus* species. This titi
may be an isolated relict
species.[357]

Dusky Titi Monkey *Callicebus moloch*

Taxonomy Disputed. Several former subspecies have been elevated to full species.[357] Much of the information included for this species may apply to the former subspecies, *C. m. brunneus*.

Distinguishing Characteristics
The head and trunk of dusky titi monkeys are buffy or grayish agouti. The sideburns, underparts, and inner sides of the limbs are a contrasting orange.[357]

Physical Characteristics Head and body length: ♀ 345mm (294–420) *[13.6in (11.6–16.5)]*, ♂ 348mm (296–450) *[13.7in (11.7–17.7)]*.[600] **Tail length:** ♀ 432mm (363–530) *[17.0in (14.3–20.9)]*, ♂ 449mm (392–500) *[17.7in (15.4–19.7)]*.[600] **Weight:** ♀ 700–1020g *[24.7–36.0oz]*, ♂ 800–1200g *[28.2–42.3oz]*.[247] **Intermembral index:** 74.[234] **Adult brain weight:** 19g *[0.7oz]*.[346]

Habitat Gallery, swamp, flooded forest. Edges and lower canopy levels are preferred.[450]

Diet Fruits, 54%; leaves, 28%; animal prey, 17%.[699] Dusky titis may supplement their diet with leaves when they are chased from fig trees by larger primates.[911] They eat spiders, ants and butterflies,

but no crickets.[140]

Life History Weaning: 8mo.[909] **Juvenile:** 4–24mo.[450] **Subadult:** 24–36mo.[450] **Sexual maturity:** 30mo.[346] **Estrus cycle:** NA. **Gestation:** 155d.[909] **Age 1st birth:** 36mo.[708] **Birth interval:** 12mo. **Life span:** 12y.[708] Mating season: Nov–Mar.[450] The male carries the infant 92% of the time, starting at birth; the female carries the infant 6% of the time, while it suckles.[909] The subadult may occasionally help with the carrying (2%) when the infant is 3 weeks old. By 8 weeks, the male carries the infant 57% of the time, and it is independent 40%. By 16 weeks, the male carries the infant only 1% of the time.[909]

Locomotion Quadrupedal walking and running.[140]

Social Structure Monogamous 1 male–1 female. Dusky titi groups are formed around a strong attachment between a single male and female. No additions to groups other than births have been observed. The male leads the group 54% of the time.[450] The female leads the group when she is lactating and the father is carrying the infant.[909] **Emigration:** Both genders emigrate after they are 3 years old.[909] **Group size:** 2–5.[453] **Home range:** 6–12ha.[826] **Day range:** 576m (315–870) *[1890ft (1034–2854)]*.[450]

Behavior Diurnal and arboreal.[198] Dusky titis have very cryptic slow movements. They blend into their surroundings and are hard to detect except in the morning, when groups announce their presence with loud calls. Both members of a pair vocalize in a coordinated fashion between 6 and 8 a.m.[826] The group stays at lower levels of the forest in tangles but feeds in the canopy for short periods.[826] All group members twine their tails together during sleep, grooming, and dueting.[450] The male shares food and plays with the infant and protects it from predators for the first year.[909] The female rarely shares solid food or plays with the infant.[909] The relationship between infant and father is

close; they remain closer than any other 2 individuals in the group, including the monogamous pair.[453] The male does most of the territorial defense.[908] The female grooms the male more often than the male grooms her.[908] In a captive study, when both members of a pair were separated, they experienced measurable stress that was relieved only by their being reunited.[708]

Association: Sympatric with collared titis (*Callicebus torquatus*)[456] and partially sympatric with ashy titi monkeys (*C. cinerascens*).[357] Dusky titis often associate with saddleback tamarins (*Saguinus fuscicollis*), for less than 1 hour per meeting; dusky titis, unlike *S. fuscicollis,* do not actively search for hidden prey,[140] and the 2 species eat only 3 species of fruit in common.[140] Dusky titis occasionally

Dusky Titi Monkey *continued*

associate with emperor tamarins *(S. imperator)*.[826]
Mating: In a captive study the female titi preferred her mate to strangers, whereas the male showed a "conflict of choice" when presented with another female.[27] Mating lasts 10–30 seconds.[909] **Scent marking:** The male has a large sternal gland, which he rubs on branches.[450] **Vocalizations:** 16.[450] The territorial long call is an antiphonal duet given before sunrise but rarely when it is raining.[450] The juvenile (1.5y old) and subadult may join in the duet.[908] The repeated calls form phrases that can be combined to form long vocal sequences.[696] One call is reported to be "turkey-like."[911] **Sleeping site:** Vine tangles.[453]

Brazil

Andean Titi Monkey *Callicebus oenanthe*

S. NASH

Taxonomy Disputed. Declared a species in 1990. Known from 6 museum specimens.[357]

Distinguishing Characteristics Andean titi monkeys have an agouti body and tail, contrasting orange underparts and side-burns, a buff frontal blaze, and a pale malar stripe.[357]

Physical Characteristics Head and body length: ♀ 315mm *[12.4in]*, ♂ 303mm (300–306) *[11.9in (11.8–12.0)]*.[357] **Tail length:** ♀ 367mm (360–375) *[14.4in (14.2–14.8)]*, ♂ 390mm (380–400) *[15.4in (15.0–15.7)]*.[357] **Weight:** NA. **Intermembral index:** NA. **Adult brain weight:** NA.

Habitat Montane forests at altitudes of 750–950m *[2461–3117ft]*.[357]

Diet NA.
Life History NA.
Locomotion NA.
Social Structure NA.

Behavior Diurnal and arboreal. **Association:** Not sympatric with any other titi monkey. Andean titis are found in the same montane habitat as yellow-tailed woolly monkeys *(Lagothrix flavicauda)* and a red-necked night monkey *(Aotus miconax—see A. nigriceps)*.[357]

Peru

Beni Titi Monkey *Callicebus olallae*

Taxonomy Disputed. Proposed in 1990. Known only from 1 museum specimen from its type locality.[315]

Distinguishing Characteristics Beni titi monkeys have an orange back, orange sides, and a dark agouti tail. The face is framed in black, and the forehead is reddish brown.[357]

Physical Characteristics Head and body length: ♂ 325mm *[12.8in]*.[357] **Tail length:** ♂ 425mm *[16.7in]*.[357] **Weight:** NA. **Intermembral index:** NA. **Adult brain weight:** NA.

Habitat Lowland forest up to 200m *[656ft]*.[357]

Diet NA.
Life History NA.
Locomotion Quadrupedal.
Social Structure NA.
Behavior Presumed to be diurnal and arboreal. No field studies had been published as of 1995.
Association: Not sympatric with

other *Callicebus* species. The Beni titi monkey may be a relict species.[357]

Peru

S. NASH

Masked Titi Monkey *Callicebus personatus*

R. A. MITTERMEIER

C. p. personatus.

Taxonomy 4 subspecies.[723] A new species with "zebra-like" stripes on its sides was proposed in 1994 but had not been recognized as of 1995.[465]

Distinguishing Characteristics Masked titis have a black forehead and sideburns, and the body is grayish to buffy to yellowish or orange. The tail is the same color as the body, but mixed with black.[357] The hands and feet are black.

Physical Characteristics Head and body length: ♀ 356mm (310–400) *[14.0in (12.2–15.7)]*, ♂ 380mm (350–420) *[15.0in (13.8–16.5)]*.[357] **Tail length:** ♀ 485mm (418–560) *[19.1in (16.5–22.0)]*, ♂ 508mm (470–550) *[20.0in (18.5–21.7)]*.[357] **Weight:** ♀ 1378g (970–1600) *[48.6oz (34.2–56.4)]*, ♂ 1270g (1050–1650) *[44.8oz (37.0–58.2)]*.[357] **Intermembral index:** 73.[234] **Adult brain weight:** NA.

Habitat Primary and secondary Atlantic coastal forest from sea level to 1000m *[3281ft]*; banana groves.[357]

Diet Fruit (27 types), 81%; leaves, 18%; flowers.[452] These monkeys use 91 plant species. No insect feeding has been observed.

Brazil

Life History NA. Birth season: Aug–Oct.[450] The male carries the infant most of the time.[453]

Locomotion Quadrupedal.

Social Structure Monogamous. **Group size:** 3.7[699] (2–6).[198] **Home range:** 4.7ha,[453] 24ha.[1006] **Day range:** 694.7m *[2279ft]*,[450] 1007m *[3304ft]*.[1006]

Behavior Diurnal and arboreal. Masked titi monkeys spend 40% of the day resting, 32% traveling, and 27% feeding.[1006] **Association:** Not sympatric with other *Callicebus* species. Masked titis occasionally feed with Geoffroy's tufted-eared marmosets (*Callithrix geoffroyi*).[450] **Vocalizations:** The dawn duet starts well after sunrise[450] with loud whoops.[198] They will call when it is raining.[450] **Sleeping site:** Branches 25–40 meters *[82–131ft]* high in large trees. The group sleeps huddled together with tails entwined.[453]

C. p. nigrifrons.

Collared Titi or Widow Monkey *Callicebus torquatus*

R. A. MITTERMEIER

Taxonomy 6 subspecies.[723]

Distinguishing Characteristics
Collared titi monkeys have buffy, yellowish, or orange hands; a dark brown body; a black tail; and a bare, black face. The neck has a white collar around it.[456]

Physical Characteristics Head and body length: ♀ 325mm (232–360) *[12.8in (9.1–14.2)]*, ♂ 331mm *[13.0in]*.[357] **Tail length:** ♀ 458mm (425–493) *[18.0in (16.7–19.4)]*, ♂ 478mm *[18.8in]*.[357] **Weight:** ♀ 1151–1462g *[40.6–51.6oz]*, ♂ 1100–1500g *[38.8–52.9oz]*.[450] **Intermembral index:** 75.[188] **Adult brain weight:** 22.4g *[0.8oz]*.[346] The male's canines are 4% larger than the female's.[450]

Habitat Primary and secondary terra firma forests; forests on white sands and black-water streams.[450]

Diet Fruit and seeds, 65%; leaves, 15%; animal prey (mostly insects), 16%; other, 4%.[452] Sixty percent of the diet is from 8[450] of the 57 plant species eaten.[188] In July the fruit of the ungurahui palm *(Jessenia polycarpa)* is the "single most important food item," making up 22% of the diet.[188]

Life History Weaning: 4.6mo. **Sexual maturity:** ♀ 24–36mo, ♂ 24–42mo.[699] **Estrus cycle:** NA. **Gestation:** NA. **Age 1st birth:** 48mo.[346] **Birth interval:** 12mo. **Life span:** NA. Mating season: Start of rainy season. Birth season: Nov–Mar.[188] Offspring: 1.

Locomotion Quadrupedal; some vertical clinging and leaping and suspensory.

Social Structure Monogamous 1 male–1 female. Groups are cohesive and territorial, with only 10% overlap in home ranges.[456] **Emigration:** Subadult males and females are forced to leave the group at 3+ years old. **Group size:** 2–6.[456] **Home range:** 4–20ha.[450] **Day range:** 820m (466–1437) *[2690ft (1529–4715)]*.[265]

Behavior Diurnal and arboreal. Collared titis suspend themselves when they feed[188] and rest in the lower levels of the forest.[456] They spend less than 1% of the time on the ground. The female leads the group.[450] The male cares for, plays with, and protects the offspring, including during agonistic encounters with its mother.[450] Each animal grooms or is groomed 135 minutes each day. The female grooms the male more than he grooms her, but he grooms the juveniles more often and longer than the female does and grooms the youngest offspring the most.[450] The infant and juvenile beg for food from the male by staring or putting hands, face, or mouth on the food.[776] In one study the infant's success rate for begging large scarce fruits, a grasshopper, and young leaves from the male was 54.5%.[776] Territories shift over time, and parents may leave part of their range to their offspring.[188] The group's day range increases with its size.[450] Activity budget: Rest, 47%; feeding, 26%; travel, 13%; other, 14%.[456] **Association:** Sympatric with dusky titis *(C. moloch)*, but the collared titi's dawn call is later (closer to dawn), less frequent, and given for a longer time.[456] Collared titis are partially sympatric with chestnut-bellied titi *(C. caligatus)* and red titis *(C. cupreus)*.[357] They occasionally associate with saddleback tamarins *(Saguinus fuscicollis)*.[456] Woodcreepers, which are insect-eating birds, have been observed to follow a troop.[450] **Mating:** These titis copulate much more often during their breeding season than do most Neotropical primates.[450] **Scent marking:** Marks are made with sternal glands.[450] **Vocalizations:** The dawn duet is given just at sunrise.[450] The male's dawn call lasts 2–7 minutes. The female group solidarity call is accompanied by a running and leaping male display.[456] The territorial morning call may be given only once per 10–14 days.[188] Frequent *chirrup* contact calls are given. **Sleeping site:** A large branch (>250mm *[>9.8in]* in diameter) slightly above the surrounding canopy[198] (25.1m *[82ft]*). These titis sleep in a different emergent tree each night. The male and female groom the most at dusk at the sleeping site.[450]

Brazil, Colombia, Venezuela

White-fronted Capuchin *Cebus albifrons*

Taxonomy 11 subspecies.[723]

Distinguishing Characteristics
The body of white-fronted capuchins varies from light to dark brown. They have a dark brown wedge-shaped cap, yellowish underparts, and a prehensile tail[251] that is dark at the base and light yellow at the tip.[198]

Physical Characteristics Head and body length: 358–460mm *[14.1–18.1in]*.[198] **Tail length:** 401–475mm *[15.8–18.7in]*.[198] **Weight:** ♀ 1400–2228g *[3.1–4.9lb]*, ♂ 1700–3260g *[3.7–7.2lb]*.[247] **Intermembral index:** 82.[234] **Adult brain weight:** 82g *[2.9oz]*.[346]

Habitat Primary deciduous, gallery, mangrove, and flooded forest up to 2000m *[6562ft]*.[898]

Diet Dry season—fruit, 53%; seeds, 42%; nectar, 3%; pith, 1%; animal prey. Wet season—fruit, 99%.[409] These capuchins use 68 plant species. Half of the animal prey is social insects: termites, ants, and wasp larvae and pupae.[409]

Life History Weaning: 9mo.[346] **Sexual maturity:** ♀ 43.1mo.[346] **Estrus cycle:** NA. **Gestation:** 162d.[870] **Age 1st birth:** 48mo.[708] **Birth interval:** 18mo. **Life span:** 44y.[708]

A young white-fronted capuchin, *C. albifrons.*

Locomotion Quadrupedal walking, running, and jumping to 4m *[13ft]*.[251]

Social Structure Multimale-multifemale groups with fewer males than females. Home ranges overlap greatly.[826] **Group size:** 10–30.[251] **Home range:** 200–300ha.[898] **Day range:** 1850m *[6070ft]*.[903]

Behavior Diurnal and arboreal. White-fronted capuchins prefer the middle strata of the forest (15–30m *[49–98ft]*)[251] but forage on the ground up to 10% of the time.[409] One male dominates the troop,[826] and social grooming involves mostly the alpha male or female and the offspring.[826] The dominant male's response to an aerial predator is to hide and not give an alarm bark until other males form a coalition to threaten the predator.[408] The alpha male is very aggressive to other groups whenever 2 groups meet.[408] Adult males associate together and cooperatively defend their group, whereas each female forages separately, avoiding other adults.[408] Feeding ecology influences male mating strategy, aggression, and cooperation.[408] Unlike tufted capuchins *(C. apella),* which get 63% of their diet in small-crowned trees (<10m *[<33ft]* in diameter), this species gets 50% of its diet in large-crowned trees (>20m *[>66ft]*). A large patch of food in a large-crowned tree cannot be monopolized, so food-related aggression is rare.[408] White-fronted males mate promiscuously and cannot be certain of paternity, so they do not come to the defense of a juvenile when it squeals.[408] **Activity budget:** Insect foraging, 39%; plant feeding, 22%; travel, 21%; rest, 18%.[826] **Association:** This species associates with tufted capuchins *(C. apella)* and displaces them at fruit sources during the dry season.[826]

White-fronted capuchins *(C. albifrons)* and tufted capuchins *(C. apella)* compete for the fruit of the *Scheelea* palm.

It associates regularly with common squirrel monkeys *(Saimiri sciureus)*[251] and reportedly associates with black-headed uacaris *(Cacajao melanocephalus).*[486] **Mating:** The dominant male follows an estrous female and sniffs her urine. Little male–male aggression is seen during estrus. **Scent marking:** Marks are made by rubbing the chest on branches.[251] **Vocalizations:** Loud vocalizations advertise a troop's location, and other groups avoid the area.[826]

Brazil, Colombia, Venezuela

Tufted or Brown Capuchin *Cebus apella*

LOWER RISK

CITES II

Taxonomy Disputed. 10 subspecies.[723] The yellow-breasted capuchin *(C. a. xanthosternos)*, which is included with this species, was recently recognized as a species,[723] but little is known about its behavior.

Distinguishing Characteristics The cap of tufted capuchins is made of short, erect black hairs that may form 2 ridges or "tufts" on either side of the crown. The shoulders are lighter than the overall body color,[198] which varies from light to dark brown. The facial pattern varies with the subspecies, except for the black sideburns. The hands and feet are always black.[509] The prehensile tail is darkest at the tip.[196]

Physical Characteristics Head and body length: 350–488mm *[13.8–19.2in]*.[198] **Tail length:** 375–488mm *[14.8–19.2in]*.[198] **Weight:** ♀ 1370–3400g *[3.0–7.5lb]*, ♂ 1300–4800g *[2.9–10.6lb]*.[247] **Intermembral index:** 82.[234] **Adult brain weight:** 71g *[2.5oz]*.[346]

Habitat Primary and secondary rain forest to semideciduous lowland and montane up to 2700m *[8859ft]*.[196]

Diet Fruit, 66%; seeds, 25%; pith, 7%; nectar, 1%; animal prey, including insects, birds, eggs, reptiles,[196] bats,[402] and mammals up to 900g *[32oz]* in body weight.[409] These capuchins eat 96 species of fruit.[408] The pith of *Scheelea* palm fronds is a keystone food during the dry season when fruit is scarce.[408]

Life History Infant: 6mo.[99] **Weaning:** 12mo.[99] **Juvenile:** 6–24mo.[99] **Subadult:** 24–42mo.[99] **Sexual maturity:** ♀ 84mo,[196] ♂ 56mo.[346] **Estrus cycle:** 18d.[251] **Gestation:** 149–158d.[698] **Age 1st birth:** 42mo.[346] **Birth interval:** 22mo.[698] **Life span:** 40y.[251] Birth seasons:

C. a. nigritus.

C. a. apella.

Oct–Jan.[698] Offspring: 1.[196] Females have no estrous swelling.[196]

Locomotion Quadrupedal; jumping to 3–4m *[10–13ft]*.[251]

Social Structure Multimale-multifemale groups with equal numbers of males and females. One male is dominant to all the others, and young males may form a socially separate subgroup.[826] **Group size:** 8–14.[826] **Home range:** 25–40ha,[196] to 355ha.[927] **Day range:** 2000m *[6562ft]*.[903]

Behavior Diurnal and arboreal. Tufted capuchins are very intelligent and curious. It has been hypothesized that intelligence is related to the way an animal searches for food. Searching for hard-to-find foods that are available only for a short time, such as insects and fruit, may require a larger brain and more energy-rich foods to maintain it. This species shows submission by a genital display and raised eyebrows.[251] The alpha male responds to aerial predators by giving loud barks and remaining visible while the rest of the troop flees.[251] When 2 groups meet at a food tree, the dominant male leads the attack; otherwise there is very little intergroup aggression.[408] The dominant male rapidly comes to the defense of juveniles that he has probably fathered,[408] but he is aggressive to juvenile males born before his tenure in the group.[408] Capuchins have been reported to hunt for and capture frogs that live in bamboo stems. Captors of frogs rarely share their prized food with infants or others in the group.[402] Allomothering is common.[847] Capuchins do not recognize themselves in a mirror.[19] These

C. a. libidinosus.

monkeys have been trained to perform tasks for quadriplegics.[889] **Association:** Tufted capuchins associate with white-fronted capucins *(C. albifrons)* and are displaced by them at fruit sources during the dry season.[826] Tufted capuchins associate with white-nosed bearded sakis *(Chiropotes albinasus)*,[850] occasionally with buffy sakis *(Pithecia albicans)*,[656] and reportedly with black-headed uacaris *(Cacajao melanocephalus)*.[238] Tufted capuchins are often followed by a troop of common squirrel monkeys *(Saimiri sciureus)*.[196] **Mating:** During the first two-thirds of estrus, females constantly follow and solicit the alpha male by using distinct calls, facial expressions, and postures. The male copulates only once a day. In the last 2 days of estrus, the dominant male "guards" the female from the subordinate males. When he stops guarding, the female copulates quickly with the other males in the group.[407] After mating, tufted capuchins display a "reverse mount" in which the "female mounts the male, clasping him around the waist with her

A former subspecies of *C. apella*, the yellow-breasted capuchin *(C. a. xanthosternos)* was proposed as a valid species in 1995. It is critically endangered.

Tufted or Brown Capuchin *continued*

Tufted capuchins often break open dead twigs searching for insects.

arms and riding his lower back."[408] **Scent marking:** Each individual maintains olfactory identity by washing its palms and feet in its own urine and scratching its fur. Females may monitor male smell to detect sexual maturity. Males do not conspicuously monitor females.[665] **Vocalizations:** Alarm calls are given at the sight of large raptors. In French Guiana the harpy eagle's second most common prey is capuchin monkeys.[251] **Sleeping site:** Palm trees are preferred.[927]

Northern and Central South America

White-throated Capuchin *Cebus capucinus*

Taxonomy 4 subspecies.[723]

Distinguishing Characteristics
White-throated capuchins have a white to yellowish throat, head and shoulders. The back and prehensile tail are black. The hair on the crown forms a V.[198]

Physical Characteristics Head and body length: 335–453mm *[13.2–17.8in].*[198] **Tail length:** 350–551mm *[13.8–21.7in].*[198] **Weight:** ♀ 2666g *[5.9lb],* ♂ 3868g *[8.5lb].*[246] **Intermembral index:** NA. **Adult brain weight:** 79.2g *[2.8oz].*[346]

Habitat Primary and secondary evergreen rain forest, mangroves, deciduous dry forest from sea level to 2100m *[6890ft].*[898] The main canopy is preferred.

Diet Fruits (95 types), 65%; leaves, 15%; berries, nuts, seeds, shoots, buds and flowers (24 species), gums, bark, and animal matter, including insects, spiders, small vertebrates (birds, infant squirrels, baby coatimundis, and lizards), eggs, crabs, and oysters, 20%.[251]

Life History Weaning: 12mo.[251] **Sexual maturity:** 36mo.[708] **Estrus cycle:** NA. **Gestation:** 157–167d.[698] **Age 1st birth:** 48mo.[708] **Birth interval:** 19mo.[698] **Life span:** 46.9y.[708] **Birth peak:** Dec–Apr.[705] Infants are born with a gray face and ears but look like the adults by 3 months.[251]

Locomotion Quadrupedal; jumping to 3–4m *[10–13ft].*[251]

Social Structure Multimale-multifemale groups with more females than males. Males defend the territory.[251] **Emigration:** Males emigrate and may remain solitary before joining another group. **Group size:** 10–20.[251] **Home range:** 32–85.5ha.[898] **Day range:** 1200[251]–2000m *[3937–6562ft].*[903]

Behavior Diurnal and arboreal.[198] White-throated capuchins, like many primates, enhance the health of the forest by pollinating some plants and dispersing the seeds of others. When they eat the buds of *Gustavia simaruba,* branching of the tree increases, enhancing fruit production.[251] Males forage near or on the ground for large vertebrates and invertebrates; females tend to forage on smaller branches for small invertebrates. This species is active before dawn and is often noisy. Allogrooming increases during the birth season and the dry season, when foraging is done on the ground and there are more ticks.[251] Jumping up and down and shaking branches are part of the threat display.[251] **Association:** This species occasionally associates with black-handed spider monkeys *(Ateles geoffroyi).*[251] **Mating:** Promiscuous.[251] A special "protruded lip face and warble vocalization" precede mating.[251] **Scent marking:** Marks are made by urinating on hands, feet, and tail

tip.[251] **Vocalizations:** 14, including an *arrawh* loud call given when an individual is out of sight of the group and wants to regain contact. A "purr" is a friendly appeasement call.[251] **Sleeping site:** This capuchin has

White-throated Capuchin *continued*

been observed to spend 47% of its nights in 1 sleeping tree in the core area of its home range. Groups of 2–4 individuals sleep together.[251]

Ecuador to
Honduras

Weeper or Wedge-capped Capuchin *Cebus olivaceus*

Taxonomy Disputed. 5 subspecies.[723] Until 1978 the name of this species was *C. nigrivittatus*.[315] In 1992 a new capuchin *(C. kaapori)* was discovered and described in Brazil. It is included here because little is known about its behavior.[678]

Distinguishing Characteristics Weeper capuchins have a tawny brown body, yellowish shoulders and upper arms, and a brownish yellow head with a V-shaped brown cap.[198] The prehensile tail is brown,[198] and the back of the head and neck is reddish.[198]

Physical Characteristics Head and body length: 374–460mm *[14.7–18.1in]*.[198] **Tail length:** 400–554mm *[15.7–21.8in]*.[198] **Weight:** ♀ 2395g *[5.3lb]*, ♂ 2974g *[6.6lb]*.[246] **Intermembral index:** NA. **Adult brain weight:** 80.8g *[2.8oz]*.[346] Females are born with an elongated clitoris that looks similar to a penis, and they have an os clitoris, a bone that is similar to a baculum.[251]

Habitat Evergreen rain forest, dry forest, and submontane forest up to 2000m *[6562ft]*.[898]

Diet Fruit, seeds, and animal prey, including snails (32%) and social insects (22%).[409]

Life History Infant: 12mo.[626] **Weaning:** 24mo.[626] **Juvenile:** 24–72mo.[626] **Subadult:** 72–144mo.[626] **Sexual maturity:** NA. **Estrus cycle:** NA. **Gestation:** NA. **Age 1st birth:** 72mo.[626] **Birth interval:** 12–24mo.[626] **Life span:** NA. Births: Year-round. Birth peak: May–Jun.[626]

H. L. QUEIROZ

The Ka'apor capuchin *(C. kaapori)* was proposed as a valid species in 1995.

Locomotion Quadrupedal walking, running, and jumping to 3m *[10ft]*.[251]

Social Structure Multimale-multifemale groups with only 1 breeding male.[626] Male hierarchical rank is by age and size.[251] Female rank is matrilineal.[626]

Emigration: Males emigrate as early as age 2.[626] Group ranges overlap, but intergroup interactions are avoided.[626] **Group size:** 10–33.[898] **Home range:** NA. **Day range:** 2300m *[7546ft]*.[903]

Behavior Diurnal and arboreal. Weeper capuchins forage on the

X. VALDERAMA (NATURAL HABITAT)

This young weeper capuchin is using its tail as a fifth limb.

Weeper or Wedge-capped Capuchin *continued*

The weeper capuchin *(C. o. apiculatus)* gets its common name from its plaintive contact call.

ground (14.5%) and in the canopy.[409] The activity budgets of females in small groups (16) are different in different seasons, unlike the activity budgets of females in large groups (32), which do not vary.[557] During the wet season, small groups spend a high percentage of time traveling and gathering food, with little rest. In the dry season, food gathering time is the same, but they travel much less and rest more.[557] The larger groups have priority access to small food patches, so smaller groups, which are displaced, must conserve energy. Females in large groups have higher fecundity and higher expected lifetime reproductive success than females in small groups.[626] "High ranking females may have an earlier age of first reproduction and a slightly shorter interbirth interval."[626] "The benefit of kinship appear[s] to be a reduction in aggression rather than increased affilia-

tion."[626] Aggressive displays include bouncing and branch shaking.[251] Social play among juveniles takes place when the adults are resting.[251] Subordinates "grin" in the presence of a dominant to appease and promote contact.[251] Allomothering has been reported.[626] **Association:** Weeper capuchins have an agonistic relationship with red howlers *(Alouatta seniculus)* when they meet in the same fig tree.[251]

Scent marking: Urine washing.[251] **Vocalizations:** 12.[347] The common name "weeper" comes from the plaintive quality of one of the contact vocalizations.[697] This species has 3 spacing calls—*huh* to maintain distance between individuals, *arrawks* to decrease distance, and *hehs* to increase distance.[697]

Guyana, French Guiana, Suriname, Brazil, Venezuela, Colombia

Bolivian Squirrel Monkey *Saimiri boliviensis*

Taxonomy Disputed. 4 subspecies.[723] Elevated from a subspecies of *S. sciureus* in 1984.[137]

Distinguishing Characteristics Bolivian squirrel monkeys are sexually dichromatic, with black on the female and gray on the male. Both genders have a narrow base of yellow on the crown, and the same color is found on the forearms, hands, and feet. Over each eye is a Roman-type (rounded) arch.[354]

Physical Characteristics Head and body length: 310mm *[12.2in]*.[354] **Tail length:** 360mm *[14.2in]*.[354] **Weight:** ♀ 700–900g *[24.7–31.7oz]*, ♂ 963–1088g *[34.0–38.4oz]*.[247] **Intermembral index:** NA. **Adult brain weight:** NA.

Habitat Primary and secondary, tropical rain forest.[826]

Diet Animal prey, particularly insects (including caterpillars), 82%; fruit and seeds, 18%.[826] During the dry season, these monkeys rely on figs.[826]

Life History Weaning: NA. **Sexual maturity:** NA. **Estrus cycle:** NA. **Gestation:** 155[891] 170d.[826] **Age 1st birth:** NA. **Birth interval:** NA. **Life span:** NA. Birth season: Yes.[891] Increased levels of testosterone in males during the breeding season cause weight gain, "fatting" of the upper torso, and an increase in testis size, which leads to sperm production.[891]

The Bolivian squirrel monkey has a rounded (Roman) arch over the eyes.

Locomotion Quadrupedal leaping.
Social Structure Multimale-multifemale groups. Males establish a

hierarchy during the breeding season.[354] Outside the breeding season, males do not relate to

LOWER RISK

CITES II

97

Bolivian Squirrel Monkey *continued*

females, and females are dominant to males.[354] **Emigration:** Males emigrate but do not move from their natal troop unless they are forced out by the more dominant males.[826] **Group size:** Up to 20.[198] **Home range:** Up to 250ha.[826] **Day range:** NA.

Behavior Diurnal and arboreal. In a captive study, Bolivian squirrel monkeys were reported to be more social, more vocal, and more active and aggressive than common squirrel monkeys *(S. sciureus)*.[354] Infants up to 6 months old spend 30% of their time with allomothers. In a study, 53% of allomothers were young females 4–6 years old and 21% were females 7–9 years old. Females whose infants died were observed to nurse unrelated infants during allomaternal care.[890] Individuals compete for fruit within groups.[570] Activity budget: Insect foraging, 50%; travel, 27%; rest, 11%; plant feeding, 11%; social grooming, <2%.[826] **Mating:** The males compete for dominance.[891] **Vocalizations:** The *girren* call is unique to this squirrel monkey.[354]

Brazil, Bolivia, Peru, Venezuela, Colombia

Red-backed Squirrel Monkey *Saimiri oerstedii*

The red-backed squirrel monkey has a pointed (Gothic) arch over the eyes.

S. BOINSKI

orange.[198] They have an olivaceous color on the hips, shoulders, and base of the tail. The tip of the tail is black. They have Gothic-type (pointed) arches over the eyes[354] and a white mask with a black muzzle.[198]

Physical Characteristics Head and body length: 270mm *[10.6in]*.[364] **Tail length:** 363mm *[14.3in]*.[364] **Weight:** ♀ 600–790g *[21.2–27.9oz]*, ♂ 750–950g *[26.5–33.5oz]*.[247] **Intermembral index:** NA. **Adult brain weight:** 25.7g *[0.9oz]*.[346] Canine length: ♀ 2.7mm *[0.1in]*, ♂ 3.9mm *[0.2in]*.[490]

Habitat Lowland forest.[198]

Diet Fruit, seeds, leaves, insects. When these squirrel monkeys capture a 20mm *[0.8in]* caterpillar, they remove the spines, head, and internal organs before eating the body.[409]

Life History NA. Mating season: Jan–Feb. Birth season: Jul.[35]

Locomotion Quadrupedal.[35]

Social Structure Multimale-multifemale groups. Group members have little social interaction, though juveniles occasionally play.[35] Females do not have a dominance hierarchy or coalitions.[570] There is no male intragroup aggression.[69] Males of the same age are socially cohesive with other males, not females. The males stay in their natal group.[69] **Emigration:** Females emigrate.[69] **Group size:** 23.[903] **Home range:** 17.4–40ha.[35] **Day range:** 3350m *[10991ft]*.[265]

Behavior Diurnal and arboreal. Red-backed squirrel monkey males are vigilant and deter potential predators and rival groups.[69] To maximize foraging, adult females try to manipulate troop movement. There is competition for insects but not for fruit. **Association:** This squirrel monkey does not associate with *Cebus capucinus* for extended periods.[67] **Vocalizations:** 4 major calls—smooth chuck, bent mask chuck, peep, and twitter.[68] Over 60% of vocalizations are 1 of the 4 subtypes of the smooth chuck.[68]

Panama, Costa Rica

Taxonomy Disputed. 2 subspecies.[723] Elevated from a subspecies of *S. sciureus* in 1984.[315] May have been introduced by humans to Central America.[198]

Distinguishing Characteristics Red-backed squirrel monkeys have a black crown, and the back, hands, and feet are golden

Common Squirrel Monkey *Saimiri sciureus*

Squirrel monkeys scent-mark by wetting their hands and feet with urine.

Taxonomy Disputed. 4 subspecies.[313] Used to include 3 other squirrel monkeys as subspecies.

Distinguishing Characteristics Common squirrel monkeys have a pink face with black around the muzzle, a gray to black crown, Gothic-type (pointed) arches over the eyes, and white hairy ear tufts.[354] The back is gray olivaceous, the underparts are light yellow, and the forearms, hands, and feet are yellow-orange.[198]

Physical Characteristics Head and body length: ♀ 312mm (275–370) *[12.3in (10.8–14.6)]*, ♂ 318mm (265–370) *[12.5in (10.4–14.6)]*.[600] **Tail length:** ♀ 405mm (380–450) *[15.9in (15.0–17.7)]*, ♂ 409mm (360–452) *[16.1in (14.2–17.8)]*.[600] **Weight:** ♀ 651–1250g *[23.0–44.1oz]*, ♂ 554–1150g *[19.5–40.6oz]*.[247] **Intermembral index:** 79.1.[425] **Adult brain weight:** 24.4g *[0.9oz]*.[346] Canine length: ♀ 2.7mm *[0.1in]*, ♂ 3.5mm *[0.1in]*.[490] Only infants have prehensile tails.

Habitat Primary and secondary moist forest, riverine forest, mangrove, swamps to 2000m *[6562ft]*;[354] not dry forest.

Diet Animal prey, including frogs, snails, and crabs; insects including flies, caterpillars, butterflies, and grasshoppers; and arachnids.[35]

Life History Infant: 11–12mo.[35] **Weaning:** 11mo.[35] **Juvenile:** 11–30mo.[35] **Subadult:** 30–60 mo.[35] **Sexual maturity:** NA. **Estrus cycle:** 18d.[346] **Gestation:** 170d.[346] **Age 1st birth:** 30mo.[708] **Birth interval:** 13.6mo.[346] **Life span:** 21y.[346] Mating season: Variable. Birth

season: Variable.[698] Offspring: 1.[35] Males gain 85–222 grams *[3.0–7.8oz]* in body weight before the breeding season.

Locomotion Quadrupedal walking and running; leaping to 7m *[23ft]*.[35]

Social Structure Multimale-multifemale groups. These squirrel monkeys are not territorial. Females remain in their natal group and have matrilineal hierarchies. They form the core of the group, traveling and resting together, and team up to dominate males. Males keep a strict dominance hierarchy, which they maintain through penile displays.[35]

Emigration: Extragroup males occasionally move together in groups of 5 or fewer.[35] **Group size:** 22–42.[698] **Home range:** 65–130ha.[35] **Day range:** 1500m *[4922ft]*.[265]

Behavior Diurnal and arboreal.[198] Squirrel monkeys are noisy and constantly on the move, often following capuchin *(Cebus)* groups. They spend less than 1% of their time on the ground.[35] Females provide all the parental care, and young females often allomother infants from about 2 weeks after birth until the infant is about 8 weeks old.[35] Males are more excitable and aggressive during mating season.[35] **Association:** These monkeys form mixed-species associations with white-fronted capuchins *(Cebus albifrons)* and tufted capuchins *(C. apella)*.[35] They associate with red uacaris *(Cacajao calvus rubicun-*

dus)[238] and reportedly associate with black-headed uacaris *(C. melanocephalus)*.[238] **Mating:** Promiscuous. These monkeys have multiple-mount matings. The mating pair is often harassed.[35]

Scent marking: Urine washing. **Vocalizations:** 20,[35] including 5 call classes. Purring/growling/spitting calls are threats; twittering/chattering/cackling calls confirm social bonds; chirping/isolation peep/squealing calls signal a desire for social contact, as well as submission or social frustration; clucking/yapping/alarm peep calls are warnings and predator alarms and signal disagreements; groaning/cawing/shrieking calls are protests.[426]

Brazil, Guyana, French Guiana, Suriname, Venezuela, Colombia

Golden-backed Squirrel Monkey *Saimiri ustus*

R. & J. SEITRE

Taxonomy Disputed. Elevated from a subspecies *S. sciureus* in 1984. Some researchers believe this species should be renamed *S. madeirae*.[137]

Distinguishing Characteristics Golden-backed squirrel monkeys have a golden back, whitish underparts, Gothic-type (pointed) arches over the eyes, and a gray crown. Females have a temporal band that males do not have. The tail is shorter than in other squirrel monkeys,[354] and the ears are less hairy.[364]

Physical Characteristics Head and body length: 310mm *[12.2in].*[364] **Tail length:** 440mm *[17.3in].*[364] **Weight:** ♀ 710–880g *[25.0–31.0oz],* ♂ 620–1200g *[21.9–42.3oz]* (wild).[247] **Intermembral index:** NA. **Adult brain weight:** NA.

Habitat NA.

Diet NA.

Life History NA.

Locomotion Quadrupedal.

Social Structure NA. **Group size:** NA. **Home range:** NA. **Day range:** NA.

Behavior Diurnal and arboreal. In a test of golden-backed squirrel monkeys for malaria, 24% tested positive. It is hypothesized that primates with a larger group size attract more mosquitoes, increasing the group's infection rate.[153] **Association:** There is evidence of hybridization with black squirrel monkeys (*S. vanzolinii*).[137]

Brazil

Black Squirrel Monkey *Saimiri vanzolinii*

Taxonomy Disputed. First described in 1985, this species may be a subspecies of *S. sciureus*[137] or *S. boliviensis*.[315] It is chromosomally different from other squirrel monkeys.[354]

Distinguishing Characteristics Black squirrel monkeys have a gray or black back with a black stripe, gray shoulders, yellow hands and forearms, and Roman-type (rounded) arches over the eyes. Males and females look the same.[354]

Physical Characteristics Head and body length: ♀ 275–295mm *[10.8–11.6in],* ♂ 278–320mm *[10.9–12.6in].*[364] **Tail length:** ♀ 415–445mm *[16.3–17.5in],* ♂ 415–430mm *[16.3–16.9in].*[364] **Weight:** ♀ 650g *[22.9oz],* ♂ 950g *[33.5oz].*[246] **Intermembral index:** NA. **Adult brain weight:** NA. The occipital lobe of the brain is "devoted to auditory processing and is proportionally the most extensive of any primate except man,"[409] perhaps the result of the complex vocalizations of squirrel monkeys.

Habitat Swamp, white water–flooded forest (varzea).[707]

Diet Fruit, animal prey.[707]

Life History NA.

Locomotion Quadrupedal.

Social Structure Multimale-multifemale.[707] **Group size:** Up to 50.[707] **Home range:** NA. **Day range:** NA.

Behavior Diurnal and arboreal. No field studies had been published as of 1995, but "all squirrel monkeys share the same feeding strategy, diet, and breeding systems."[137] Social organization is different "only in conjunction with changes in forest structure and the distribution of food resources."[137] This species has the smallest distribution of any squirrel monkey—only 950 sq km *[367 sq mi].*[707] **Association:** There is evidence of hybridization with golden-backed squirrel monkeys (*S. ustus*).[137]

L. C. MARIGO

This black squirrel monkey is eating a large praying mantis.

Brazil

Equatorial Saki *Pithecia aequatorialis*

R. C. MITTERMEIER

Taxonomy Disputed. Monotypic.[723] Elevated from a subspecies of *P. monachus* in 1987.[356]

Distinguishing Characteristics Equatorial sakis have a blackish body with white highlights and long, shaggy fur. The underparts are orange or brown, and the face has a white ring around it.[198]

Physical Characteristics Head and body length: NA. **Tail length:** NA. **Weight:** NA. **Intermembral index:** NA. **Adult brain weight:** NA.

Habitat Primary lowland, riverine, and white water–flooded forest (varzea).[198]

Diet Fruit, seeds, leaves.[198]

Life History NA.

Locomotion Quadrupedal; leaping.[198]

Social Structure NA. **Group size:** NA. **Home range:** NA. **Day range:** NA.

Behavior Diurnal and arboreal. Equatorial sakis hide in vegetation and can be very quiet and cryptic.[198] They reportedly sit in trees along rivers, where "they look like big black balls of fur with bushy tails."[198]

Ecuador, Peru

Buffy Saki *Pithecia albicans*

L. C. MARIGO

Taxonomy Disputed. Monotypic.[723] Elevated from a subspecies of *P. monachus* in 1979.[315] This species is the largest and most derived in the genus.[656]

Distinguishing Characteristics Buffy sakis are buffy to reddish on the head, sides, and underparts.[352] The back and tail are blackish.[352] Neonates have a dark brown coat and a naked tail.[416]

Physical Characteristics Head and body length: ♀ 380mm (365–405) *[15.0in (14.4–15.9)]*, ♂ 404–410mm *[15.9–16.1in]*.[352] **Tail length:** ♀ 428mm (405–455) *[16.9in (15.9–17.9)]*, ♂ 416–440mm *[16.4–17.3in]*.[352] **Weight:** 3.0kg *[6.6lb]*.[656] **Intermembral index:** NA. **Adult brain weight:** NA.

Habitat Primary flooded and terra firma forest.[656]

Diet Seeds, 46.2%; fruit, 33.7%; young leaves, 9.5%; flowers, 6.5%; gums, 0.8%; nectar, 1.5%; bark, 1.1%; spiders, 0.4%.[656] Buffy sakis have been described as sclerocarpic (seed-eating) frugivores.[656] They use 81 species of plants.[656]

Life History Infant: 0–6 mo.[416] **Juvenile:** 6–12mo.[416] **Sexual maturity:** NA. **Estrus cycle:** NA. **Gestation:** NA. **Age 1st birth:** NA. **Birth interval:** NA. **Life span:** NA. Birth season: No. At 6 months the offspring travel independently. By 11 months they are difficult to distinguish from adults.[416]

Locomotion Quadrupedal.[198]

Social Structure NA. Uncohesive groups with overlapping home ranges.[656] **Group size:** 4.6 (3–7).[656] **Home range:** 147–204ha.[656] **Day range:** NA.

Behavior Diurnal and arboreal. Buffy sakis prefer large branches in the canopy and subcanopy, at heights of 15–25 meters *[49–82ft]*.[416] They avoid emergent trees and low undergrowth and have never been seen on the ground.[416] They are seed predators and specialize in unripened

Brazil

L. C. MARIGO

fruit, which is abundant and more available year-round than ripe fruit.[656] Buffy sakis have a rapid rate of travel but are cryptic in response to humans. The male has a distraction display to lead humans away from the other troop members.[416] Humans hunt sakis for meat and use the tail as a dust mop or for cleaning shotguns.[416] **Association:** Buffy sakis occasionally associate with tufted capuchins *(Cebus apella)*.[656]

Vocalizations: The alarm call is the most notable vocalization.

Sleeping site: Trees. Buffy sakis have no routine sleeping site.[416]

Bald-faced Saki *Pithecia irrorata*

Taxonomy Disputed. 2 subspecies.[723] This species may be a subspecies of *P. monachus*.[315]

Distinguishing Characteristics Bald-faced sakis are blackish with white, giving them a silvery gray appearance. They have white around their face.[198]

Physical Characteristics Head and body length: ♀ 375–420mm *[14.8–16.5in]*, ♂ 390–410mm *[15.4–16.1in]*.[364] **Tail length:** ♀ 470–545mm *[18.5–21.5in]*, ♂ 465–500mm *[18.3–19.7in]*.[364] **Weight:** ♀ 2160g *[4.8lb]*, ♂ 2920g *[6.4lb]*.[356] **Intermembral index:** NA. **Adult brain weight:** NA.

Habitat Lowland and flooded forest.[198]

Diet NA.

Life History NA.

Locomotion Quadrupedal.

Social Structure 1 male–1 female monogamous family groups. **Group size:** 3.3. **Home range:** NA. **Day range:** NA.

Brazil, Bolivia

Behavior Diurnal and arboreal. Bald-faced sakis prefer the canopy and have been observed at an average height of 23.6 meters (6–37m) *[77ft (20–121)]*.[656] Only 2% of tested individuals had malaria.[153]

Vocalizations: Bald-faced sakis make loud growl-grunts at human observers.[364]

Monk Saki *Pithecia monachus*

Taxonomy Disputed. 2 subspecies.[723] This species was named *P. hirsuta* until 1987.[315]

Distinguishing Characteristics Monk sakis have a blackish or gray body, a reddish beard and underparts, and pale hands and feet. Males are distinguished by buffy crown hairs.[87] The face has a white stripe down each side from below the eyes. Infants are born almost black.[198] The tail is not prehensile.

Physical Characteristics Head and body length: 370–480mm *[14.6–18.9in]*.[198] **Tail length:** 404–500mm *[15.9–19.7in]*.[198] **Weight:** ♀ 1.3–2.5kg *[2.9–5.5lb]*, ♂ 2.5–3.1kg *[5.5–6.8lb]*.[247] **Intermembral index:** 77.[234] **Adult brain weight:** 38.1g *[1.3oz]*.[346]

R. A. MITTERMEIER

Habitat Primary, humid, lowland terra firma and some white water–flooded forest (varzea).[898]

Diet Fruit, 55%; seeds, 38%; leaves, 4%; flowers, 3%.[355] Ants were found in the stomachs of monk sakis that were shot for a morphological study in the early 1970s.[364]

Life History NA.

Locomotion Quadrupedal branch running, bipedal hopping.[87]

Social Structure 1 male–1 female monogamous groups. Monk sakis often forage alone or in pairs. **Group size:** 2.9 (2–6).[87] **Home range:** NA. **Day range:** NA.

Behavior Diurnal and arboreal.[656] Monk sakis are shy, cryptic,[198] and quiet, making them difficult to detect.[79] They prefer the canopy at an average height of 20.9 meters (7–35m) *[69ft (23–115)]*[656] and feed at 15–24 meters *[49–79ft]* 89% of the time.[656] All group members groom each other. Their several facial expressions include frowning when frustrated, teeth baring when hostile, and brow raising when frightened.[87] In Colombia they are known as flying monkeys.[87] **Association:** Monk sakis associate with Spix's black-mantled tamarins *(Saguinus nigricollis)*.[403] They have been observed following woolly monkeys *(Lagothrix lagotricha)* and bald uacaris *(Cacajao calvus)*.[87]

Sleeping site: Large branches in the upper canopy.[198]

R. A. MITTERMEIER

Brazil, Peru, Colombia, Ecuador

White-faced Saki *Pithecia pithecia*

Female.

Taxonomy 2 subspecies.[723]

Distinguishing Characteristics White-faced sakis are sexually dichromatic, a rare trait in New World monkeys. Males are black, with a white face and a black nose; females are agouti brown and have white stripes along the sides of the nose.[196] The tail is not prehensile.[87]

Physical Characteristics Head and body length: ♀ 335, 343mm *[13.2, 13.5in]*; ♂ 350mm (330–375) *[13.8in (13.0–14.8)]*.[352] **Tail length:** ♀ 342, 435mm *[13.5, 17.1in]*; ♂ 397mm (348–445) *[15.6in (13.7–17.5)]*.[352] **Weight:** ♀ 779–1750g *[27.5–61.7oz]*, ♂ 964–2500g *[34.0–88.2oz]*.[247] **Intermembral index:** 75.[234] **Adult brain weight:** 31.7g *[1.1oz]*.[346]

Habitat Primary evergreen,[196] coastal, secondary, liana,[898] savanna,[87] palm,[898] and gallery forest from lowland to moderate elevations; rarely in flooded forest.[87]

Diet Fruit, 59.5%; seeds, 33%; flowers, 6.7%;[455] leaves and animal prey, including small birds and bats.[87] White-faced sakis are seed predators[87] and crack hard nuts with their canine teeth.[196] Only 2 fruit species account for 49% of their fruit intake. They feed on termite nests, which are high in iron.[455]

Life History Weaning: 4mo.[87] **Juvenile:** 4–9mo.[87] **Sexual maturity:** 24–36mo. **Estrus cycle:** 16d.[1012] **Gestation:** 163–176d.[87] **Age 1st birth:** NA. **Birth interval:** 12mo.[346] **Life span:** 13.7y.[346] Birth season: Dec–Apr. Offspring: 1.[196] At 2 months, male infants begin to show the dimorphic adult coloration.[87]

Locomotion Quadrupedal walking, climbing, and leaping; bipedal hopping.[87]

Social Structure 1 male–1 female family groups.[87] Solitary individuals have been observed.[87] **Group size:** 2.6 (2–5).[898] **Home range:** 4–10ha.[87] **Day range:** NA.

Behavior Diurnal and arboreal. White-faced sakis prefer to use solid supports in the lower to middle canopy levels (3–15m *[10–49ft]*).[656] They are occasionally found on the ground or in emergent trees.[87] They may use lower levels of forest than other *Pithecia* species because of competition with bearded sakis *(Chiropotes satanas),* which are also seed predators.[656] Allogrooming is common among the

whole social group.[87] The rapid bipedal hopping of these sakis has earned them the nickname "flying jacks" in Guyana.[87]

Association: Sympatric with bearded sakis *(Chiropotes satanas).*[452] White-faced sakis reportedly feed and travel alone but have been seen with both tamarins *(Saguinus)* and bearded sakis *(Chiropotes* spp.).[87]

Vocalizations: These sakis have a loud call for territorial spacing that is "similar to [that of] the howler monkey, though lower in amplitude."[196] An aggressive display starts with a body shake, an arched posture, and a growl; then they shake a branch with their whole body.[87]

Brazil, Suriname, Guyana, French Guiana

Infants of both sexes are born the same color as the females. This juvenile male has begun to change.

Male.

White-nosed Bearded Saki *Chiropotes albinasus*

Taxonomy Monotypic.[723]

Distinguishing Characteristics

White-nosed bearded sakis have an all-black body and tail, a thick beard, and 2 tufts of hair on the crown. The upper lip and the nose are pink and covered with some yellowish white hair. The scrotum is white. Infants are born with a prehensile tail used to cling to their mother; the tail is no longer prehensile after 2 months.[198]

Physical Characteristics Head and body length: 427mm (412–470) *[16.8in (16.2–18.5)]*.[355] **Tail length:** 412mm (380–450) *[16.2in (15.0–17.7)]*.[355] **Weight:** ♀ 2220–2800g *[4.9–6.2lb]*, ♂ 2720–3320g *[6.0–7.3lb]*.[247]

Intermembral index: NA. **Adult brain weight:** NA. Males are slightly larger than females.[198]

Habitat Primary nonflooded forest (terra firma) and black water–flooded forest.[898]

Diet Fruit, 54%; seeds, 36%; flowers, 3%; no insects.[851]

Life History Weaning: NA. **Sexual maturity:** NA. **Estrus cycle:** NA. **Gestation:** 152–162d.[698] **Age 1st birth:** NA. **Birth interval:** NA. **Life span:** NA. Births: Year-round. Birth peaks: Feb–Mar, Aug–Sep.[698] Estrus is signaled by reddening of the anogenital region.[79] A female's teats and anogenital region are red throughout pregnancy.[851] In a captive study of

Brazil

development, an infant slept most of its first 24 days. At 3.5 weeks it was able to climb on its mother's back. By day 37 it ate its first solid food. By day 54 it attempted a suspensory posture of hanging by its feet. After 3 months it ate mostly solid food.[851]

Locomotion Quadrupedal walking and climbing.

Social Structure Multimale-multifemale groups.[851] **Group size:** 2–26.[198] **Home range:** 100ha.[198] **Day range:** 2000–5000m *[6562–16,405ft]*.[198]

Behavior Diurnal and arboreal. White-nosed bearded sakis use the middle and upper levels of forest.[198] Tail wagging is an expression of excitement and may be a silent warning to other troop members.[851] **Association:** These sakis associate with tufted capuchins *(Cebus apella)*.[851] **Mating:** The female presents to the male by lying down and lifting her tail to show her red estrous coloration. Both males and females make a purring vocalization while mating.[851] **Vocalizations:** A high-pitched whistling contact call distinguishes these sakis in the forest.[851] They are said to produce a weak chirp when feeding.[851] They reportedly vocalize "profusely" at human observers and may approach them rather than flee.[79]

Bearded Saki *Chiropotes satanas*

Taxonomy 3 subspecies.[723]

Distinguishing Characteristics

Bearded sakis have an all-black face, a prominent black beard, 2 "bulbous temporal swellings" on the top of the head, a thick bushy nonprehensile tail, and hind limbs that are noticeably longer than the forelimbs.[851] The back and shoulders vary from dark brown to pale yellow brown to ocherous, depending on the subspecies. Males have a pink scrotum.[851] The infant's tail is prehensile for 2 months.[851]

Physical Characteristics Head and body length: ♀ 397mm (380–410) *[15.6in (15.0–16.1)]*, ♂ 422mm (400–480) *[16.6in (15.7–18.9)]*.[600] **Tail length:** ♀ 389mm (370–420) *[15.3in (14.6–16.5)]*, ♂ 393mm

(395–420) *[15.5in (15.6–16.5)]*.[600] **Weight:** ♀ 1.9–3.3kg *[4.2–7.3lb]*, ♂ 2.2–4.0kg *[4.8–8.8lb]*.[247] **Intermembral index:** 83.[234] **Adult brain weight:** 53g *[1.9oz]*.[346]

Habitat High rain forest, mountain savanna forest, black-water swamp forest (igapo), and terra firma forest.[851]

Diet Seeds, 66%; fruit, 30%; flowers, 3%; animal prey, 0%.[452] These sakis use 86 plant species. They are seed predators on 52 species and seed dispersers of 7 species.[851] Some researchers report that insects, including ants, beetles, and caterpillars, are eaten.[575] Specialized teeth enable this species to ingest less-ripe fruit with a hard pericarp, thus avoiding competition with spider monkeys *(Ateles)*, which eat softer, ripe fruit.[455] This species eats smaller amounts of leaves and harder fruit than white-faced sakis *(Pithecia pithecia)*, which have a similar diet and are sympatric.

Life History Weaning: NA. **Sexual maturity:** NA. **Estrus cycle:** NA. **Gestation:** 152–167d.[699] **Age 1st birth:** NA. **Birth interval:** NA. **Life span:** 15y.[346] Birth season: Dec–Jan.[851] Offspring: 1. During estrus the female's anogenital region is red.[851] The nipples and anogenital area are a deep red throughout pregnancy.

Bearded Saki *continued*

Locomotion Quadrupedal walking, bounding, galloping, climbing, and leaping. The tail is nonprehensile but may be draped over a branch for support when the monkey suspends itself by its hind limbs to feed.[246]

Social Structure Multimale-multifemale groups. One researcher reported monogamous pairs within the large group.[699] These sakis travel in a cohesive group and spread out when they reach feeding trees.[1013] **Group size:** 10–30.[851] **Home range:** 100ha[196] to 200–250ha.[851] **Day range:** 2500–3000m *[8203–9843ft]*.[851]

Behavior Diurnal and arboreal. Bearded sakis prefer the upper canopy levels[851] for feeding and the lower levels for traveling.[858] They rarely come to the ground.[452] Their feeding sites are far apart, and they travel very fast. When an individual is left behind, it may temporarily join a troop of spider monkeys *(Ateles)* or tufted capuchins *(Cebus*

apella).[455] When excited, these sakis wag their tails.[851] When sitting, they wag their tail below the branch. When standing, they wag their tail over their back. The genus got the name *Chiropotes,* meaning "hand drinker," because it may dip its hands into water to drink.[851] **Association:** Sympatric with white-faced sakis *(Pithecia pithecia)*.[452] Black spider monkeys *(Ateles paniscus)* tolerate bearded sakis feeding at the same food tree.[850] **Vocalizations:** "High pitched whistles rising in crescendo and abruptly cut off"[198] are believed to be a contact call.[851] The alarm call is a more intense whistle.[246] In Suriname, they are named Bisa, after the sound of their vocalizations.

Sleeping site: Variable from night to night.[851]

Brazil, Suriname, French Guiana, Guyana, Venezuela

Bald Uacari *Cacajao calvus*

Taxonomy Disputed. 4 subspecies.[723] *C. c. rubicundus* may be a distinct species.[315]

Distinguishing Characteristics Bald uacaris have a pink to scarlet face and no hair on the top of the head. The tail is short, and the hair is long and shaggy.[198] The coat color varies from white *(C. c. calvus)* to chestnut red *(C. c. rubicundus)*.[238]

Physical Characteristics Head and body length: ♀ 555mm *(540–570) [21.9in (21.3–22.4)]*, ♂ 550mm *(540–560) [21.7in (21.3–22.0)]*.[600] **Tail length:** 155mm *(150–160) [6.1in (5.9–6.3)]*.[600] **Weight:** ♀ 2880g *[6.3lb]*, ♂ 3450g *[7.6lb]*.[246] **Intermembral index:** 83.[234] **Adult brain weight:** 73.3g *[2.6oz]*.[346]

Habitat *C. c. calvus*—white water–flooded forest (varzea); *C. c. rubicundus*—black-water swamp forest (igapo).[898]

Diet Seeds, 67%; fruit, 18%; flowers, 6%; animal prey, 5%; buds.[452] Insects are not actively sought but are eaten.[238]

Life History Infant: 1–15mo.[238] **Weaning:** 15–21mo.[238] **Juvenile:** 15–36mo.[238] **Subadult:** 36–60mo.[238] **Sexual maturity:** ♀ 43mo, ♂ 66mo.[699] **Estrus cycle:** 14–48d.[238] **Gestation:** NA. **Age 1st birth:** 42mo.[238] **Birth interval:** NA. **Life span:** 20.1y.[708] **Mating season:** Oct–May.[238] **Birth peak:** May–Oct.[699] At 15 months, infants feed independently.[238] The infant's face is gray, not red.[238] At 3–12 months, red uacaris *(C. c. rubicundus)* develop gray scalp hair, which lasts until the subadult stage.

Locomotion Quadrupedal walking, running, galloping, leaping, and occasionally swinging.[238]

Social Structure Multimale-multifemale groups.[238] **Group size:** 5–30,[238] to 100.[238] **Home range:**

Red uacari *(C. c. rubicundus)*.

500–600ha.[33] **Day range:** NA.
Behavior Diurnal and arboreal.[198] Bald uacaris come to the ground to feed on seeds and shoots in the dry season.[198] Groups may split

up when feeding or when disturbed.[454] When excited, bald uacaris wag their tail.[198] Communication includes 10 graded facial expressions.[358] In captivity

Bald Uacari *continued*

White uacari *(C. c. calvus)*.

they express dominance by noisy fighting.[238] Allogrooming is reciprocal among females, but males do not groom females as much.[238] Healthier individuals have redder faces.[328] **Association:** Red uacaris *(C. c. rubicundus)*

associate with common squirrel monkeys *(Saimiri sciureus)*.[238] White uacaris *(C. c. calvus)* have been observed in the same tree with monk sakis *(Pithecia monachus)*.[238] **Mating:** Olfactory signals from the female stimulate

the male to initiate mating.[238] The male mounts several times.[169] **Vocalizations:** Bald uacaris have 12 vocalizations, including a continuous barking vocalization[358] and a short sonorous *hic* vocalization, which is used by all age groups except adult males. The response to a predator is to make *chick* alarm calls, wag the tail, and erect the hair.[238] In captivity this species responds to squirrel monkey *(Saimiri)* alarm calls.[238] **Sleeping site:** Thin branches in the highest part of large trees.[198] Bald uacaris sleep separately, except for mothers and infants.[238]

Brazil, Peru

Black-headed Uacari *Cacajao melanocephalus*

Taxonomy 2 subspecies.[723]

Distinguishing Characteristics
Black-headed uacaris have a black face and head and varying amounts of light brown on the hind limbs, back, and tail. The tail is short (one-third the body length).[198]

Physical Characteristics Head and body length: 300–500mm [11.8–19.7in].[198] **Tail length:** 125–210mm [4.9–8.3in].[198] **Weight:** 2.4–4.0kg [5.3–8.8lb].[198] **Intermembral index:** NA. **Adult brain weight:** NA.

Habitat Black-water swamp forest (igapo), caatinga forest on white sands, terra firma forest, and

mountain slopes up to 600m [1969ft].[76]

Diet Seeds, fruit, leaves, and animal prey,[76] including amphibians.[94] Seeds of unripe fruit appear to be an important dietary component, as for all pitheciines. Parts of uaraba fruit *(Swartzia polyphylla)*, which are 180mm [7in] long and weigh 290g [10oz], are eaten.[76]

Life History NA. Offspring: 1.[196]

Locomotion Quadrupedal walking and running;[898] leaping to 2.5m [8ft]; some suspensory.[486]

Social Structure Variable. 1 male–multifemale or multimale-multifemale groups. **Group size:** 15–25,[196] up to 100.[76] **Home range:** NA. **Day range:** NA.

Behavior Diurnal and arboreal.[198] Black-headed uacaris forage in the terra firma forest during the dry season. In the wet season (August–November), they move to the forests of flooded black-water rivers (igapo) and feed in tall trees. They travel just above water level in vines and bushes[76] and suspend themselves by their hind limbs when feeding.[486] When disturbed, they wag their tail 2–3 beats per second.[76] Human hunting pressure rather than habitat destruction is the major threat to these uacaris.[486] In the flood season many monkeys are drawn to

the large concentrations of fruit[162] and are easy targets for hunters in canoes.[76] **Association:** This species reportedly associates with common squirrel monkeys *(Saimiri sciureus)*[238] and white-fronted and tufted capuchins *(Cebus albifrons,*[486] *C. apella*[238]). **Vocalizations:** The high-pitched alarm call has been compared to the call of an Amazon parrot.[76] Hunters locate groups of uacaris by listening for their contact calls and the sound of fruit falling into water.[76]

Brazil, Venezuela

Red-handed Howler *Alouatta belzebul*

Taxonomy 4 subspecies.[723]

Distinguishing Characteristics
Red-handed howlers vary from all black to brownish to completely red.[736] The hands are yellow to reddish,[509] and the male's scrotum is a rusty red.[198] The fur color variation has been observed within the same troop in the wild.[30]

Physical Characteristics Head and body length: ♀ 508mm (456–570) *[20.0in (18.0–22.4)]*, ♂ 590mm (580–600) *[23.2in (22.8–23.6)]*.[600] **Tail length:** ♀ 630mm (580–690) *[24.8in (22.8–27.2)]*, ♂ 607mm (555–660) *[23.9in (21.9–26.0)]*.[600] **Weight:** ♀ 4850–6200g *[10.7–13.7lb]*, ♂ 6540–8000g

[14.4–17.6lb].[247] **Intermembral index:** NA. **Adult brain weight:** NA. Diploid chromosome number: ♀ 50, ♂ 49.[30]

Habitat Primary and secondary, terra firma, and flooded forest.[198]

Diet NA.

Life History NA.

Locomotion Quadrupedal.

Social Structure 1 male–multifemale groups.[30] **Group size:** 2–8. **Home range:** NA. **Day range:** NA.

Behavior Diurnal and arboreal.[198] Red-handed howlers are the least well studied of the howlers. A genetic study of 2 populations found that the eastern population had more diversity of coat color but almost half the heterozygosity genetically.[736] Of individuals tested, 14% were positive for malaria.

Brazil

R. A. MITTERMEIER (DIGITALLY COMPOSITED)

Black-and-gold Howler *Alouatta caraya*

Taxonomy Monotypic.[723]

Distinguishing Characteristics
Adult male black-and-gold howlers are black. Adult females are golden light to dark yellow.[615] The tail is prehensile.

Physical Characteristics Head and body length: 420–550mm *[16.5–21.7in]*.[198] **Tail length:** 530–650mm *[20.9–25.6in]*.[198] **Weight:** ♀ 3800–5410g *[8.4–11.9lb]*, ♂ 5000–8280g *[11.0–18.2lb]*.[247] **Intermembral index:** 97.[234] **Adult brain weight:** 56.7g *[2.0oz]*.[346] The female's weight is 68% of the male's

weight.[615]

Habitat Primary, riparian, dry deciduous, and broadleaf forest, as well as semiarid caatinga.[898]

Diet Leaves, 76%; fruit, 24%; flowers, buds.[790]

Life History Weaning: NA. **Sexual maturity:** ♀ 35–42mo, ♂ 24–37mo.[141] **Estrus cycle:** 20d.[615] **Gestation:** 187d.[141] **Age 1st birth:** 44.4mo.[708] **Birth interval:** NA. **Life span:** NA. Birth season: No. Birth peak: Mid–dry season. Offspring: 1, but twins reported once.[56] Both genders are the same color until the male turns black at around

Brazil, Paraguay, Argentina

the age of 2.5 years.[615]

Locomotion
Quadrupedal; climbing; suspensory feeding.[615]

Social Structure Variable. 1 male[55] or multimale-multifemale groups.[615] The most males reported in a group have been 4, with up to 4 females.[615] In the female dominance hierarchy, younger adults rank higher than older ones.[615] **Group size:** 7.2–8.9.[141] **Home range:** NA. **Day range:** NA.

Behavior Diurnal and arboreal.[198] Black-and-gold howlers may come to the ground to drink from ponds. They also drink water from arboreal sources by wetting and then licking their hands. They drink water more when there is a lack of young leaves.[55] Allogrooming is seen mostly between adult females and immatures. A dominant male is groomed by the adult females.[615] **Mating:** Tongue flicking is a ritualized display of sexual solicitation. It is especially distinctive in this species because the tongues are pink, bordered with black.[615] **Vocalizations:** The vocalizations are lower pitched than those of mantled howlers (*A. palliata*) but similar to those of red howlers (*A. seniculus*).[836]

Howlers have an enlarged hyoid bone in their throat that enhances their territorial call.

The male is black and almost twice the weight of the brownish gold female.

Coiba Island Howler *Alouatta coibensis*

CRITICALLY ENDANGERED IUCN

Taxonomy Disputed. 2 subspecies.[723] Elevated from a subspecies of mantled howlers *(A. palliata)* in 1987. The dermal ridges of the hands and feet are different from the mantled howler's and were the basis for considering this a separate species.[253]

Distinguishing Characteristics Coiba Island howlers have a walnut-colored back and long, golden hair on the flanks.[834] The skull dimensions are more sexually dimorphic than in any other howler, the male's being larger than the female's. The tail is prehensile.[253]

Physical Characteristics Head and body length: 560mm *[22.0in]*.[834] **Tail length:** 580mm *[22.8in]*.[834] **Weight:** NA. **Intermembral index:** NA. **Adult brain weight:** NA. Infants are difficult to sex because the genitalia of both genders are very similar in appearance until a later age.[252]

Habitat Primary forest, mangrove swamp.[578]

Diet Leaves, fruit.[578]

Life History NA.

Locomotion Quadrupedal; climbing.

Social Structure Variable. 1 male–multifemale or multimale-multifemale groups. Solitary males and up to 3 males in a group have been observed.[615] **Group size:** 6.[615] **Home range:** NA. **Day range:** NA.

Behavior Diurnal and arboreal. Coiba Island howlers reportedly use the upper canopy and emergent trees. They are not listed by CITES yet but are very threatened by habitat destruction and hunting.[578]

S. NASH

Panama

Brown Howler *Alouatta fusca*

CRITICALLY ENDANGERED IUCN

NATURAL HABITAT

Taxonomy Disputed. 2 subspecies.[723] The correct name may be *A. guariba*.[315]

Distinguishing Characteristics Male brown howlers are brown, dark red, or black, with a yellow or gold beard and brown to orange-brown underparts. Females are paler. The tail is prehensile.[198]

Physical Characteristics Head and body length: ♀ 468mm (450–490) *[18.4in (17.7–19.3)]*, ♂ 559mm (535–585) *[22.0in (21.1–23.0)]*.[600] **Tail length:** 532mm (515–570) *[20.9in (20.3–22.4)]*, 572mm (485–670) *[22.5in (19.1–26.4)]*.[600] **Weight:** ♀ 4100–5000g *[9.0–11.0lb]*, ♂ 5300–7150g *[11.7–15.8lb]*.[247] **Intermembral index:** NA. **Adult brain weight:** NA.

Habitat Primary and secondary forest and in Paraná pines *(Araucaria angustifolia)*.[198] Brown howlers prefer to be near streams.[898]

Diet Depending on the season, leaves, 66–77%; fruit, 2–29%; buds, 0.7–11%; flowers, 5–9%.[615] Figs and fig leaves are eaten.[675]

Life History NA.

Locomotion Quadrupedal walking, climbing, and suspensory feeding.[615]

Social Structure Variable. 1 male–multifemale or 2 male–multifemale groups.[615] **Group size:** 6, up to 11.[615] **Home range:** 5.8ha.[615] **Day range:** 225m (197–540) *[738ft (646–1772)]*.[615]

Behavior Diurnal and arboreal.[198] Brown howlers descend from trees to drink from streams.[711] All members of a troop groom one another, but females do the most grooming.[615] Brown howlers never recline while feeding; they sit or hang suspended by 3 limbs to feed, rarely by 1 limb. They rest in sitting position on larger branches. **Mating:** Tongue flicking is a sexual solicitation display. **Association:** Brown howlers are sympatric with woolly spider monkeys *(Brachyteles arachnoides)* but eat more leaves and leaf buds and less fruit.[615] **Vocalizations:** The peak time for brown howlers to make their territorial roar is

Brown Howler *continued*

Howlers use their prehensile tail to help suspend themselves while feeding.

late afternoon between 4 and 6 p.m.[615] **Sleeping site:** The preferred resting and sleeping site is the Paraná pine tree, whose brown dead leaves provide camouflage. Brown howlers sleep huddled together on branches in small subgroups of 2 or 3, their tails wrapped around a branch.[198]

Brazil

Mantled Howler *Alouatta palliata*

Taxonomy 3 subspecies.[723]

Distinguishing Characteristics
Mantled howlers are black except for a fringe of long, gold to buff hairs on the sides. The tail is prehensile.[198]

Physical Characteristics Head and body length: ♀ 520mm (481–632) *[20.5in (18.9–24.9)]*, ♂ 561mm (508–675) *[22.1in (20.0–26.6)]*.[600] **Tail length:** ♀ 609mm (563–655) *[24.0in (22.2–25.8)]*, ♂ 583mm (545–605) *[23.0in (21.5–23.8)]*.[600] **Weight:** ♀ 3.1–7.6kg *[6.8–16.8lb]*, ♂ 4.5–9.8kg *[9.9–21.6lb]*.[247] **Intermembral index:** 98.[234] **Adult brain weight:** 55.1g *[1.9oz]*.[346] The hyoid bone is smaller than in other howlers.[878]

Habitat Evergreen rain forest, cloud forest, dry lowland deciduous forest, and coastal mangrove forest up to 2000m *[6562ft]*.[898] The upper third of the canopy is preferred.[615]

Diet Young leaves, 44.2%; mature leaves, 19.4%; flowers, 18.2%; fruit, 12.5%; petioles, 5.7%.[615] Mantled howlers use more than 100 plant species, but only 19.9% of the individual trees (331 of 1665 in a study) in the home range are used for food. Mature fruit is eaten only from May to August.[205] Mature leaves of trees are eaten more in the wet season.[615]

Life History Infant: 0–6mo.[615] **Weaning:** 21mo.[346] **Juvenile:** 6–20mo.[142] **Subadult:** 30–48mo.[615] **Sexual maturity:** ♀ 36mo, ♂ 42mo.[141] **Estrus cycle:** 16.3d (11–24).[615] **Gestation:** 186d (180–194).[615] **Age 1st birth:** 48mo.[123] **Birth interval:** 22.5mo (18–24).[123] **Life span:** 20y.[708] Births: Year-round.[123] When

males reach adulthood, the scrotum turns white. In females the color of the genital area changes from white to pink during estrus.[615]

Locomotion Quadrupedal walking, climbing, and suspensory feeding.[615]

Social Structure Multimale–multifemale groups.[615] Females form a hierarchy in which the youngest adult is the most dominant, although it may have less reproductive success than the older females.[615] The females of midrank have the greatest reproductive success.[123] An adult female's rank decreases with age. The tenure of the alpha male is 2.5–3 years, during which time he may sire 18 offspring, or more than 2 times the average female's 8 offspring.[123] It is to a female's reproductive advantage to have male offspring during periods of social instability or drought and female infants during stable periods with sufficient food.[123] Mantled howlers do not maintain exclusive territories; overlap is up to 100%. They defend the place where they are rather than an exclusive area.[615] **Emigration:** Both genders emigrate as juveniles or subadults.[615] **Group size:** 4.2–21 in 12 different studies. **Home range:** 9.9–60ha.[205] **Day range:** 100–800m *[328–2625ft]*.[615]

Behavior Diurnal and arboreal.[198] Mantled howlers will occasionally travel on the ground, where they can outrun a human; they can also swim.[205] They are subject to high levels of botfly larvae infestation in May–October, which may increase adult and infant mortality.[142] Mothers give infants little

grooming. Males reportedly allomother infants by carrying them.[531] More female than male infants survive to the age of 1 year.[123] Infanticide is the cause of 40% of infant mortality. Infants are often killed after a new male takes over a group.[123] Other females attracted to an infant may often prevent a new mother from feeding and sleeping.[123] All age groups have been reported to fall out of trees.[123] Activity budget: Rest, 74%; feeding, 15–22%; social activity, 4%.[615] **Mating:** Dominant males have "priority access to estrous females."[615] **Scent marking:** Mantled howlers scent-mark by rubbing their throat on branches and by urine washing.[615] **Vocalizations:** The long-distance territorial roar is produced mostly when exhaling; other howlers vocalize on inhalation and exhalation.[878] Infants have a play vocalization.[531]

Mexico to Equador

LOWER RISK

CITES II

Black Howler *Alouatta pigra*

Taxonomy Disputed. Monotypic.[723] Elevated from a subspecies of *A. palliata*.[898] This species' name was *A. villosa* until 1988.[315]

Distinguishing Characteristics Both genders of the black howler are all black, with long hair and a prehensile tail.[198]

Physical Characteristics Head and body length: 521–639mm *[20.5–25.2in]*.[198] **Tail length:** 590–690mm *[23.2–27.2in]*.[198] **Weight:** ♀ 6434g *[14.2lb]*, ♂ 11,352g *[25.0lb]*.[246] **Intermembral index:** NA. **Adult brain weight:** NA. Canine length: ♀ 6.2mm *[0.2in]*, ♂ 7.6mm *[0.3in]*.[490] Males are 1.76 times the size of females.[246]

Habitat Primary and secondary semideciduous and lowland rain forest up to 250m *[820ft]*.[898]

Diet Leaves, fruit, flowers. During some seasons, 86% of the diet is from 1 tree *(Brosimum alicastrum)*.[615]

Life History Weaning: NA. **Sexual maturity:** ♀ 48mo, ♂ 72–96mo.[944] **Estrus cycle:** NA. **Gestation:** NA. **Age 1st birth:** 48–60mo.[944] **Birth interval:** NA. **Life span:** 20y.[944] Infants are brown when born; at 4 months, males develop a white scrotum.[198]

Locomotion Quadrupedal walking, climbing, and suspensory feeding.[615]

The black howler can live in secondary forests near farms. They are nicknamed baboons in the local Creole dialect.

NATURAL HABITAT

Social Structure Variable. 1 male–multifemale or 2 male–multifemale cohesive groups.[615] **Emigration:** Males emigrate. Females rarely leave their natal troop.[944] **Group size:** 7.[615] **Home range:** 3–25ha.[944]

Day range: 250m (40–700) *[820ft (131–2297)]*.[734]

Behavior Diurnal and arboreal. Black howlers rarely groom one another.[615] Males interact with infants infrequently.[688] A yellow fever epidemic in the late 1960s caused a sharp decline in black howler populations.[734] Activity budget: Rest, 66%; feeding, 22%; travel, 12%.[734] **Mating:** Males sniff the urine of estrous females.[386] The female tongue-flicks at the male before mating. The male holds the female's shoulders during mating, which lasts 30–60 seconds. One female was observed to cross her troop's territorial border to consort with a male from another troop.[386] **Scent marking:** Males scent-mark branches with their chin and throat. They catch urine with their hands and feet to mark branches as they travel.[944] **Vocalizations:** Black howlers usually roar within 100 meters *[328ft]* of their home range border at both dawn and dusk. During the rainy season, there is more roaring during midday.[387]

Mexico, Belize, Guatemala

Bolivian Red Howler *Alouatta sara*

Taxonomy Disputed. Elevated from a subspecies of *A. seniculus* in 1985, based on genetic analysis.[315]

Distinguishing Characteristics Bolivian red howlers have a dark maroon head and a uniformly golden orange body, with slightly darker limbs and yellowish underparts. They have a red beard and a black fringe around the face. The tail is prehensile.[365]

Physical Characteristics Head and body length: ♀ 535mm *[21.1in]*.[365] **Tail length:** ♀ 590mm *[23.2in]*.[365] **Weight:** NA. **Intermembral index:** NA. **Adult brain weight:** NA.

Habitat NA.

Diet NA.

Life History NA.

Locomotion Quadrupedal.

Social Structure NA. **Group size:** NA. **Home range:** NA.

Day range: NA.

Behavior Diurnal and arboreal. Little had been published about this species by 1995.

Bolivia

Male.

Red Howler *Alouatta seniculus*

Northern South America

C. CROCKETT (NATURAL HABITAT)

A troop of red howlers roars in unison.

Taxonomy Disputed. 8 subspecies. *A. s. arctoidea* has been proposed as a full species.[723]

Distinguishing Characteristics Red howlers have a dark red back with paler sides. Adult males have a long, blackish beard.[198]

Physical Characteristics Head and body length: ♀ 527mm (475–570) *[20.7in (18.7–22.4)]*, ♂ 581mm (510–630) *[22.9in (20.1–24.8)]*.[600] **Tail length:** ♀ 602mm (522–680) *[23.7in (20.6–26.8)]*, ♂ 612mm (565–680) *[24.1in (22.2–26.8)]*.[600] **Weight:** ♀ 4.2–7.0kg *[9.3–15.4lb]*, ♂ 5.4–9.0kg *[11.9–19.8lb]*.[247] **Intermembral index:** NA. **Adult brain weight:** 57.9g *[2.0oz]*.[346] The female's weight is 69% of the male's.[615]

Habitat Gallery forest, mangrove swamps, semideciduous forest, savanna, woodlands, and secondary forest[615] below 1000m *[3281ft]*.[898] The forest canopy above 20 meters *[66ft]* is preferred.[898]

Diet Young leaves, 44.5%; fruit, 42.3%; mature leaves, 7.5%; flowers, 5.4%;[615] petioles, 0.1%.[404] Red howlers prefer ripe fruit to unripe.[270] They use 195 plant species[424] and consume 1.23kg *[2.7lb]* of fresh food per day.[270]

Life History Infant: 10mo.[615] **Juvenile:** 10–30mo.[615] **Subadult:** 30–53mo.[615] **Sexual maturity:** ♀ 43–54mo, ♂ 58–66mo. **Estrus cycle:** 17d.[143] **Gestation:** 191d (184–194).[143] **Age 1st birth:** 62mo.[142] **Birth interval:** 16.6mo.[615] **Life span:** 25y.[708] Birth season: No. Two births have been observed in the afternoon. Delivery took 3 minutes, and the placenta was eaten.[615] In infancy the female's genitalia mimic the male's.[198]

Locomotion Quadrupedal walking, 80%; leaping, 4%.[615]

Social Structure Variable. 1 male–multifemale[758] and multimale-multifemale groups.[743] Male tenure is 5–7.5 years.[142] Solitary males have been observed.[270] In the adult female hierarchy, younger females are dominant to older ones.[615] A male's rank in the hierarchy depends on size.[615] Home ranges overlap 31.9–63%.[743] **Emigration:** Both sexes migrate from their natal group.[270] **Group size:** 4.6[758] (4–17).[743] **Home range:** 8ha,[743] 23ha.[615] **Day range:** 370m (20–840) *[1214ft (66–2756)]*.[743]

Behavior Diurnal and arboreal.[198] Red howlers travel on the ground 28% of the time.[582] They feed more on high-energy food sources early in the day (6–7 a.m.) and high-protein sources later (3–5 p.m.).[270] They come to the ground to eat soil on 30 of 203 days.[404] Eating leaves makes up 50% of the daily activity.[270] Competition for food is greater when food is scarcer and found in smaller patches.[744] A 1-male group may become a multimale troop after being invaded by an all-male bachelor troop. After taking over a troop, a new male may increase his reproductive success by killing the young infants to shorten the females' interbirth interval.[615] In the 9 takeovers that were observed in a study, 15 of 20 infants under 9 months of age were killed.[615] Agonistic interactions are very rare between established members of the same group.[615] In 3 cases of females' adopting infants, 2 infants were related to the adopter, and 1 was not.[826] Adult males do not interact with infants.[615] When alarmed, red howlers may urinate and defecate on the observer. Researchers can detect a group's recent departure by the strong stablelike scent of copious droppings.[198] Activity budget: Rest, 78.5%; travel, 5.6%; feeding, 12.7%.[270] **Association:** Red howlers compete with and usually displace black spider monkeys *(Ateles paniscus)* in fruiting *Bagassa guianensis* trees.[758] They have an agonistic relationship with weeper capuchins *(Cebus olivaceus)* when the 2 species meet in the same fig tree.[251] **Vocalizations:** Both males and females howl. The female's howl is higher pitched than the male's. Males respond more to male howls, and females respond to the howls of females from other groups.[744] During the full moon, the predawn chorus begins almost 1 hour earlier than usual.[743] "Interacting troops come closer to each other and roar more during the dry season, when they often meet around fruiting figs."[744] Before moving, red howlers are believed to call in a way that does not indicate their direction of travel.[743] They do not respond to roars of other troops that are more than 100 meters *[328ft]* away. The alarm call is a few soft grunts.[198] **Sleeping site:** *Albizia caribea,* a tree that is a food source.[743]

Juvenile.

White-bellied Spider Monkey *Ateles belzebuth*

Taxonomy Disputed. 3 subspecies.[723] Used to include *A. marginatus,* which is now accepted as a full species.[720]

Distinguishing Characteristics *A. b. belzebuth* has a black back and pale buffy underparts.[850] *A. b. hybridus* has a pale brown back. Both subspecies may have a white triangular patch on their brow[198] (30% do not have a patch[461]). These spider monkeys are individually variable and recognizable.[461]

Physical Characteristics Head and body length: 416–582mm *[16.4–22.9in].*[198] **Tail length:** 680–899mm *[26.8–35.4in].*[198] **Weight:** ♀ 7491–10,400g *[16.5–22.9lb],* ♂ 7264–9800g *[16.0–21.6lb].*[247] **Intermembral index:** 109.[234] **Adult brain weight:** 106.6g *[3.8oz].*[346] These spider monkeys have a sternal gland; females have an elongated clitoris.[461]

Habitat High tropical rain forest and riverine, marsh, and semideciduous forest up to 600m *[1969ft].*[850]

Diet Fruit, 83%; decaying wood, 10%; leaves, 5%; epiphytes, 2%; flowers, 1%.[461] Ripe fruit from 33 species is eaten.[462] These spider monkeys eat the decaying wood and mud used by termites (*Constrictotermes* sp.) to build a tree nest, but they do not eat the termites.[461] Why they include this wood in their diet is currently a mystery.[461] Palm fruits are important to these spider monkeys but not to howlers. An individ-

Female spider monkeys have an elongated clitoris for scent marking.

Variegated spider monkey (*A. b. hybridus*).

ual white-bellied spider monkey ate 100 fruits (20mm *[0.8in]* in size) of the *Pouteria* tree in 7 minutes.[462]

Life History Infant: 0–12mo.[461] **Juvenile:** 12–48mo.[461] **Subadult:** 48–72mo.[461] **Sexual maturity:** 72mo.[461] **Estrus cycle:** NA. **Gestation:** NA. **Age 1st birth:** NA. **Birth interval:** 36mo.[698] **Life span:** 20y.[89] Birth season: No. Neonates have a reddish face and dark fur. Development: Infant first independent, 2.5mo;[461] first play with another infant, 5mo;[461] coat change to adult color, 12mo.[461] Juveniles eat adult food but still nurse.[461] Subadult males have canines and visible scrotums.[461]

Locomotion Brachiation, climbing, and suspensory.[850]

Social Structure Fission-fusion communities.[850] Communities break into subgroups of 3–4 individuals to forage.[462] Adult males often travel together; they travel faster than females and infants.[461] **Group size:** 3–22.[850] **Home range:** 260–390ha.[850] **Day range:** 1000–2400m,[461] to 4000[850] *[3281–7874ft, to 13,124].*

Behavior Diurnal and arboreal. White-bellied spider monkeys feed at 15–35 meters *[49–115ft]* above the ground. Up to 30 individuals have been observed to congregate at "salados" (salt lick sites) on the ground.[720] When 2 subgroups reunite, there is an excited greeting display that includes vocalizing, chasing, hugging with tails wrapped around each other, and sniffing of the pectoral gland.[850] The subgroups disperse quietly without a

display.[461] The largest group sizes are reported during the wet season's peak, when fruit is most available.[347] Activity budget: Rest, 62%; feeding, 22.2%; travel, 14.8%;[462] allogrooming, 1%.[850] **Mating:** Duration is 5–10 minutes, with the male sitting on a branch behind the female.[169] **Scent marking:** White-bellied spider monkeys rub the chest to deposit secretions from the pectoral glands.[850] The pendulous clitoris may retain urinary droplets and scent.[461] **Vocalizations:** "Ts chookis," whoops, and wails are used to locate other subgroups that have been heard a short distance away (5–30m *[16–98ft]*).[461] These spider monkeys "whinny" during 86% of reunions.[461] They shake branches at potential predators. They have been observed to vocalize "ook-sob" at a coatimundi and to give growls, "ookbarks," and whines at a tayra (*Eira barbara*), This spider monkey has been observed to shake a vine occupied by a tayra, to cause the predator to fall.[461] Various screams and barks have also been des-cribed.[198] **Sleeping site:** Adults sleep in separate crotches of tall trees 20–30 meters *[66–98ft]* above the ground.[461]

White-bellied spider monkey (*A. b. belzebuth*).

R. A. MITTERMEIER

Colombia, Ecuador, Peru, Brazil

Black-faced Black Spider Monkey *Ateles chamek*

L. C. MARIGO

Taxonomy Disputed. Elevated from a subspecies of *A. paniscus* in 1989 from genetic studies.[725]

Distinguishing Characteristics Black-faced black spider monkeys have a black face and body. They are smaller than black spider monkeys *(A. paniscus),* and their fur is shorter in length.[850] Infants have whitish skin around the eyes,[850] pink feet, and sparse fur that appears to be gray.[815]

Physical Characteristics Head and body length:
♀ 403–520mm [15.9–20.5in], ♂ 450mm [17.7in].[365]
Tail length: ♀ 800–880mm [31.5–34.6in], ♂ 820mm [32.3in].[365] **Weight:** 7.0kg [15.4lb].[247] **Intermembral**

index: NA. **Adult brain weight:** NA.
Habitat Primary tropical rain forest up to 1424m [4672ft].

Diet Fruit and seeds, 80%; young leaves, 17%; flowers, 2%;[877] honey, wood, caterpillars, termites.[814] The seeds of 17 species are ingested whole and probably dispersed.[877]

Life History Infant: 15–18mo.[815] **Juvenile:** 18–48mo.[815] **Subadult:** 48–72mo.[815] **Sexual maturity:** NA. **Estrus cycle:** NA. **Gestation:** 225d.[815] **Age 1st birth:** NA. **Birth interval:** 34.5mo (25–42).[815] **Life span:** NA. Infant mortality is 33%; 5 of 15 infants observed in one study did not survive to 1 year of age.[815]

Locomotion Brachiation, climbing, leaping, and bipedal walking on horizontal branches.[826]

Social Structure Fission-fusion communities. These monkeys have discrete territories with only 10–15% overlap.[815] Females use a core area of only 22–33% (50ha) of the home range of the total group and do not travel in cohesive groups.[815] Males range over the entire group territory. **Emigration:** Females emigrate.[815] **Group size:** Mean 4.5 at higher elevations (1424m [4672ft]) to 18.5 at lower elevations (275m

[902ft]).[850] **Home range:** 150–231ha.[815] **Day range:** 1927m (465–4070) [6322ft (1526–13,354)].[815]

Behavior Diurnal and arboreal. This species has been observed to use its tail to obtain water from a deep tree hole (850mm [33.5in]) in the dry season.[220] The amount of fruit (density) and the size of the fruit tree (patch size) affect group size, feeding time, and day range.[814] The size of the foraging party increases from 2 to 6 when more fruit is available.[814] The passage time (ingestion to defecation) is 4–8 hours.[877] Activity budgets of the genders differ. Males feed 22.5% of the day and travel 30.0% of the day; females with infants feed 30.7% and travel only 20.9%.[815] Males are more vigilant and reportedly cooperate in mobbing any predators they detect.[813]

Vocalizations: Male coalitions vocalize and chase other groups near the boundaries of their territories.[815]

Brazil, Peru, Bolivia

Brown-headed Spider Monkey *Ateles fusciceps*

Taxonomy 2 subspecies.[723]

Distinguishing Characteristics
One subspecies *(A. f. fusciceps)* has a black to brownish body with a brownish head. The other *(A. f. robustus)* is all black except for a few white hairs on the chin.[850] Infants have a pink face and pink ears.

Life History Head and body length:
393–538mm [15.5–21.2in].[850] **Tail length:** 710–855mm [28.0–33.7in].[850] **Weight:** ♀ 8800g [19.4lb], ♂ 8890g [19.6lb].[246] **Intermembral index:** 103.[234] **Adult brain weight:**
114.7g [4.0oz].[346]

Habitat
Lowland to lower montane rain forest.[198]

Panama to Ecuador

Diet Fruit and leaves.[723]

Life History Weaning: 20mo.[195]
Sexual maturity: ♀ 51mo,

♂ 56mo.[698] **Estrus cycle:** 26d.[346] **Gestation:** 226–232d.[195] **Age 1st birth:** 51mo.[195] **Birth interval:** 36mo.[346] **Life span:** 24y.[708] The infant first leaves its mother to play at 10 weeks. It rides on the mother's back at 16 weeks.[850]

Locomotion Semibrachiation, climbing.

Social Structure Fission-fusion communities.[89] **Group size:** NA. **Home range:** NA. **Day range:** NA.

Behavior Diurnal and arboreal.[198] When brown-headed spider monkeys display aggressively, they can piloerect the hair on their neck, shoulders, and tail base.[850] In captivity an adult male groomed with a young male more than the females did.[850] **Mating:** Females reportedly consort with a male for up to 3 days or mate promiscuously.[89] Mating is face-to-face[13] for 5–10 minutes,[169] up to 25 minutes.[195]

Vocalizations: 9.[347] High-frequency whinnies and sobs are friendly or feeding vocalizations.[850] Other calls such as trills, twitters, and squeaks are given in situations of social discomfort.[850]

Black-handed Spider Monkey *Ateles geoffroyi*

Taxonomy 9 subspecies.[850]

Distinguishing Characteristics The coat color of black-handed spider monkeys varies from light buff to black, depending on the subspecies. The hands and feet are usually black.[850] Infants are born black; in some subspecies the body color lightens appreciably during the first 5 months.[850]

Physical Characteristics Head and body length: 305–630mm [12.0–24.8in].[198] **Tail length:** 635–840mm [25.0–33.1in].[198] **Weight:** ♀ 6000–8912g [13.2–19.6lb], ♂ 7420–9000g [16.4–19.8lb].[247] **Intermembral index:** 105.[234] **Adult brain weight:** 110.9g [3.9oz].[346]

Habitat Evergreen rain forest, semideciduous forest, mangrove forest.[850]

Diet Fruit, 77.7%; seeds, 11.1%; flowers, 9.8%; young leaves, 7.3%; buds, 2.6%; mature leaves, 1.2%; animal prey, 1.2%.[106]

Life History Infant: 0–24mo.[560] **Sexual maturity:** ♀ 48–

60mo, ♂ 60mo.[698] **Estrus cycle:** 26d.[346] **Gestation:** 226–232d.[698] **Age 1st birth:** 60–90mo.[560] **Birth interval:** 17–45mo.[850] **Life span:** 27y.[560]

Locomotion Quadrupedal walking and running.[850]

Social Structure Fission-fusion communities. The ratio of males to females in a community is 1:1.8. The community divides into subgroups when resources are scarce.[107] **Group size:** 4.5, to 35.[198] **Home range:** 62.4ha,[214] to 100–115.[850] **Day range:** 350–3000m [1148–9843ft].[688]

Behavior Diurnal and arboreal.[198] Female black-handed spider monkeys seek out trees with large amounts of fruit per visit to minimize the cost of searching for food. Males spend less time feeding at each site and travel farther, thus increasing their chances of encountering different females. Home range size varies with gender. Males have the largest territories; females with infants have smaller territories.[214] Male juveniles 24–36 months old affiliate more with subadult and adult males.[560] The male's affinitive interactions are directed to other males (85%).[792] Activity budget: Rest, 54%; travel, 27.6%; feeding, 10.8%; interaction, 6.9%; vocalization, 0.41%. **Association:** This species occasionally associates with white-throated capuchins (*Cebus capucinus*).[688] **Mating:**

These spider monkeys mate in a sitting position, the male behind the female, "with his legs over and between the female's thighs, locking her legs outward." Mating lasts 8–25 minutes.[850]

Vocalizations: Barks, whinnies, and screams.[198]

Mexico to Panama

White-whiskered Spider Monkey *Ateles marginatus*

Taxonomy Disputed. Elevated from a subspecies of *A. belzebuth* in 1989.[315]

Distinguishing Characteristics White-whiskered spider monkeys are all black except for a white brow and a white fringe of hair on the face from ears to chin.[198]

Physical Characteristics Head and body length: ♀ 344–505mm [13.5–19.9in], ♂ 500mm [19.7in].[365] **Tail length:** ♀ 613–770mm [24.1–30.3in], ♂ 750mm [29.5in].[365] **Weight:** ♀ 5824g [12.8lb].[247] **Intermembral index:** NA. **Adult brain weight:** NA.

Habitat Tall primary terra firma forest.[530]

Diet Fruit, seeds.[530]

Life History NA.

Locomotion Semibrachiation, climbing.

Social Structure NA. **Group size:** NA. **Home range:** NA. **Day range:** NA.

Behavior Diurnal and arboreal. To get water during the dry season, white-whiskered spider monkeys dip their hands in tree holes, then lick the fur and fingers.[220] As of 1995 they had not been the subject of any long-term study. This species is threatened by dam projects and colonization.[530]

L. C. MARIGO

Brazil

Black Spider Monkey *Ateles paniscus*

Taxonomy Disputed. Monotypic.[723] Included *A. chamek* as a subspecies until 1989.[723]

Distinguishing Characteristics Black spider monkeys have long, glossy black hair and a pinkish face.[850] The prehensile tail has thicker hair one-third of its length from the base.[198] Infants have dark pigmentation around the eyes and muzzle.[850]

Physical Characteristics Head and body length: ♀ 540mm (490–620) *[21.3in (19.3–24.4)]*, ♂ 545mm (515–580) *[21.5in (20.3–22.8)]*.[600] **Tail length:** 814mm (640–930) *[32.0in (25.2–36.6)]*, 807mm (720–852) *[31.8in (28.3–33.5)]*.[600] **Weight:** ♀ 6500–11,000g *[14.3–24.2lb]*, ♂ 5470–9200g *[12.1–20.3lb]*.[247] **Intermembral index:** 105.[234] **Adult brain weight:** 109.9g *[3.9oz]*.[346]

Habitat Mature rain forest, pina swamp.[850]

Diet Fruits, including figs, 82.9%; leaves, 6%; flowers, 6.4%; bark, 1.7%; some caterpillars and termites. These monkeys eat 171 species of fruit.[850] They are seed predators on 23 tree species but may disperse the seeds of more than 135 tree species.[850] Most feeding bouts are 1–15 minutes long.[850]

Life History Infant: 0–12mo.[850] **Juvenile:** 36–50mo.[850] **Subadult:** 50–65mo.[850] **Sexual maturity:** 50–65mo. **Estrus cycle:** 26–27d.[850] **Gestation:** NA. **Age 1st birth:** 60mo.[708] **Birth interval:** 24mo,[826] to 46–50.[850] **Life span:** 33y.[708] Birth peak: Nov–Jan (end of dry season).[850] An infant rides on its mother's back when 6–15 months old.[850] Infants are not entirely independent until 36 months old.[850]

Locomotion Quadrupedal, suspensory, climbing, leaping, and bipedal walking on horizontal branches.[850]

Social Structure Fission-fusion communities. Males are often solitary or in small groups. Males have a hierarchy determined by age[850] and are dominant over females.[247] Female and offspring often forage alone.[850] **Group size:** 18; 3 when foraging.[903] **Home range:** 225ha.[758] **Day range:** 2700m[9] (500–5000)[850] *[8859ft (1641–16,405)]*.

Behavior Diurnal and arboreal.[198] Black spider monkeys only rarely forage below 20 meters in the canopy *[66ft]*.[850] Travel through the canopy is noisy because of branch-shaking leaps.[198] Males associate with other males twice as often as with females.[792] Males cooperate in territorial patrols and perform aggressive displays at their borders.[850] These monkeys threaten observers by shaking vegetation and growling.[198] Hunters prefer them because of their large body size. **Association:** Subgroups of spider monkeys give way to red howlers (*Alouatta seniculus*)[758] but tolerate the presence of bearded sakis (*Chiropotes satanas*) feeding at the same food tree. **Mating:** A male and female may consort for up to 72 hours. The female initiates mat-

Brazil, Guyana, Suriname, French Guiana

ing.[850] **Scent marking:** Both sexes sniff and embrace when greeting.[850] **Vocalizations:** These monkeys have a variety of loud calls, audible for 800–1000 meters *[2625– 3281ft]* on the ground and 2000 meters *[6562ft]* above the canopy. Males and females give the "ook-barking" call to threaten ground predators or humans.[850] Greeting calls are horselike whinnies. Only males whoop.[850] Juveniles develop their long calls by trial and error.[850]

Yellow-tailed Woolly Monkey *Lagothrix flavicauda*

Taxonomy Monotypic.[723] Until rediscovered in 1974, this species was known only from museum specimens and thought to be extinct.

Distinguishing Characteristics Yellow-tailed woolly monkeys are deep mahogany,[508] with yellow on the tail's underside and a yellow scrotal tuft of hair.[684] The face is "brown with a pale yellow triangular patch covering mouth and nose, [and] the point of the triangle is between the eyes."[198]

Physical Characteristics Head and body length: 515–535mm *[20.3–21.1in]*.[198] **Tail length:** 560–610mm *[22.0–24.0in]*.[198] **Weight:** 10kg *[22lb]*.[246] **Intermembral index:** NA. **Adult brain weight:** NA.

Habitat Humid montane cloud forest at

Peru

500–3000m *[1641–9843ft]*[898] and 4–25°C *[39–77°F]*.

Diet Fruit, leaves, animal prey, flowers, buds, epiphyte roots.[508]

Life History Weaning: 12mo.[89] **Sexual maturity:** 48mo.[507] **Estrus cycle:** NA. **Gestation:** NA. **Age 1st birth:** NA. **Birth interval:** NA. **Life span:** NA.

Locomotion Quadrupedal and suspensory.

Social Structure Multimale-multifemale groups.[698] There is little competition among troop members.[442] **Group size:** NA. **Home range:** NA. **Day range:** NA.

Behavior Diurnal and arboreal.[198] The largest male in a group threatens humans "by shaking and dropping branches and by urinating and defecating."[198] Yellow-tailed woolly monkeys are threatened with habitat destruction and hunting by humans.[508] **Association:** These monkeys associate with spider monkeys

A. YOUNG

(*Ateles*).[507] **Vocalizations:** A sharp puppylike bark,[198] as well as a loud call and soft contact calls.[307]

Woolly Monkey *Lagothrix lagotricha*

Colombian woolly monkey *(L. l. lugens)*.

[26.1in (24.3–28.3)], ♂ 674mm (532–770) [26.5in (20.9–30.3)].[600]

Weight: ♀ 3.5–6.5kg [7.7–14.3lb], ♂ 3.6–10kg [7.9–22.0lb].[247]

Intermembral index: 97.9.[425] **Adult brain weight:** 96.4g [3.4oz].[346]

Habitat Humid, primary flooded and nonflooded forest;[515] up to 3000m [9843ft] in Colombia.[515]

Diet Fruit, 67.5%; leaves, 14.4%; seeds, 7.1%; gums, 6.2%; flowers, 3.1%;[657] animal prey, including mammals.[94] There are 225 plant species in the diet.[657] They reportedly eat as much as 32% of their body weight in a day.[657]

Life History Weaning: 20mo.[684] **Sexual maturity:** NA. **Estrus cycle:** 21d. **Gestation:** 223d.[698] **Age 1st birth:** 60mo.[708] **Birth interval:** 18–24mo.[698] **Life span:** 25.9y.[708] Births: Year-round. Development: Infant clings to mother's belly, week 1; rides on mother's back, week 2;[684] travels independently most of day and eats solid food, week 24.[684]

(R. A. MITTERMEIER)

Humboldt's woolly monkey *(L. l. lagotricha)*.

Geoffroy's woolly monkey *(L. l. cana)*.

Taxonomy 4 subspecies,[723] some of which may be valid species with further study.[574]

Distinguishing Characteristics
The fur color of woolly monkeys is pale, dark brown or gray to blackish and varies with individuals and with subspecies. Woolly monkeys have a round head, a black face, prominent suborbital ridges, a strong prehensile tail, and a potbelly.[684] Females have a pendulous clitoris longer than the male's penis.[89] Infants are straw colored;[198] the natal coat is lighter than the adult coat.[684]

Physical Characteristics Head and body length: ♀ 490mm (461–580) [19.3in (18.1–22.8)], ♂ 526mm (460–650) [20.7in (18.1–25.6)].[600]

Tail length: ♀ 663mm (617–718)

Locomotion Quadrupedal climbing and suspensory. Woolly monkeys do not travel as fast as spider monkeys *(Ateles)* and rarely leap.[684] They can hang by their tails only.[684]

Social Structure Multimale-multifemale groups. Within the larger community, subgroups forage independently.[198] Woolly monkeys have overlapping territories with little intergroup hostility.[684] Males have a dominance hierarchy determined by age.[684] **Group size:** 33[684] (5–43, to 70).[898] **Home range:** 400– 1100ha.[898] **Day range:** 1000[3]– 3000m [3281–9843ft].[790]

Behavior Diurnal and arboreal.[198] Woolly monkeys prefer the upper canopy.[684] In a captive study, males had higher rates of social behavior than females and formed male coalitions.[515] Females interact predominantly with close female relatives and males. The dominant male protects females with new infants from the rest of the group for 1–2 months.[684] Adult males perform hostile displays by rubbing their chests.[684] These monkeys have a submissive display in which their hand shields their eyes while they make a sobbing vocalization.[684] The threat display includes shaking branches.[684] Less than 2% of the daily activity is spent on grooming.[684] Woolly monkeys are the number one target of human hunters for meat and pets.[684] They are the first monkey to disappear when extensive forests are fragmented by agriculture or

logging.[53] **Association:** Monk sakis *(Pithecia monachus)* have been observed following woolly monkeys. **Mating:** Females lip-smack when in estrus. Males sit down, place their legs over the females' thighs, and grip branches with their feet, thus locking females into a snug dorsoventral contact.[489] Mating lasts 4 minutes.[489] Females mate with more than 1 male.[684] **Vocalizations:** 14, including loud and descending trills, barks and screams. When alarmed, the whole group choruses with *yoohk-yoohk*.[198]

Amazon Basin

(R. A. MITTERMEIER)
Poeppig's woolly monkey *(L. l. poeppigii)*.

Woolly Spider Monkey or Muriqui *Brachyteles arachnoides*

Taxonomy Disputed. In 1995 the 2 subspecies were recognized as 2 species:[723] the northern *B. hypoxanthus* and the southern *B. arachnoides,* which has thumbs, sexually dimorphic canines, and a different social structure and genetic makeup.[489]

Distinguishing Characteristics Woolly spider monkeys are the largest New World primate.[621] They have gray, yellow, or brown fur; a protruding belly and large testicles;[198] and a bare patch of skin (volar skin) on their prehensile tail.[621] The southern species has a black face; the northern species' face is individually mottled.

Physical Characteristics Head and body length:
♀ 573mm (545–600) *[22.6in (21.5–23.6)]*, ♂ 595mm (580–610) *[23.4in (22.8–24.0)]*.[600] **Tail length:**
♀ 791mm (740–840) *[31.1in (29.1–33.1)]*, F 680mm (670–690) *[26.8in (26.4–27.2)]*.[600] **Weight:** ♀ 9450g *[20.8lb]*, ♂ 12,125g *[26.7lb]*.[246] **Intermembral index:** 104.[234] **Adult brain weight:** 120.4g *[4.2oz]*.[346]

Habitat Moist rain forest, seasonal semideciduous forest. Primary and secondary forests are used equally.[787]

Diet Leaves, 51%; fruit, 32%;[790] nectar, seeds.[621]

Life History Infant: 0–12mo.[788] **Weaning:** 30mo.[788] **Juvenile:** 12–36mo.[788] **Subadult:** 36–84mo.[788] **Sexual maturity:** NA. **Estrus cycle:** NA. **Gestation:** 210–255d.[790] **Age 1st birth:** 90mo.[788] **Birth interval:** 33.8mo.[788] **Life span:** 30y.[790] Birth season: Jun–Aug, during the dry season. The female's nipples are under the armpits. An infant rides on its mother's side until it is 6 months old; then it rides on her back.

Locomotion Brachiation; suspensory; quadrupedal walking and running on large branches; leaping to 8m *[26ft]*. This species will cross open ground quadrupedally.[621]

Social Structure Variable. Multimale-multifemale and fission-fusion.[621] Males tend to associate with males, and females with females. Territorial disputes are settled by vocalization rather than a physical fight.
Emigration: Males remain in their natal group. Females emigrate voluntarily at age 5–6.5 years but have difficulty immigrating into another group because resident females are hostile to newcomers. They may be accepted and helped into the group by subadult males.[788] **Group size:** 26 (8–45);[790] 5–24 when foraging.[787] **Home range:** 70[625]–168ha.[790] **Day range:** 1283m[790] (350–1400)[621] *[4210ft (1148–4593)]*.

Behavior Diurnal and arboreal. Woolly spider monkeys are reported to be one of the most nonaggressive primate species. They do very little social grooming, but individuals embrace each other for reassurance, and a greeting hug is used when meeting a friend of the same gender.[787] They alarm-call at large feline predators, but because of their large size, they do not react to aerial predators. They travel 1 hour per day and forage "from tree to tree eating what's available in each tree rather than [traveling] a long distance to a specific food tree as do howlers."[559] They feed 3 hours per day and rest 50% of each day.[559] Less than 1% of their time is spent in social interaction with individuals other than dependent offspring.[559] **Association:** Sympatric with brown howlers *(Alouatta fusca).*[615]
Mating: Males compete via sperm competition rather than overt aggression.[621] An estrous female mates promiscuously, and several males may line up to

mate with her.[787] After intromission the male remains motionless for 2–7 minutes; then there is 10–28 seconds of thrusting.[169] The male's ejaculate hardens to form a vaginal plug, which is later removed by the female or other individuals and dropped.[787] **Scent marking:** Adults "deposit urine on their hands."[559] **Vocalizations:** 8. A loud long-distance call is given before sleeping.[621] An aggressive barklike call is used to threaten a human observer.[621] Chuckles and warbles convey reassurance and excitement. Neighs have variable meanings besides being a contact call.[789] Females produce a mating twitter.[559] **Sleeping site:** Canopy. Several individuals sleep in a huddled cluster.[621]

NATURAL HABITAT
A male eating bark.

NATURAL HABITAT
The northern muriqui *(B. hypoxanthus)* has a mottled face that is individually distinct.

Brazil

The southern muriqui *(B. arachnoides)* has a black face.

NATURAL HABITAT
This muriqui is hanging upside down by its tail to feed.

Drill *Mandrillus leucophaeus*

Old World Monkeys

The anthropoid monkeys and apes of Africa and Asia are all included in the infraorder Catarrhini. The name "Catarrhini" refers to the shape of their nose. In catarrhines the nostrils face downward and are narrow; the Platyrrhini of the New World have round nostrils facing to the side. The skull and ear structures of these two infraorders have small differences. The dental formula of the larger platyrrhines includes 3 premolars, whereas catarrhines, from which humans evolved, have only 2 premolars, for a total of 32 teeth.[234]

$$\frac{2.1.2.3.}{2.1.2.3.} \times 2 = 32$$

Old World primates are diurnal and generally larger than Platyrrhini primates, and many are partly terrestrial. All catarrhines have flattened nails on their digits, and most have pads on their buttocks (ischial callosities). The Catarrhini are divided into two superfamilies—the Cercopithecoidea (Old World monkeys) and the Hominoidea (apes).

Macaques, Baboons, Guenons, and Colobines　　　　　　*Family:* Cercopithecidae

All of the Old World monkeys belong to the same family, the Cercopithecidae. Members share several skeletal features. The hind limbs are longer than the forelimbs. The tail when present is not prehensile. The nose and palate are narrow. Each of the molars has four cusps, which form two distinct ridges or lophs referred to as bilophodont. This makes the molars of the Cercopithecidae easy to distinguish from the simple molars of the Hominoidea.

Anatomical features divide the Cercopithecidae into two subfamilies. These features are related to dietary adaptations. The Cercopithecinae have low-cusp molars, eat mostly fruit, and have pouches in their cheeks to store food for short periods. The Colobinae have high-cusp molars, specialized stomachs, eat more leaves and seeds, and have no cheek pouches.[234] The Cercopithecinae range in size from the arboreal dwarf guenon *(Miopithecus talapoin),* which weighs about 1 kilogram (about 2 pounds), to the terrestrial olive baboon *(Papio hamadryas anubis),* which can weigh over 37 kilograms (over 80 pounds).

Macaques have a rigid dominance hierarchy, and agonistic interactions are common.
These are Assamese macaques *(Macaca assamensis).*

119

Cheek Pouch Monkeys

Cheek Pouch Monkeys   *Subfamily:* Cercopithecinae

All of the genera of cheek pouch monkeys are found in Africa. Only 1 species of the genus *Macaca* is found in Africa, however, and the other 15 *Macaca* species are found in Asia. Macaques can presently be found in more climates and habitats than any other primates except humans. The geographic ranges that the macaques inhabit are from as far north and east as Japan (Japanese macaques, *M. fuscata*) to as far west as Morocco (Barbary macaques, *M. sylvanus*). Macaques are adaptable frugivores. Rhesus macaques *(M. mulatta)* have even adapted to life in cities. Macaques are the most common monkey used in biomedical research. Over 6000 were legally imported in 1993 to be used in biomedical or laboratory research.

Baboons of the genus *Papio* are the largest and most terrestrial of the cercopithecines. They have long, doglike snouts and impressive canines, their main weapons of defense. Most species live in dry savanna woodlands and are well studied in their natural habitat.

The genus *Mandrillus* consists of two species that live in a few remaining forests in western Africa. They have never been successfully studied in their natural habitat because they live in dense forest and are wary of humans, who hunt them.

The gelada baboon *(Theropithecus gelada)* is the only living member of a large genus known

from fossils. It lives only on the grassland plateaus in the highlands of Ethiopia.

The mangabeys are forest monkeys that have recently been separated into two genera. Although the species look similar, they are differentiated by their behavior and ecology. The three species of *Cercocebus* travel and feed on the ground and understory, and the two species of *Lophocebus* are predominantly arboreal, rarely coming to the ground.

There are three monotypic genera of monkeys that are closely related to the guenons *(Cercopithecus)*. Allen's swamp monkey *(Allenopithecus)* has not been well studied because it inhabits lowland swamp forests in the Zaire basin. The dwarf guenon *(Miopithecus)* is the smallest member of the Old World monkeys. The patas monkey *(Erythrocebus)* is a terrestrially adapted monkey that inhabits savanna woodlands from Senegal to Ethiopia.

The genus *Chlorocebus,* which was recently separated from *Cercopithecus,* currently includes only one species of vervet, or green monkey *(Chlorocebus aethiops).*

The genus *Cercopithecus,* which has the largest number of species, is the most colorful and varied genus of the subfamily. It is covered in more detail on page 153.

Stump-tailed Macaque *Macaca arctoides*

Infant stump-tailed macaques are white.

NATURAL HABITAT

Taxonomy Monotypic. The name was changed from
M. speciosa in 1976.[315]

Distinguishing Characteristics Stump-tailed
macaques are dark brown, with a short, nearly hairless tail. The face is hairless and mottled with varying
amounts of black and red skin.[509]

Physical Characteristics Head and body length:
♀ 485–585mm *[19.1–23.0in]*, ♂ 517–650mm
[20.4–25.6in].[211] **Tail length:** ♀ 14.5–69mm
[0.6–2.7in], ♂ 3.2–12.5mm *[0.1–0.5in]*.[211]
Weight: ♀ 7.5–9.1kg *[16.5–20.1lb]*, ♂ 9.9–10.2kg
[21.8–22.5lb].[211] **Intermembral index:** 98.[234] **Adult brain
weight:** 104.1g *[3.7oz]*.[346]

Habitat Lowland forest, monsoon forest, dry forest, and
montane forest up to 2000m *[65ft]*. These macaques
prefer dense forests and are occasionally found near
human settlements and temples.[898]

Diet Fruit, seeds, young leaves, flowers, buds, and animal prey, including insects, birds, and eggs.[693]

Life History Infant: 6–12mo.[650] **Weaning:** 9–18mo.[54]
Juvenile: 12–48mo.[650] **Subadult:** 48–96mo.[650] **Sexual
maturity:** NA. **Estrus cycle:** 29d.[346] **Gestation:** 178.2d[709]
(166–185).[650] **Age 1st birth:** 45.6[708]–56.4mo.[709] **Birth
interval:** 19mo (12–24).[709] **Life span:** 30y.[708] Females
have no visual sign of estrus.[156] Neonates are
creamy white.[650]

Locomotion Quadrupedal.

Social Structure Multimale-multifemale groups.
Matrilineal hierarchies are very strong.
Group size: 5–40.[898] **Home range:** NA. **Day range:** 400–
3000m *[1312–9843ft]*.[54]

Behavior Diurnal, arboreal, and terrestrial. Stump-tailed
macaques travel on the forest floor and along the
banks of streams.[693] They do not swim.[54] They have
been the subject of many captive behavioral studies
but few field studies. After a conflict, they have a ritual
reconciliation behavior in which the subordinate presents its hindquarters to the dominant, which clasps
the presenter by the rump.[159] This eases the tension
between individuals. Grins, teeth chattering, and lip-
smacking are other signs of submission.[159] The subordinate may also offer a hand for a mock bite. The
dominant animal may present to a low-ranking individual to reassure or pacify it.[54] Grooming is a form of
social interaction that promotes appeasement and
group cohesion. These macaques may groom for a
few seconds to more than an hour.[54] In a captive
study, wounded individuals were groomed more than
usual.[54] The dominant male stops female fights and

protects infants.[159] In a captive
study, 66% of all social behavior
involved touching and hud-
dling.[627] Some 50% of the agonistic behavior toward mating
pairs was directed by adult
females.[627] The tail position indicates an individual's intentions:
tail down means submission or
fright; tail curled up indicates
excitement; tail straight up is an
assertion of dominance.[54] The
teeth-chattering face is a greeting
or appeasement signal. When
attacked by a high-ranking animal, a stumptail may redirect its
aggression by attacking a nearby
subordinate.[54] "In Assam these
monkeys are feared and reputed
to attack people if they are disturbed."[693] **Mating:** Dominant
males copulate with high-ranking
females throughout their estrus
cycles.[782] Mating lasts 12–20
minutes,[782] after which partners
remain attached and are often
harassed by other members of
the group.[156] Of all the observed
matings, 92% had some form of
harassment.[782] High-ranking
males are the most likely to be
harassed, usually by adult
females and juvenile males and
females. When low-ranking males
mate, they are often interrupted
by the dominant male; to avoid
interruption, they mate while he is
mating with another female.[782] A

NATURAL HABITAT

This male stump-tailed macaque is showing
his uneasiness about the nearness of the
photographer.

male can copulate 10 times a
day. Males and females make an
orgasm face.[156] A crescent-
shaped vaginal plug is formed by
semen. The male's penis fits lock-
and-key with the female's specialized reproductive tract and may
function as a reproductive isolation method.[169] **Vocalizations:** 17
graded calls.[347] The most common vocalization is a *coo* used
when approaching other group
members to avoid aggression
and initiate grooming or other
friendly interactions.[38]

NATURAL HABITAT

Southern China and
Southeast Asia

Assamese Macaque *Macaca assamensis*

A juvenile climbing quadrupedally.

Taxonomy 2 subspecies.[243]

Distinguishing Characteristics The coat of Assamese macaques varies from yellowish to dark brown, the face is hairless, and the skin is red in adults.[509]

Physical Characteristics Head and body length: ♀ 431–587mm *[17.0–23.1in]*, ♂ 538–730mm *[21.2–28.7in]*.[211] **Tail length:** ♀ 204–292mm *[8.0–11.5in]*, ♂ 140–460mm *[5.5–18.1in]*.[211] **Weight:** ♀ 4.9–8.6kg *[10.8– 19.0lb]*, ♂ 7.9–15.0kg *[17.4– 33.1lb]*.[211] **Intermembral index:** NA. **Adult brain**

weight: NA.

Habitat Monsoon, montane, evergreen, bamboo, and deciduous dry forest at 300–3500m *[984–11,484ft]* elevation.[898]

Diet Fruit,[693] young leaves, insects, crops, and mammal prey.[509] Assamese macaques have been observed in fig trees, but not much information from the wild had been published by 1995.[207]

Life History NA.

Locomotion Quadrupedal.

Social Structure Multimale-multifemale. **Group size**: 10–50.[898] **Home range:** NA. **Day range:** NA.

Behavior Diurnal, arboreal, and terrestrial. These macaques have been observed frequenting the

A submissive grin is given to appease a dominant that has threatened this female.

high canopy.[693] **Association:** These macaques displace golden langurs (*Trachypithecus geei*) from feeding sites.[588] **Sleeping site:** Trees and rock outcrops. This species is nicknamed rock monkey by Hmong and Karen hill tribes in western Thailand.[207]

Nepal to Viet Nam to Southern China

Formosan Rock Macaque *Macaca cyclopis*

Taxonomy Monotypic.[898]

Distinguishing Characteristics Formosan rock macaques have a dark brown coat and "moderately long tail." They are endemic to Taiwan.[210]

Physical Characteristics Head and body length: ♀ 400–500mm *[15.7–19.7in]*, ♂ 450–550mm *[17.7– 21.7in]*.[650] **Tail length:** 260–450mm *[10.2– 17.7in]*.[650]

Weight: ♀ 4945g *[10.9lb]*, ♂ 6000g *[13.2lb]*.[490] **Intermembral index:** NA. **Adult brain weight:** NA.

Habitat Mixed coniferous-hardwood temperate forest, as well as bamboo[206] and grassland[693] at 100–3600m *[328–11,812ft]*.[918]

Diet Fruit, leaves, animal prey, buds,[650] young shoots.[693] These macaques reportedly raid crops.[898]

Life History Weaning: NA. **Sexual maturity:** ♀ 30mo.[917] **Estrus cycle:** 165d.[528] **Gestation:** NA. **Age 1st birth:** 48–60mo.[917] **Birth interval:** 15.4mo (10.5–24).[917] **Life span:** NA. Mating season: Nov–Jan. Birth season: Apr–Jun.[917] The mating season coincides with the peak of fruit availability.[917] Estrous swelling is pronounced.[650] Females 5–9 years old usually give birth every other year; older females give birth every year.[917]

Locomotion Quadrupedal.

Social Structure Variable. 1 male–multifemale or multimale-multifemale.[918] Troops have 2–8 males, with a ratio of 1.25 males to 1.5 females.[918] Territories overlap partially.[918] **Emigration:** Males emigrate and are solitary or form bachelor troops.[918] **Group size:** 10–36,[918] to 100.[482] **Home range:** 140–1400ha.[918] **Day range:** NA. Small troops with only 1 male have been observed to have an influx of bachelor males during the breeding season.[917]

Behavior Diurnal, arboreal, and terrestrial. Formosan rock macaques rest in forest and forage in grassland.[918] High-ranking matrilines have more reproductive success.[917] Habitat destruction rather than hunting is the greatest risk to the population.[536]

Taiwan

Long-tailed or Crab-eating Macaque *Macaca fascicularis*

Long-tailed macaque infants are born black and gradually change. This fat female is one of the dominant members of a troop that is provisioned by humans.

Southern Indochina, Burma, Indonesia, Philippines, Nicobar Islands (India)

An adult male.

Taxonomy 10 subspecies.[957] Synonymous with *M. cynomolgus* and *M. irus,* names that were formerly used for this species.

Distinguishing Characteristics The body of long-tailed macaques varies from gray to reddish brown, with lighter underparts. The hair on the crown of the head grows directly backward, often resulting in a pointed crest. The face is pinkish. Males have cheek whiskers and a mustache; females have a beard.[210] Infants are born black.

Physical Characteristics Head and body length: ♀ 385–503mm *[15.2–19.8in],* ♂ 412–648mm *[16.2–25.5in].*[211] **Tail length:** ♀ 400–550mm *[15.7–21.7in],* ♂ 435–655mm *[17.1–25.8in].*[211] **Weight:** ♀ 2.5–5.7kg *[5.5–12.6lb],* ♂ 4.7–8.3kg

[10.4–18.3lb]. **Intermembral index:** 93.[903] **Adult brain weight:** 69.2g *[2.4oz].*[346]

Habitat Primary, secondary, coastal, mangrove, swamp, and riverine forest up to 2000m *[6562ft].* These macaques are tolerant of humans and may be found near villages.[898]

Diet Fruit, 64%; seeds, buds, leaves, other plant parts, and animal prey such as insects, frogs, and crabs.[654] These macaques can be crop raiders.[898]

Life History Infant: 6–12mo.[704] **Weaning:** 14mo.[346] **Juvenile:** 12–42mo.[704] **Subadult:** 42–54mo.[704] **Sexual maturity:** ♀ 51.6mo, ♂ 50.4mo.[346] **Estrus cycle:** 28d.[346] **Gestation:** 160–170d.[625] **Age 1st birth:** 46mo.[346] **Birth interval:** 13mo (12–24).[709] **Life span:** 37.1y.[555;708] Mating and births: Year-round.[555] Infant mortality: 20%. Juvenile mortality: 15–20%.[849]

Locomotion Quadrupedal; leaping to 5m *[16ft].*

Social Structure Multimale-multifemale groups,[704] with 2.5 females to 1 male in the average troop. Groups often split into subgroups.[704] **Emigration:** All juvenile males emigrate by age 7, most at ages 4–5.[849] **Group size:** 10–48, up to 100.[898] **Home range:** 25–200ha.[898] **Day range:** 150–1500m,[654] to 1900[903] *[492–4922ft, to 6234].*

Behavior Diurnal and arboreal. Long-tailed macaques swim well

and jump into water from nearby trees. The male dominance hierarchy is less marked than in other macaques. High-ranking individuals lead the group. Infants play with other infants of the same age group, although infants of high rank play together more often. The play behavior of 1-to-2-year-old juveniles appears to be shaped by gender; males play more with males, and females more with females.[849] Tension after an aggressive interaction is indicated by increased levels of self-grooming, body shaking, and scratching.[136] Tension-reducing reconciliation between individuals consists of the dominant one approaching with raised eyebrows, while the opponent stares into the eyes of the dominant, lip-smacks, and touches the other's genitals.[136] **Association:** In South Borneo these macaques compete with and displace proboscis monkeys *(Nasalis larvatus).*[924] **Mating:** In a captive study, single-mount (47%) and multiple-mount matings (53%) were observed.[751] **Vocalizations:** 15.[347] In the forest these macaques can often be located by their *kraa* calls.[654] The contact call is a *coo,* the warning call is a "raucous rattle," and the threat call is a "shrill scream ending in rattle." The juvenile alarm call is a short scream.[704]

Japanese Macaque *Macaca fuscata*

Taxonomy 2 subspecies.[315]

Distinguishing Characteristics Japanese macaques are brown to gray and have a red face and bottom and a short tail.[210]

Physical Characteristics Head and body length: ♀ 472–601mm *[18.6–23.7in]*, ♂ 535–607mm *[21.1–23.9in]*.[211] **Tail length:** ♀ 72–103mm *[2.8–4.1in]*, ♂ 81–124mm *[3.2–4.9in]*.[211] **Weight:** ♀ 8.3–18.0kg *[18.3–39.7lb]*, ♂ 11.0–18.0kg *[24.2–39.7lb]*.[211] **Intermembral index:** NA. **Adult brain weight:** 109.1g *[3.8oz]*.[346]

Habitat Subtropical to subalpine, deciduous, broadleaf, and evergreen forest of Japan below 1500m *[4922ft]*.[898] This macaque lives at the northernmost latitude of any nonhuman primate. An introduced free-ranging population has lived in Texas (USA) since 1972.[959]

Diet Fruit, seeds, leaves, bark, fungi, bird eggs, and invertebrates such as snails and crayfish.[650] In the wild in Japan, 192 plant species are consumed.[959]

Life History Weaning: NA. **Sexual maturity:** ♀ 42mo.[551] **Estrus cycle:** 29.4d.[896] **Gestation:** 173d.[896] **Age**

A dispute between two females will often escalate into a fight between two matrilines.

1st birth: 60[709]–65mo. **Birth interval:** 14mo.[709] **Life span:** 33y.[708] Birth season: May–Sep.[551]

Locomotion Quadrupedal.

Social Structure Multimale-multifemale groups.[898] Females outnumber males by 3.4 to 1.[551] Females have a rigid hierarchy in which an infant inherits its mother's rank. Males have a dominance hierarchy, but it is less fixed because they change troops. **Emigration:** Males emigrate to a new troop every 2–4 years, usually during mating season.[958] **Group size:** 40–194.[898] **Home range:** 101–530ha.[551] **Day range:** NA.

Behavior Diurnal, arboreal, and terrestrial. Japanese macaques are good swimmers. They have been studied in the field since 1952. Researchers documented the first case of cultural innovation in nonhuman primates with this species. A female named Imo learned to wash sand off of provisioned sweet potatoes and then clean sand off of wheat by putting it in water. This habit spread from her to her kin and the troop but was never learned by adult males of that generation. This species often invades human territories, including hot thermal baths in the winter. **Mating:** Matings

A female grooming a male.

Several macaque species are good swimmers.

The Arashiyama West troop in Texas now numbers more than 400 individuals, up from 150 when they were first brought from Japan in 1972.

Japanese Macaque *continued*

The female infant ranks above her older sisters in the matrilineal hierarchy.

Drying off after a swim.

female orgasm. Female masturbation to orgasm has been observed and may be a learned behavior, as it occurs in some matrilines but not others.[896] Female–female homosexual interactions have been documented during the breeding season and after conception in captivity.[896] The alpha male may interrupt the matings of other males.[751] **Vocalizations:** 40,[347] from contact-call coos to submissive squeals. The macaques in Texas have invented a special alarm call for rattlesnakes that is not a part of the vocabulary of macaques in Japan.[974]

Japan

are multiple-mount, with an average of 18 mountings, and last an average 13.22 minutes.[896] A

female's reaching back to clutch the male and stare at his face is reported to be an indication of

This adult male has a dark spot on his naturally red face. The spot is an identifying tattoo so that researchers can study the behavior of each individual in the troop.

Since their move to Texas, Japanese macaques have developed a taste for native plants.

Celebes Moor Macaque *Macaca maura*

Taxonomy Disputed. Elevated from a subspecies of Celebes black macaque *(Macaca nigra)*.[315]

Distinguishing Characteristics Celebes moor macaques have brown to dark brownish black body fur and a pale brownish gray rump patch.[239] The ischial callosities are pink.[210]

Physical Characteristics Head and body length:
♀ 500–585mm *[19.7–23.0in]*, ♂ 640–690mm *[25.2–27.2in]*.[211] **Tail length:** NA. **Weight:** NA. **Intermembral index:** NA. **Adult brain weight:** NA.

Habitat Tropical primary karst forest.

Diet Figs, other fruit.[210]

Life History Weaning: NA. **Sexual maturity:** NA. **Estrus**

Estrous female.

Sulawesi
(Indonesia)

NATURAL HABITAT

Figs are a major part of the Celebes moor macaque's diet.

NATURAL HABITAT

cycle: 36.2d.[538] **Gestation:** 175–176d.[538] **Age 1st birth:** NA. **Birth interval:** NA. **Life span:** 28y.[708] **Birth season:** No.[538] Estrous swellings have not been observed after conception.[538] **Locomotion** Quadrupedal.[210] **Social Structure** Multimale-multifemale. **Group size:** NA. **Home range:** NA. **Day range:** NA. **Behavior** Diurnal and arboreal.

Mating: Matings are multiple-mount.[169] Estrous female Celebes moor macaques present more to the alpha male. Subordinate males are able to mate only when dominant males are out of sight.[538]

Rhesus Macaque *Macaca mulatta*

Taxonomy 3 subspecies.[240]

Distinguishing Characteristics Rhesus macaques are brown, and adults have a red face and rump. The underparts are lighter brown.[210] The tail is medium length, and the hair on the top of the head is short.

Physical Characteristics Head and body length: ♀ 470–531mm *[18.5–20.9in]*, ♂ 483–635mm *[19.0–25.0in]*.[650] **Tail length:** ♀ 189–284mm *[7.4–11.2in]*, ♂ 203–305mm *[8.0–12.0in]*.[650] **Weight:** ♀ 4.4–10.9kg *[9.7–24.0lb]*, ♂ 5.6–10.9kg *[12.3–24.0lb]*.[211] **Intermembral index:** 93.[234] **Adult brain weight:** 95g *[3.4oz]*.[346] **Canine length:** ♀ 6mm *[0.2in]*, ♂ 8.9mm *[0.4in]*.[490]

Habitat Semidesert, dry deciduous, mixed deciduous and bamboo, and temperate cedar-oak forest[898] to tropical woodland and swamps,[745] from sea level to 3000m *[9843ft]*.[650] Water is a lim-

iting factor in the range.[898]

Diet Fruit, seeds, leaves, gums, buds, grass, clover, roots, bark, resin, and small invertebrates.[693] Up to 92 plant species are used.[491] Rhesus macaques raid crops and are fed by humans near Hindu temples in India.[693] In the dry season, they drink 3–4 times per day.[521]

Life History Infant: 0–12mo. **Juvenile:** 12–36mo.[745] **Subadult:** 48–72mo. **Sexual maturity:** 42–48mo.[551] **Estrus cycle:** 29d.[346] **Gestation:** 164d.[528] **Age 1st birth:** 54mo.[708] **Birth interval:** 12–24mo.[709] **Life span:** 29y.[708] Mating season: Variable, usually Mar–Jun.[551] The female's bright red bottom indicates estrus. Subadult males have a pink scrotum; adult males have a red scrotum.

Locomotion Quadrupedal.

Social Structure Multimale-multifemale. Rhesus macaques

Male macaques change troops every few years.

Rhesus Macaque *Continued*

A low-ranking female giving a submissive grin.

have a 2-tiered class system. A female associates with and supports the family and class in a strict female-bonded matrilineal hierarchy. Males are dominant to females but are peripheral to the group and change groups every few years. If a group splits into 2 groups, the split occurs between different matrilines, with mothers and their offspring staying together.[650] **Group size:** 10–50, up to 90.[745] Near temples there can be up to 242 per group.[745] **Home range:** 0.01–400ha.[745] Temple troops have smaller ranges than troops in a natural forest habitat.[745] **Day range:** 1428m (830–1895) *[4685ft (2723–6217)]*.[347]

Behavior Diurnal, mostly terrestrial, and partly arboreal.[745] Rhesus macaques are well known as a belligerent species. In a captive study, an average of 18 aggressive acts by each monkey during a 10-hour period were recorded. Their most common threat is a wide-open mouth with staring eyes, which is given by a dominant animal. Submission is signaled by screaming and baring of the teeth because of fear. A subordinate will give a silent bared-teeth display as a ritualized signal of submission to a more dominant-ranked individual.[160] Rhesus macaques often reconcile after an aggressive interaction by lip smacking and embracing. Often when threatened by a dominant, subordinates redirect their aggression by threatening low-ranking bystanders.[159] Adult males have vigilance behaviors, including protecting individuals, investigating, and scanning.[491] Rhesus macaques have been observed to mob a tiger.[491] Adult males are tolerant of juveniles until the latter reach age 4.[160] These macaques spend about 10–13% of their daily activity on grooming.[521] They feed at heights to 10 meters *[33ft]* above the ground.[491] Dominant individuals feed on higher-quality foods such as ripe fruit;

subordinates eat lower-quality foods.[745] In Pakistan some 45% of the day is spent feeding.[297] **Association:** Rhesus macaques may associate with Hanuman langurs *(Semnopithecus entellus);* the two compete for human offerings around temples.[711] These macaques displace golden langurs *(Trachypithecus geei)* from feeding sites.[588] **Mating:** Females have a "head ducking" behavior to initiate mating; during mating they have a lip-smacking facial expression.[443] They are multiple-mount maters. The amount of sexual interest a male has for a female depends as much on their previous social and sexual experience as on olfactory cues.[295] "Photoperiod, among other environmental factors, helps to synchronize and enhance the annual changes in hormone levels as well as those in the sexual and aggressive behavior of both males and females."[555] "In the male, testosterone clearly facilitates sexual behavior and appears to enhance aggressiveness."[555] "The data [indicate] that estradiol is the primary hormone

responsible for the female's sexual motivation and for her sexual attractiveness to males [and that] attractiveness depends on her proceptive behavior, partly on olfactory signals and partly on tactile information from the vagina."[555] **Vocalizations:** 16.[347] Rhesus macaques grunt to a newborn baby as a sign of friendliness.[156]

Afghanistan and India to Thailand and Southern China

Pig-tailed Macaque *Macaca nemestrina*

The Mentawai Island macaque *(M. n. pagensis)* is recognized by some taxonomists as a valid species with two subspecies.

Taxonomy Disputed. 4 subspecies. Mentawai Island macaques *(M. n. pagensis)*[315] are considered by some to be a separate species with 2 subspecies, because they look different, are genetically distinct,[550] and live only on the Mentawai Islands.[102]

Distinguishing Characteristics Pig-tailed macaques are olive brown above, with white underparts. The top of the head is dark brown.[654] The tail is short, slender, and thinly furred or naked.[240]

Physical Characteristics Head and body length: ♀ 467–564mm *[18.4–22.2in]*, ♂ 495–564mm *[19.5–22.2in]*.[211] **Tail length:** ♀ 130–253mm *[5.1–10.0in]*, ♂ 160–245mm *[6.3–9.6in]*.[211] **Weight:** ♀ 4.7–10.9kg *[10.4–24.0lb]*, ♂ 6.2–14.5kg *[13.7–32.0lb]*.[211] **Intermembral**

index: 92.[234] **Adult brain weight:** 106g *[3.7oz]*.[346] Canine length: ♀ 7.3mm *[0.3in]*, ♂ 11.7mm *[0.5in]*.[490]

Habitat Lowland primary and secondary forest[693] and coastal, swamp, dry land, and montane forest up to 1700m *[5578ft]*.[102]

Diet Fruit and seeds, 73.8%; animal prey (including insects, nestling birds,[693] termite eggs and larvae, and river crabs[884]), 12.2%; leaves, 5.4%; buds, 3%; flowers, 1.1%; other plant matter, including fungus.[102] More than 160 plants are used. In Sumatra these macaques may raid ripe corn crops and oil palms.[898]

Life History Weaning: 12mo.[346] **Sexual maturity:** ♀ 35mo.[102] **Estrus cycle:** 30–35d.[102] **Gestation:** 171d.[102] **Age 1st birth:** 44–47mo.[102] **Birth interval:** 12–24mo.[708] **Life span:** 26.3y.[550] Mating: Year-round. Mating peak: Jan–May.[102]

Locomotion Quadrupedal; some suspensory behavior.[102]

Social Structure Multimale-multifemale groups, with a ratio of 1 male to 5–8 females.[102] Females have a matrilineal dominance hierarchy.[643] **Emigration:** Males emigrate[643] and remain solitary or peripheral to a group.[102] **Group size:** 15–40.[102] **Home range:** 62–828ha.[102] **Day range:** 2000m *[6562ft]*.[903]

Behavior Diurnal, arboreal, and terrestrial. Pig-tailed macaques are on the ground 8.4% of the time, in the lower canopy 33.8%, in the middle canopy 47.4%, and in the upper canopy 10.4%.[102] They have a unique facial expression called a pucker, which has variable meanings but usually implies aggressiveness.[102] Males groom only estrous females and, rarely, other males.[643] Female–female grooming is kept within rank and probably kinship.[643] Female–female mounting is seen after agonistic interactions with the dominant macaque mounting the subordinate.[643] This is unlike male–male mounting, in which the subordinate mounts the dominant after a subordinate presents. This reconciliation display expresses the dominant male's tolerance.[643] Kisses have been observed from dominant females to subordinates.[643] Females may form coalitions to attack a male.[643] This species is currently in demand by medical laboratories for HIV (AIDS virus) research.

Association: White-handed gibbons *(Hylobates lar)*

Adult male.

compete with pig-tailed macaques for food and harass them.[682] *M. n. pagensis* often associates with Mentawai Island leaf monkeys *(Presbytis potenziani)*.[973] **Mating:** Pig-tailed macaques have multiple-mount matings and courtships that last up to 3 days.[102] The alpha male interferes with the matings of other males.[751] **Vocalizations:** 15[347]–32.[379] The most notable are a loud, harsh bark given by the male at dawn; several contact calls; and high-pitched vocalizations used in agonistic situations.[102] **Sleeping site:** Often dipterocarp trees.

Burma to Malay Peninsula and Sumatra

Pig-tailed macaques are trained to harvest ripe coconuts on plantations.

D. FINFLAY

Celebes or Crested Black Macaque *Macaca nigra*

Taxonomy Disputed. 2 subspecies, including *M. n. nigrescens*.[315] This species was considered 1 species with 7 subspecies until 1976.[240]

Distinguishing Characteristics Celebes black macaques are all black, with a short tail, long hair that forms a pointed crest on the head, and high bony cheek ridges.[650]

Physical Characteristics Head and body length: ♀ 445–550mm *[17.5–21.7in]*, ♂ 520–570mm *[20.5–22.4in]*.[239] **Tail length:** 25mm *[1.0in]*.[239] **Weight:** NA. **Intermembral index:** 94.[234] **Adult brain weight:** 94.4g *[3.3oz]*.[346]

Habitat Primary and secondary tropical forest.[650]

Diet Fruit, buds, sprouts, and insects, including caterpillars.[650] These macaques use 120 plant species.[449]

Life History Weaning: NA. **Sexual maturity:** ♀ 49mo.[346] **Estrus cycle:** 36d.[346] **Gestation:** 174,[449] 196d.[346] **Age 1st birth:** 65mo.[708] **Birth interval:** 18mo.[449] **Life span:** 18y.[449] **Infant mortality:** 21%.[449] Females have pronounced sexual swellings that are bright pink to red.[210]

Locomotion Quadrupedal.

Social Structure Multimale-multifemale groups with a ratio of 1 male to 3.4 females.[449] **Emigration:** Males disperse.[449]

M. KINNAIRD (NATURAL HABITAT)

Celebes black macaque *(M. n. nigra)*.

Group size: 47 (27–97).[449] **Home range:** 114–320ha.[449] **Day range:** 6000m *[19,686ft]*.[449]

Behavior Diurnal, terrestrial, and arboreal. Black macaques show less agonistic behavior (3.6 agonistic interactions/hour) than stump-tailed macaques *(M. arctoides)*.[52] During aggressive encounters, biting is rare and not performed by adult males. In a captive study, coalitions were rarely observed.[50] Females have a mutual embrace in which they meet head to tail and sniff each other's genitals like dogs.[52] These macaques huddle more during rainy weather, and females groom more than males. Black macaque populations have decreased 60% in the last 10 years. They are hunted with snares and are considered a Christmas delicacy by Christians. Muslims consider monkeys to be unclean and may hunt but not eat them.[485] **Mating:** Matings are multiple-mount. The female presents, looks at the male, and lip-smacks.[751]

Sulawesi (Indonesia)

This macaque was once misnamed the Celebes black ape.

The Dumoga-Bone macaque *(M. n. nigrescens)* is recognized by some taxonomists as a valid species. It is much more arboreal than the Celebes black macaque *(M. n. nigra)*.

ENDANGERED

IUCN

Booted Macaque *Macaca ochreata*

Taxonomy Disputed. 2 subspecies, including *M. o. brunnescens*.[315] Elevated from a subspecies of *M. nigra*.

Distinguishing Characteristics Booted macaques have ocherous gray on the forearms and hind limbs. The body is black, with a gray rump patch. Infants are brown and change to black gradually.[239]

Physical Characteristics Head and body length: ♀ 500mm *[19.7in]*, ♂ 590mm *[23.2in]*.[211] **Tail length:** ♀ 35–40mm *[1.4–1.6in]*, ♂ 35mm *[1.4in]*.[211] **Weight:** NA. **Intermembral index:** 100.[234] **Adult brain weight:** NA.

Habitat Tropical forest.

Diet Fruit. Booted macaques raid cacao crops.[203]

Life History NA. Females have estrous swellings.[211]

Locomotion Quadrupedal.

Social Structure NA. **Group size:** NA. **Home range:** NA. **Day range:** NA.

Behavior Diurnal and arboreal. Poisoned bait is used to kill booted macaques that raid cacao plantations.[203] **Association:** Booted macaques may hybridize with Tonkean macaques *(M. tonkeana)* where their ranges overlap.[746]

Sulawesi (Indonesia)

Estrous female.

C. KUNTUNIDISZ

J. FOODEN

Bonnet Macaque *Macaca radiata*

Taxonomy 2 subspecies.[242]

Distinguishing Characteristics Bonnet macaques have a grayish brown back and a well-defined circular cap. The tail is two-thirds the length of the body.[315]

Physical Characteristics Head and body length: ♀ 375–480mm *[14.8–18.9in]*, ♂ 450–590mm *[17.7–23.2in]*.[211] **Tail length:** ♀ 330–566mm *[13.0–22.3in]*, ♂ 498–639mm *[19.6–25.2in]*. **Weight:** ♀ 3900–4440g *[8.6–9.8lb]*, ♂ 5400–8850g *[11.9–19.5lb]*.[211] **Intermembral index:** NA. **Adult brain weight:** 76.8g *[3.0oz]*.[346]

Habitat Wet lowland to dry deciduous forest up to 2134m *[7002ft]*. Bonnet macaques also live near urban areas and temples.[898]

Diet Fruit, 47–53%; seeds, leaves, flowers, and animal prey, including insects, lizards, and frogs.[693] Bonnet macaques eat 39 plant species,[471] as well as raid crops and eat what humans offer at temples. Juveniles eat small soft fruit and adults eat harder fruits.[477] Subadults and juveniles feed on fruit in the smaller branches of trees that will not hold the weight of adults.[477]

Life History Weaning: NA. **Sexual maturity:** ♀ 48mo, ♂ 72mo. **Estrus cycle:** NA. **Gestation:** 162d.[709] **Age 1st birth:** 49mo.[709] **Birth interval:** 15.6mo.[709] **Life span:** 30y.[708] Mating season: Oct–Nov. Birth season: Jan–May. Females have genital reddening during estrus but no perineal swelling.[210]

Locomotion Quadrupedal.

Social Structure Multimale-multifemale groups with up to 12 males and 15 females.[477] This species defends territories and has a dominant hierarchy.[760] A troop moves as a cohesive group in forest.[760] **Emigration:** Both males and females emigrate.[477] **Group size:** 15–40,[898] to 95.[477] **Home range:** 40[898]–200ha.[477] **Day range:** 1500–2000m *[4922–6562ft]*.[477]

Behavior Diurnal, arboreal, and terrestrial. Bonnet macaques are good swimmers.[693] They search the ground for insects and chase flying grasshoppers;[898] adult males spend more time on the ground.[477] These macaques sit in contact with others or huddle together when they rest.[760] In a captive study, when access to food was limited to 1 individual at a time (clumped, rather than

Bonnet Macaque *continued*

evenly distributed or patchy), each individual's dominance status was strictly followed. The alpha male, which had the highest rank, ate first. The experiment resulted in an increase in agonism and a decrease in play.[61] Subordinate males that were holding infants were not the subjects of aggression by dominant males. Male infants 25–84 weeks old were used intentionally as buffers.[756] Dominant females have been known to kidnap other females' infants. This behavior is not allomothering but a form of competition between females of different matrilines, and often the infant is injured in the process.[754] Males form coalitions with other males with whom they groom. Low-ranking males appear to prefer to groom high-ranking males, which then support the low-ranking males in agonistic interactions with other males.[755] **Association:** Sympatric with lion-tailed macaques *(M. silenus)*.[704] **Mating:** Matings are single-mount.[751] Females have a vocalization made only during mating, but when a female mates with a low-ranking male, she is often quiet so as to avoid being interrupted by other troop members. During the mating season peak, males mate up to 9 times.[373] When mating was studied in captivity, the youngest and lowest-ranking adult males copulated most frequently. **Vocalizations:** 25.[373] Males and females make the same types of vocalizations.[375] "Contact rattles" make up 29.3% of all calls; threat

After a fight over some food a juvenile gives a submissive grin.

rattles, 13.8%; and growls, 12%.[375] Two vocalizations are made only by infants. Individuals make more diverse vocalizations as they mature. The chuckle call is given by juveniles and subadults in agonistic interactions within the troop and toward other troops.[373] **Sleeping site:** Fig trees near human settlements,[693] as well as tall trees with dense foliage.[477] The troop separates into subgroups of 2–3 individuals that huddle together to sleep.[477]

India

A male gives a threat display by showing his long canine teeth.

Lion-tailed Macaque *Macaca silenus*

Taxonomy Monotypic.[960]

Distinguishing Characteristics
Unlike any other macaque, lion-tailed macaques have a long, brownish gray mane around the face. They have a black body and a black tail with a slight tuft of hair on the tip.[240]

Physical Characteristics Head and body length: ♀ 460mm *[18.1in]*, ♂ 510–610mm *[20.1–24.0in]*.[211]
Tail length: ♀ 254–320mm *[10.0–12.6in]*, ♂ 254–386mm *[10.0–15.2in]*.[211] **Weight:** ♀ 3.0–6.0kg *[6.6–13.2lb]*, ♂ 5.0–10.0kg *[11.0–22.0lb]*.[211]
Intermembral index: NA. **Adult brain weight:** 85g *[3.0oz]*.[346] Females have perineal estrous swellings.[210]

Habitat Evergreen, broadleaf forest in hilly country[704] up to 1500m *[4922ft]*.[898]

Diet Fruits, seeds, mushrooms, flowers, young buds, and animal prey, including snails and giant squirrel infants.[476]

Life History Weaning: 12–18 mo.[650]
Juvenile: 18–48mo.[379] **Subadult:** ♂ 48–96mo.[650] **Sexual maturity:** 30–48mo.[650] **Estrus cycle:** NA. **Gestation:** 162–186d.[650] **Age 1st birth:** 59mo.[708] **Birth interval:** NA. **Life span:** 38y.[708] **Mating peak:** Jan–Feb.[704] **Births:** Year-round.[704] Neonates have brown hair and pale pink skin.[704]

Locomotion Quadrupedal; rarely jumps.

Social Structure Variable. 1 male–multifemale or 2 male–multifemale groups with overlapping territories.[704] Troops may temporarily split into subgroups.[704]
Group size: 4–30.[704] **Home range:** 131ha[476] (100–500).[967] **Day**

131

Lion-tailed Macaque *continued*

range: 750–2500m,[476] to 4000[967] *[2461–8203ft, to 13,124].*

Behavior Diurnal and arboreal. Lion-tailed macaques use the middle and upper levels of canopy (93%) and rarely come to the ground.[476] They forage for insects under bark and by breaking dead branches.[476] They are good swimmers.[704] An individual's presentation of its hindquarters to another individual is a sign of submission.[704] After a fight, 60% of the reconciliations are initiated by the aggressor, and 40% by the victim.[121] In captivity some individual macaques have used tools to extract honey from a simulated termite mound.[876] A 1993 population and habitat viability analysis

(PHVA) report estimated that there were 3660 lion-tailed macaques in the wild and 570 in captivity.[967] Activity budget: Feeding and foraging, 50%; rest, 33%; moving, 15%.[476]

Association: Sympatric with Nilgiri langurs *(Trachypithecus johnii)* and bonnet macaques *(M. radiata).*[704] **Mating:** These macaques have multiple-mount matings. The female's copulation call varies with her stage of estrus and may be another indication, along with sexual swellings, of female receptivity.[379] **Vocalizations:** 21.[379] The loud call is produced at the best frequency for optimum sound transmission through the rain forest in which this species lives.[379]

Researchers can tell that a female has not had an infant (is nulliparous) if she does not have elongated nipples, which results from nursing.

Liontails are the only macaque with a male loud call that is a territorial spacing call. The *whoo* call is the most frequent contact call heard (60%)[375] and occurs most often when the group is traveling.[379] They have an alarm call for aerial predators but not for terrestrial predators. These are the only macaques with both male and female copulation calls.[347] The calls of liontails are similar in structure to the calls of Nilgiri langurs *(T. johnii),* which live in the same forest. The similarity is thought to be a case of convergent evolution for the most effective communication within the same environment.[347]

An estrous female being groomed by another female.

India

Toque Macaque *Macaca sinica*

Taxonomy 2 subspecies.[968]

Distinguishing Characteristics
Toque macaques have a well-formed caplike whorl of hair radiating outward from the center of the head. They have a dusky brown to golden yellow body, black ears, and a black lower lip. The female's face is red.[172]

Physical Characteristics Head and body length: ♀ 432–452mm *[17.0–17.8in]*, ♂ 442–533mm *[17.4–21.0in].*[210] **Tail length:** ♀ 465–569mm *[18.3–22.4in]*, ♂ 549–622mm *[21.6–24.5in].*[210] **Weight:** ♀ 3.4–4.3kg *[7.5–9.5lb]*, ♂ 4.4–8.4kg *[9.7–18.5lb].*[210] **Intermembral index:** NA. **Adult brain weight:** 69.9g *[2.5oz].*[346]

Habitat Lowland, gallery, and semideciduous forest near permanent water[172] up to 1524m *[5000ft].*[898]

Diet Fruit and seeds, 75%; other plant parts, 23%;[102] animal prey

(including reptiles, birds, and mammals), 2%.[94] Toque macaques raid crops, garbage dumps, and rice mills.[693]

Life History Infant: 12mo.[172] **Juvenile:** 12–60mo. **Subadult:** 60–84mo.[172] **Sexual maturity:** ♂ 60–84mo,

J. M. WALKER (NATURAL HABITAT)

132

Toque Macaque *continued*

♀ 54–66mo.[551] **Estrus cycle:** NA. **Gestation:** NA. **Age 1st birth:** 56–60mo.[172] **Birth interval:** 18mo.[709] **Life span:** 30y.[709] Mating season: Yes.[555] Birth season: Feb–Apr.[551] Male infants have a higher mortality rate.[172]

Locomotion Quadrupedal.

Social Structure Multimale-multifemale troops with 2.3 females per male.[551] Toque macaques have overlapping home ranges[172] and a clear male dominance hierarchy.[751] "Competition between troops, in part, determines the amount of food available and thereby contributes to setting the upper limit to troop size."[173] **Emigration:** Males emigrate on average every 5.83 years;[172] females remain in their natal group. Males are most likely to change troops when they lose a high position in the hierarchy.[172] All-male subgroups are peripheral to the main troop.[172] **Group size:** 24.8 (8–43).[551] **Home range:** 17–115ha.[172] **Day range:** NA.

Behavior Diurnal, arboreal, and terrestrial. The competition between toque macaque troops with overlapping home ranges means that the harsher the environment is, the fewer young and old animals will survive in the troop.[172] In the dominance hierarchy, adults dominate juveniles, juveniles dominate infants, and males dominate females. Juvenile females fare the worst because mothers protect only their youngest offspring.[173] The majority of all threats (81.5%) have been recorded during foraging, and 36% of the time, food was taken from the one threatened.[173] Dogs are the main predators of this species.[172] **Mating:** Matings are single-mount. High-ranking males are observed to mate more often.[751]

Sri Lanka

Barbary Macaque *Macaca sylvanus*

The male juveniles sleep with an adult male or other juveniles.

R. SEITRE

Taxonomy Monotypic.[240]

Distinguishing Characteristics
Barbary macaques are yellowish gray to grayish brown.[210] The underparts are paler[210] and the face is dark. They have no tail.

Physical Characteristics Head and body length: ♀ 450mm [17.7in],
♂ 550–600mm [21.7–23.6in].[211]
Tail length: Vestigial.[211] **Weight:**
♀ 10.2–11.0kg [22.5–24.2lb],
♂ 15.3–17.0kg [33.7–37.5lb].[211]
Intermembral index: NA. **Adult brain weight:** 93.2g [3.3oz].[346]

Habitat Mixed cedar and holm oak or cork oak forest up to 1600–2160m [5250–7087ft].[898]

Diet Acorns; bark, cones, and needles of cedar trees;[693] mushrooms; bulbs; animal prey, including insects, other invertebrates, and amphibians. Barbary macaques eat 100 of 195 known plants in the Ghormaran region of Morocco.[544] The cedar tree is a keystone food species during the snowy winter. The diet changes seasonally. In winter these macaques are arboreal folivores; in spring they are folivorous, using all forest levels; in summer they are terrestrial frugivores and graminivores and eat tadpoles from streams; in autumn they are more arboreal but still are frugivore-graminivores.[544]

Life History Infant: 0–12mo.[545] **Juvenile:** 12–36mo.[545] **Subadult:** ♀ 36–48mo, ♂ 36–60mo.[545] **Sexual maturity:** ♀ 46mo.[346] **Estrus cycle:** NA. **Gestation:** 164.7d.[709] **Age 1st birth:** 46mo.[346] **Birth interval:** 13[709]–31mo.[346] **Life span:** 22y.[708] Mating season: Nov–Dec. Birth season: Apr–Jun. Estrous females have a dark reddish gray perineal swelling.[210] Infant mortality: 23–27%. The prepartum phase of birth lasts 38 minutes for an experienced mother. The infant shrieks after birth. As many mammals do, the mother consumes the placenta.[332] Reproductive senescence and menopause have been observed in the last 5 years of female life in captive study.[653]

Locomotion Quadrupedal.

Social Structure Multimale-multifemale groups with matrilineal hierarchies.[751] **Emigration:** Males rarely emigrate.[751] Home ranges overlap 80%.[545] In a study in Algeria the group size doubled (from 37 to 76) in 5 years, and the group fissioned into 3 groups along matrilineal lines. The split was led by females during mating season. An influx of nonresident males took place after the fission.[552] **Group size:** 24 (12–59).[545] **Home range:** 7200ha (3000–9000).[545] **Day range:** NA.

Behavior Diurnal, arboreal, and terrestrial. Barbary macaques forage in trees, on the ground, and even under rocks.[693] Males associate with infants soon after birth. The males "do not associate preferentially with related infants and are almost surely not able to identify their own offspring."[652] Allomothering is usually done by females related to the infant.[652] There is little male aggression.[751] Barbary macaques, which were introduced to Gibraltar, are the only monkey in

Barbary Macaque continued

Europe. Their natural habitat is threatened by intensive logging. **Mating:** Matings are single-mount and last an average of 8.7 seconds.[821] One male was observed to mate 3 times within 37 minutes.[821]
Sleeping site: Barbary macaques sleep in clusters of 2–3 individuals. An adult female sleeps with her infant and her other female juveniles. Male juveniles sleep with an adult male or other juveniles. Adult males and females have not been observed sharing a sleeping cluster.[25]

Morocco, Algeria,
Gibraltar
(introduced)

Adult male.

Tibetan Macaque *Macaca thibetana*

QI-KUM ZHAO

Taxonomy Disputed. Monotypic. Formerly included as a subspecies of *M. arctoides*.[244]

Distinguishing Characteristics
Tibetan macaques have long, dense, grayish brown fur.[650] The whiskers and beard are lighter than the top of the head.[240] Infants are blackish with silver, which changes to yellow at age 2.[929]

Physical Characteristics Head and body length: ♀ 613–710mm *[24.1–28.0in]*, ♂ 507–630mm *[20.0–24.8in]*.[650] **Tail length:** ♀ 56–80mm *[2.2–3.1in]*, ♂ 55–65mm *[2.2–2.6in]*.[650] **Weight:** ♂ 14.2–17.5kg *[31.3–38.6lb]*,[650] ♀ 13.0kg *[28.7lb]*.[929] **Intermembral index:** 95.[234] **Adult brain weight:** NA.

Habitat Subtropical evergreen forest to mixed deciduous temperate forest[693] at 800–2000m *[2625–6562ft]* elevation.[650]

Diet Leaves, shoots, fruits, roots, mushrooms, and animal prey, including eggs, birds, snakes, and invertebrates.[693] These macaques are fed by humans near temples.[693]

Life History Infant: 0–12mo.[929] **Juvenile:** 12–36mo.[929] **Subadult:** 60–84mo.[929] **Sexual maturity:** NA. **Estrus cycle:** NA. **Gestation:** NA. **Age 1st birth:** 5mo.[929] **Birth interval:** NA. **Life span:** 20y.[89] Mating season: Yes. Birth season: Yes.[928] Estrus is signaled by a slight swelling between the anus and the tail. Subadult males have fully erupted canines.[931]

Locomotion Quadrupedal.[650]

Social Structure Multimale-multifemale.[928] **Group size:** NA. **Home range:** 300ha.[928] **Day range:** NA.

Behavior Diurnal and primarily terrestrial. Tibetan macaque males are reported to have a "favorite" infant that they hold and groom. Subordinate males recognize this care and carry the favorite infant to the dominant males in order to interact affiliatively with them.[642] **Mating:** During mating season, competition and transfers cause a group's composition to change. Matings are multiple-mount. While mating, dominant males vocalize, and a "bared teeth chatter is usually performed by both partners."[931] To avoid being harassed by more-dominant males, low-ranking males mate out of sight of the group and make no vocalizations.[931] **Sleeping site:** Cliffs with sheltering overhangs, trees on cliff tops, or tall trees in the forest.[928]

China

Tonkean Macaque *Macaca tonkeana*

Adult male.

males.[532] **Association:** Tonkean macaques may hybridize with booted macaques *(M. ochreata)* where their ranges overlap.[746] **Mating:** Matings are multiple-mount. The mating pair is often harassed by the immature of the female involved, and the harassment is directed at the male mounter, which is rarely aggressive in return.[831] **Vocalizations:** A loud call is given only by the alpha male.[532] The alarm call is similar to that of rhesus macaques *(M. mulatta)*.[532] Tonkean macaques have a specific affiliation call.[532] Females have a call that is given at the beginning of estrus and while mating.[532] During agonistic interactions these macaques make rattle and chuckle calls.[532]

Sulawesi
(Indonesia)

Taxonomy Disputed. 2 subspecies, including *M. t. hecki*.[315] Considered a subspecies of *M. nigra* until 1976. The taxonomic status of the Togian Island macaque as a valid species or subspecies has not been resolved.

Distinguishing Characteristics
Tonkean macaques are dark brown to black, with prominent brownish gray to buff cheek tufts. Their large, buff rump patch contrasts with the darker body.[239]

Physical Characteristics Head and body length: ♀ 500–565mm *[19.7–22.2in]*, ♂ 575–675mm *[22.6–26.6in]*.[239] **Tail length:** ♀ 28–56mm *[1.1–2.2in]*, ♂ 40–70mm *[1.6–2.8in]*.[239] **Weight:** ♀ 8636g *[19.0lb]*, ♂ 10,454g *[23.0lb]*.[239] **Intermembral index:** 95.[234] **Adult brain weight:** NA.

Habitat Tropical rain forest that grows on the ultrabasic soil of northern Sulawesi.[98]

Diet NA. Tonkean macaques reportedly raid crops.[98]

Life History NA. No mating season or birth season has been recorded in captivity.[532] Females have estrous swellings.

Locomotion Quadrupedal.

Social Structure Multimale-multifemale. These macaques have no formal hierarchy; in captivity they have an open social system.[661] **Group size:** NA. **Home range:** NA. **Day range:** NA.

Behavior Diurnal, arboreal, and terrestrial. Tonkean macaques have been studied in captivity more than in their natural habitat. Grooming among females is not linked to kinship or dominance relationships.[833] A third party has been recorded to intervene nonaggressively to stop a conflict by clasping one of the opponents.[661] Although nonaggressive intervention is uncommon, it is usually directed at juveniles and adult females, not adult males.[661] Baring the teeth, a submissive signal in many other macaques, is used by Tonkean macaques as an affiliative signal and is used during play.[832] Females groom more than

Olive Baboon *Papio hamadryas anubis*

NATURAL HABITAT

Olive baboons *(P. h. anubis)* in the Awash River valley naturally hybridize with hamadryas baboons *(P. h. hamadryas)*.

NATURAL HABITAT

Baboons often sleep on cliffs to avoid nocturnal predators
such as leopards.

Taxonomy Disputed. Formerly considered a valid species.

Distinguishing Characteristics
Olive baboons have a greenish olive agouti coat.[509] Adult males have longer hair on the shoulders than the females do. Both genders have a black face and ruffs on the cheeks.[601]

Physical Characteristics Head and body length: ♀ 597mm (535–700) *[23.5in (21.1–27.6)]*, ♂ 741mm (600–857) *[29.2in (23.6–33.7)]*.[601] **Tail length:** ♀ 434mm (380–533) *[17.1in (15.0–21.0)]*, ♂ 499mm (410–584) *[19.6in (16.1–23.0)]*.[601] **Weight:** ♀ 14,744g (14,520–14,968) *[32.5lb (32.0–33.0)]*, ♂ 28,414g (22,000–37,200) *[62.6lb (48.5–82.0)]*.[601] **Intermembral index:** 97.[234] **Adult brain weight:** 175.1g *[6.2oz]*.[346]

Habitat Semidesert, thorn scrub, savanna, woodland, gallery, and rain forest up to 4500m *[14,765ft]*. Water must be available.[898]

Diet Fruits, seeds, tubers, roots, leaves, flowers, and animal prey, including invertebrates, reptiles, birds, and mammals.[94]

Life History Infant: 7mo. **Weaning:** 14mo.[346] **Juvenile:** 7–48mo. **Subadult:** 84–120mo. **Sexual maturity:** ♀ 54mo, ♂ 60–84mo.[551] **Estrus cycle:** 31[346]–40d.[741] **Gestation:** 180d.[551] **Age 1st birth:** NA. **Birth interval:** 18–34mo. **Life span:** 30–45y. Births: Year-round.[712] Females have prominent estrous swellings. In a study in Tanzania more males were born if there was less rain during January, and more females if rainfall was high.[204]

Locomotion Quadrupedal.

Social Structure Multimale-multifemale groups.[551] Olive baboons have a rigid dominance hierarchy. Males determine the group's direction. **Emigration:** Males first emigrate when they are 4 years old and change troops every few years.[204] Females remain in their natal troop.[741] **Group size:** 50, up to 140.[898] **Home range:** 1968ha[551] (74.5–4014).[898] **Day range:** 5000m *[16,405ft]*.[551]

Olive Baboon *continued*

Behavior Diurnal, mostly terrestrial, and partly arboreal. Olive baboons have complex social relations, including "friendships" between males and females and between males and infants.[204] The most socially adept adult males have the least stress, as measured by low basal cortical concentrations,[686] and are generally more successful at mating than the more aggressively dominant males.[807] Victims of aggression were found to have lower lymphocyte counts and higher basal cortical concentrations, which are physical reactions to intragroup stress.[9] Males use infants as social buffers in dominance struggles.[807] Daughters born early in the birth peak survive better than later infants.[804] Sons of high-ranking females are more likely to survive.[804]
Association: Along the Awash River in Ethiopia, olive baboons hybridize with hamadryas baboons *(P. h. hamadryas).*[596]
Mating: Olive baboons have single-mount matings.[741] Females mate with 1–5 males per cycle.[971] Adolescent females sexually present to males of all ages when

Baboons catch and eat mammals such as this spring hare. Unlike chimpanzees, baboons do not share their catch.

they begin to cycle[741] but do not consort with adult males until after their 6th estrus cycle.[741]
Vocalizations: The most audible vocalization is an alarm bark, and the grating roar of the adult male

is often heard at night when the group is disturbed.[204] Lip smacking is a means of social communication associated with affiliative behaviors.[189] **Sleeping site:** Trees or cliff faces.

Equatorial Africa

A young female presenting to a male.

Yellow Baboon *Papio hamadryas cynocephalus*

Taxonomy Disputed. Formerly considered a valid species.

Distinguishing Characteristics Yellow baboons are yellowish brown to yellowish gray. The cheek hair is lighter than the hair on the top of the head. Adult males have a mane. The nostrils do not protrude beyond the upper lip.[601] Infants have a black coat.[16]

Physical Characteristics Head and body length: ♀ 616mm (550–685) *[24.3in (21.7–27.0)]*, ♂ 727mm (620–840) *[28.6in (24.4–33.1)]*.[601] **Tail length:** ♀ 502mm (380–565) *[19.8in (15.0–22.2)]*, ♂ 601mm (450–660) *[23.7in (17.7–26.0)]*.[601] **Weight:** ♀ 12,300g *[27.1lb]*, ♂ 24,893g (22,793–28,300) *[54.9lb (50.2–62.4)]*.[601] **Intermembral index:** 96.[234] **Adult brain weight:** 169.1g *[6.0oz]*.[346]

Habitat Thorn scrub, savanna, woodland, and gallery forest up to 1000m *[3281ft]*. Water must be nearby.[898]

Diet Fruit, seeds, leaves, flowers, roots, tubers, bulbs, animal prey (invertebrates, reptiles, amphibians, birds, and mammals).[94] A total of 180 plant species are eaten.[969]

Life History Weaning: 17mo. **Sexual maturity:** ♀ 51mo, ♂ 73.8mo.[346] **Estrus cycle:** 32.5d.[502] **Gestation:** 175d.[166] **Age 1st birth:** 66mo.[708] **Birth interval:** 20mo.[166] **Life span:** 40y.[708] **Births:** Year-round.[551] Offspring: 1.[204] Conceptions occur more during the season of most food abundance. Females have prominent estrous swellings. Pregnancy is marked by red around the callosities.[204]

Locomotion Quadrupedal.

Social Structure Multimale-multifemale groups with a distinct dominance hierarchy and overlapping territories.[204] A juvenile female inherits her mother's rank. **Emigration:** Males emigrate at age 4. **Group size:** 28 (7–198).[898] **Home range:** 518–2409ha.[898] **Day range:** 5900m *[19,358ft]*.[551]

Behavior Diurnal, mostly terrestrial, and partly arboreal. In one study an alpha male inseminated 90% of estrous females, and the beta male only 10%. High-ranking females have more female infants and a lower infant mortality rate. Low-ranking females have more male offspring and a higher infant mortality rate.[123] Daughters of high-ranking females enhance their reproductive success by conceiving earlier than daughters of low-ranking females.[716] Activity budget: Dry season—feeding, 52–55%; travel, 25%. Wet season—feeding, 40–42%; travel, 17–19%. The resistance of baboons to HIV (AIDS virus) has led to several well-publicized experiments. In 1992 the liver of a yellow baboon was transplanted into a 35-year-old human male who had hepatitis B and HIV. The human died of an infection from the surgery.[779] **Mating:** Matings are single-mount. **Vocalizations:** At least 10. The most audible is a 2-phase bark *wahoo* given by adult males at the sight of a predator or during aggressive encounters.[204] **Sleeping site:** Trees or cliff faces.

Male baboons protect the infants of their female friends.

Southern Equatorial and East Africa

Hamadryas Baboon *Papio hamadryas hamadryas*

Adult male.

Taxonomy Disputed. 5 subspecies, all of which were formerly considered full species.[315] Because these taxa are well known and well studied, each subspecies is given a separate treatment in this book.

Distinguishing Characteristics The male hamadryas baboon is gray, with a long shoulder cape; the female is olive brown, without a cape. The skin on the face and bottom is pink. Infants are black.[601]

Physical Characteristics Head and body length: ♂ 750mm *[29.5in]*.[601] **Tail length:** ♂ 550mm *[21.7in]*.[601] **Weight:** ♀ 12.0kg *[26.4lb]*, ♂ 21.3kg *[46.9lb]*.[234] **Intermembral index:** 94.[234] **Adult brain weight:** 142.5g *[5.0oz]*.[346] **Canine length:** ♂ 83mm *[3.3in]*.[475]

Habitat Arid subdesert and savanna woodland up to 2600m *[8531ft]*.[475]

Diet Grass seeds, roots, tubers, leaves, and animal prey, including invertebrates (termites) and small vertebrates. Hamadryas baboons may raid crops and garbage dumps.[898]

Life History Infant: 8–15mo.[753] **Weaning:** NA. **Juvenile:**

Hamadryas Baboon *continued*

A juvenile riding on its mother, who is in estrus.

15–50mo.[753] **Subadult:** 50–102mo.[753] **Sexual maturity:** ♀ 51.5mo, ♂ 57.5–81.5mo.[769] **Estrus cycle:** 30d.[502] **Gestation:** 165–174d.[769] **Age 1st birth:** 73mo.[753] **Birth interval:** 22mo.[475] **Life span:** 35.6y.[708] Birth peaks: May–Jun, Nov–Dec.[769] Males grow their first long shoulder cape at age 10.[475]

Locomotion Quadrupedal.

Social Structure Fission-fusion community.[473] Hamadryas baboons have a complex 4-level social structure unlike that of any other mammal.[475] The basic unit is 1 male with multiple females, and 2–4 of these units with bachelor followers make up a clan. Several clans make up a band of 60 individuals. Several bands may form a troop, and several troops share a cliff face sleeping site.[475] **Emigration:** Males stay within their natal clan; females transfer to other clans or bands by age 3.5.[753] **Group size:** Foraging group, 25–38;[752] troops to 750.[474] **Home range:** 2800ha.[752] **Day range:** 6.5–19.1km *[4.0–11.9mi]*.[898]

Behavior Diurnal and terrestrial. A hamadryas baboon male starts his own troop by adopting and mothering subadult females. The adopted females come into estrus 1–2 years earlier than other baboons but do not conceive until age 5.[650] The male protects and

All the females of the troop are attracted to a newborn infant.

herds the females and teaches them to follow him by chasing them and giving them a ritual neck bite.[473] The study of a natural hybrid zone for *P. h. hamadryas* and *P. h. anubis* has proven that male herding behavior is a genetically linked trait.[475] The group is centered around the male, and the unrelated females groom the male more than they groom other females.[474] These baboons have the largest day range recorded for any primate.[475] They will dig for water in dry streambeds.[475] The ancient

Egyptians considered them to be the sacred attendants of Thoth, the scribe to the gods.[475]

Association: Along the Awash River in Ethiopia, hamadryas baboons hybridize with olive baboons (*P. h. anubis*).[596]

Sleeping site: Vertical cliffs, 15–25 meters *[49–82ft]* high.[474]

Somalia, Ethiopia, Saudi Arabia, Yemen

High-ranking baboon mothers are more permissive parents than low-ranking females, which often keep their infants under tight control.

NATURAL HABITAT

Male hamadryas baboons have long hair on their shoulders that looks like a cape.

Guinea Baboon *Papio hamadryas papio*

An adult male following an adult estrous female.

Taxonomy Disputed. Formerly considered a valid species.[315]

Distinguishing Characteristics The coat of the Guinea baboon is an overall reddish brown. The adult male does not have as long a mane as those found in other baboon species.[601]

Physical Characteristics Head and body length: ♂ 687mm [27.0in].[601] **Tail length:** ♂ 560mm [22.0in].[601] **Weight:** 17,596g [38.8lb].[708] **Intermembral index:** NA. **Adult brain weight:** 165.3g [5.8oz].[346]

Habitat Evergreen gallery forest and woodland savanna.[65] Guinea baboons avoid tall grass.[898]

Diet Fruit, seeds, flowers, and animal prey, including mammals.[94] These baboons will raid crops.[898]

Life History Weaning: NA. **Sexual maturity:** NA. **Estrus cycle:** NA. **Gestation:** 184d.[346] **Age 1st birth:** NA. **Birth interval:** 14.1mo.[346] **Life span:** 40y.[708]

Locomotion Quadrupedal.

Social Structure Multimale-multifemale groups.[65] Guinea baboons may have a rudimentary fission-fusion social structure in which 1 male and 3–4 females in a subgroup forage separately and several subgroups coalesce at night at the sleeping site.[65] **Group size:** 40–200.[898] **Home range:** 1036–1554ha.[898] **Day range:** NA.

Behavior Diurnal, mostly terrestrial, and partly arboreal. Male Guinea baboons have a herding behavior toward females that includes neck biting and rapid grooming.[65] **Mating:** Males usually mate only with females of their subgroup.[65] Females will mate with young males outside the subgroup. **Sleeping site:** Palm trees and kapok trees. These baboons sleep in subgroups of 1–11 individuals.[182]

Senegal, Mauritania, Guinea, Sierra Leone

Chacma Baboon *Papio hamadryas ursinus*

Taxonomy Disputed. Formerly considered a full species.[315]

Distinguishing Characteristics Chacma baboons are dark yellowish gray to dark brown to almost black.[601] The face is black, with white hair below the eyes on the muzzle.

Physical Characteristics Head and body length: ♀ 587mm [23.1in], ♂ 765mm (710–820) [30.1in (28.0–32.3)].[601] **Tail length:** ♀ 589mm [23.2in], ♂ 640mm (530–840) [25.2in (20.9–33.1)].[601] **Weight:** ♀ 16.8kg [37.0lb], ♂ 20.4kg [45.0lb].[346] **Intermembral index:** NA. **Adult brain weight:** 214.4g [7.6oz].[346]

Habitat Woodland, grassland, acacia scrub, and semidesert habitats, including small hills (kopjes), seaside cliffs, and mountains up to 2980m [9777ft]. Water must be nearby.[898]

Diet Fruit, seeds, leaves, flowers, and animal prey, including reptiles, birds, and mammals. Baboons living near the sea eat crabs, mussels, and limpets.[601] Chacma baboons raid farms[898] and beg food from tourists.

Life History Weaning: NA. **Sexual maturity:** ♀ 38mo, ♂ 60mo.[346] **Estrus cycle:** 35.6d.[601] **Gestation:** 187d (173–193).[601] **Age 1st birth:** 44.4mo.[708] **Birth interval:** 18–24mo.[551] **Life span:** 45y.[708] Infants, which weigh 600–800 grams [21.2–28.2oz], are born with black hair that gradually changes to the adult coloration by age 1. When females are pregnant, the black naked skin over the hips gradually changes to red.[601]

Locomotion Quadrupedal.

Social Structure Variable. Multimale-multifemale groups,[18] 1 male–multifemale groups.[972] The environment in which the troop lives may influence its social structure.[972] **Group size:** 20–50, to 128.[898] **Home range:** 210–3367ha.[898] **Day range:** 4670–10,460m,[903] to 14,500[551] [2.9–6.5mi, to 9.0].

Behavior Diurnal, mostly terrestrial, and partly arboreal. Chacma baboon juveniles are much more susceptible to drought conditions than adults. In South Africa, high infant mortality was found to be due to tick infestation and kidnapping by adult females. Adult males often die of wounds received in fights with other males.[78] Cannibalistic infanticide has been observed.[367] **Mating:** Matings are multiple-mount and last 3–11 minutes. Females give

R. & J. SEITRE

a grunting call during mating.[204] **Sleeping site:** Sites that are most secure from predators, such as steep cliff faces, emergent trees above the canopy, and open woodland trees.[331]

Southern Africa

Drill *Mandrillus leucophaeus*

Taxonomy Disputed. 2 subspecies.[313] Included in *Papio* until 1989.[315]

Distinguishing Characteristics the drill is dark brown, with a black face and a white fringe of hair. The muzzle is long and has lateral ridges. The naked rump is blue to purple, and there is red on the inner thighs.[509]

Physical Characteristics Head and body length: ♀ 660mm *[26.0in]*, ♂ 700mm *[27.6in]*.[601] **Tail length:** ♀ 81mm *[3.2in]*, ♂ 120mm *[4.7in]*.[601] **Weight:** ♀ 1.0kg *[22.0lb]*, ♂ 17.0kg *[37.5lb]*.[346] **Intermembral index:** 92–100.[601] **Adult brain weight:** NA.

Habitat Gallery, lowland rain forest to montane forest. Drills have never been observed outside forest boundaries.[898]

Diet Fruit, seeds, roots, fungus, small vertebrates, insects.[509]

Life History Weaning: NA. **Sexual maturity:** ♀ 42mo.[346] **Estrus cycle:** 35d.[502] **Gestation:** 174d.[346] **Age 1st birth:** 60mo.[708] **Birth interval:** 15mo.[346] **Life span:** 28.6y.[708]

Locomotion Quadrupedal.

Social Structure 1 male–multifemale groups, with 1 male having up to 20 females.[269] During the dry season several 1-male groups may form into a group as large as 200.[269] **Emigration:** Males emigrate and may remain solitary. **Group size:** 14–179.[269] **Home range:** 4000–5000ha. **Day range:** NA.

Behavior Diurnal, arboreal, and largely terrestrial. Male drills are more terrestrial than females and juveniles. Males forage to the rear of the group, and subadult males have been reported to herd the juveniles away from a researcher.[269] In Nigeria, hunters use dogs to hunt drills, which are the preferred game because they stand their ground and several may be killed with shotguns before they flee. **Vocalizations:** Drills have a crowing call used to contact other subgroups that are

An adult male in a threatening stance.

foraging separately.[269] A deep grunt initiates group cohesion and movement.[269] Juveniles give the most alarm barks.[269]

Nigeria, Cameroon

A juvenile stuffing his cheek pouches.

Estrous female.

Mandrill *Mandrillus sphinx*

An adult female presents to a male that is almost twice her size.

Cameroon, Gabon

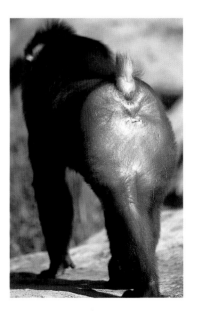

Taxonomy Disputed. Monotypic. Included in *Papio* until 1989.[315]

Distinguishing Characteristics Mandrill males have an orange yellow beard and an unmistakable bright red-and-blue snout and rump. Females and juveniles have a duller blue snout and a buffy beard.

Physical Characteristics Head and body length: ♀ 560mm *[22.0in]*, ♂ 810mm *[31.9in]*.[601] **Tail length:** 70mm *[2.8in]*.[601] **Weight:** ♀ 11.5kg *[25.3lb]*, ♂ 26.9kg *[59.3lb]*.[234] **Intermembral index:** 88–96.[234] **Adult brain weight:** 159.4g *[5.6oz]*.[346] This is the largest cercopithecine monkey.[479]

Habitat Primary and secondary, dense rain forest, as well as gallery and coastal forests. The savanna is used only rarely.[898]

Diet Fruit and seeds, 92%; bark, leaves, stems,[479] pith of plants, palm nuts,[480] and animal prey, including ants, termites,[388] dung beetles, spiders, tortoises,[480] duikers, birds, mice, frogs,[388]

crabs, bird eggs, and nestlings. In all, 113 plant species are eaten.[388] Mandrills are seed predators and feed continuously throughout day.[388] They raid manioc crops, oil palm plantations, and, during the dry season, banana plantations.[898] They eat ants consistently throughout the year and termites seasonally.[388]

Life History Weaning: NA. **Sexual maturity:** ♀ 39mo.[970] **Estrus cycle:** 33d.[601] **Gestation:** 220d.[601] **Age 1st birth:** 48[7]–60mo.[346] **Birth interval:** 17.3mo.[970] **Life span:** 46.3y.[708] **Mating season:** Jul–Oct (dry season in Gabon).[346] Birth season: Dec–Apr (Rio Muni).[708] The female's estrous swelling is small and between the callosities.[601]

Locomotion Quadrupedal; climbing.

Social Structure Multimale-multifemale groups.[215] Larger groups may contain smaller 1-male subgroups.[601] Males have a distinct hierarchy. **Group size:** Variable; 95 (2–250).[898] **Home range:** 1000–5000ha.[898] **Day range:** 1500–4500m *[4922–14,765ft]*.[898]

Behavior Diurnal, arboreal, and terrestrial. Mandrills prefer to forage along streams and in mature or secondary forest without dense undergrowth. They travel more during the major fruiting season (4.5 versus 2.5km/d *[14,765 vs. 8203ft/d]*), going rapidly from one fruit source to the next.[388] Dominant males are more colorful than subordinates. A bobbing of the head is the most common threat. Yawning without a direct stare is a sign of tension.[549] When one mandrill greets another, the facial expression is said to be a "smile."[549] **Mating:** In a captive study in a 6-hectare enclosure, the 2 top-ranking males that guarded females and mated during their maximum estrous swellings fathered the most infants, according to a DNA fingerprinting study of infant paternity.[177] **Scent marking:** Mandrills are one of the few Old World monkeys to have cutaneous glands. The sternal gland is "in the triangular area in the middle of the chest and

is covered with modified hairs."[215] Males older than 7 years scent-mark, but alpha males scent-mark most frequently.[215]

Vocalizations: 5.[347] Males have a 2-phase grunt or roar that mobilizes the group for traveling, prompting subadult males to round up the juveniles.[601]

Sleeping site: Trees at a different site each night.[388]

Gelada Baboon *Theropithecus gelada*

NATURAL HABITAT

Taxonomy Disputed. Included in *Papio* until 1979.[315]

Distinguishing Characteristics Adult male geladas have a long, heavy, dark cape. Both genders have an "hour glass shaped area of naked bright pink skin on neck and chest."[601] Estrus is signaled by swollen bumps on the skin of the female's chest. Geladas have thumbs that are more opposite the fingers than in any other Old World primate. Males are much larger than females.[601]

Physical Characteristics Head and body length:
♀ 500–650mm *[19.7–25.6in]* ♂ 690–740mm *[27.2–29.1in].*[598] **Tail length:** ♀ 325–400mm *[12.8–15.7in]*, ♂ 460–500mm *[15.7–19.7in].*[598] **Weight:** ♀ 11.7kg *[25.8lb]*, ♂ 20.0kg *[44.1lb].*[234] **Intermembral index:** 100.[234] **Adult brain weight:** NA.

Habitat Montane grassland with no tall trees, only at altitudes of 1400–4400m *[4593–14,436ft].*[898]

Diet Grass, 90%; seeds, leaves, bulbs,[183] animal prey (insects, mammals).[423] Crops are raided.[898]

Life History Weaning: 15mo.[346] **Sexual maturity:** ♀ 49.5mo.[346] **Estrus cycle:** 35.7d (32–36d).[601] **Gestation:** 150–180d.[769] **Age 1st birth:** 48[708]–54mo.[346] **Birth interval:** 24mo (12–30).[769] **Life span:** 19.2y.[708] Births: Year-round. Birth peaks: Jun–July, Nov–Dec.[769] The female's perineal skin changes from pink to scarlet during weeks 4–8 of pregnancy.[601]

Locomotion Quadrupedal.[234]

Social Structure Multimale-multifemale. The basic reproductive unit consists of 1 male and several related females (3–20).[898] Several 1-male units and all-male units of 3–13 individuals[601] form a band, and several bands may form temporary herds of up to 600 individuals.[898] **Emigration:** Males emigrate; females stay in their natal group.[601] **Group size:** Reproductive unit, 3–20;[601] band, 30–300.[769] **Home range:** 344ha.[898] **Day range:** 2500m (1500–3500m) *[8203ft (4922–11,484)].*[183]

Behavior Diurnal and terrestrial. Gelada baboons sit on tough ischial callosities to pluck grass with their hands.[423] In the morning they spend 2 hours in social activity, particularly grooming.[183] They feed periodically throughout the day (50–60% of activity budget). In captivity, infanticide has been observed after a new male

Adult males are much larger than females and have a cape.

leader was introduced to a group.[583] The "lip flip" is a fear grimace in which the upper lip is inverted to show the teeth and gums, and the lip completely covers the nostrils.[601] Dominance or threat is expressed by means of the eyelid flash signal. The scalp is retracted and the pale eyelids are flashed.[601] Geladas are the only survivors from a widespread radiation of Theropithecus baboons during the Pleistocene and Pliocene epochs. They were hunted by early hominids.[423]

Vocalizations: 22.[347] Contact calls make up half of the vocalizations; the other 11 calls are related to aggression and defense. One call is given only by the highest-ranking female in the group.[5] **Sleeping site:** Rock cliff faces.[898]

Ethiopia

NATURAL HABITAT

Gelada baboons live only in in the high plateau of Ethiopia and sleep on cliffs.

Agile Mangabey *Cercocebus agilis*

Taxonomy Separated from *C. galeritus* in 1978. Includes the subspecies *C. a. chrysogaster* and may include *C. a. sanje*, a subspecies first described in 1981 from the Uzungwa Mountains of Tanzania.[315]

Distinguishing Characteristics Agile mangabeys have a dark brown back and yellowish gold hair on the chest and abdomen.[571]

Physical Characteristics Head and body length: NA. **Tail length:** NA. **Weight:** ♀ 4700g *[165.8oz]*, ♂ 9250g *[326.2oz]*.[315] **Intermembral index:** NA. **Adult brain weight:** NA.

Habitat Tropical rain forest.

Diet NA.

Life History NA.

Social Structure NA. **Group size:** NA. **Home range:** NA.

Day range: NA.

Locomotion Quadrupedal.

Behavior Diurnal and arboreal. In captivity, agile mangabey males display significantly more aggressive behaviors than females, which are more active groomers and more vocal.[571] **Vocalizations:** Males have a species-specific loud call that is usually given in the morning and answered by neighboring males. It carries up to 600m *[1969ft]* and is believed to be a spacing mechanism[679] that identifies the caller.[864] Other vocalizations show only minor differences from those of Tana River mangabeys *(C. galeritus)*.[864]

Equatorial Guinea, Gabon, Cameroon, Congo Republic, Zaire

C. a. chrysogaster.

Tana River Mangabey *Cercocebus galeritus*

This mangabey is eating palm fruits.

M. KINNAIRD (NATURAL HABITAT)

Taxonomy Disputed. Included *C. agilis* as a subspecies until 1978.[315]

Distinguishing Characteristics The Tana River mangabey has a yellowish brown back, white underparts, and dark gray hands, feet, and tail.[509]

Physical Characteristics Head and body length: ♀ 483mm (440–525) *[19.0in (17.3–20.7)]*,[601] ♂ 549mm (500–625) *[21.6in (19.7–24.6)]*.[601] **Tail length:** ♀ 570mm (40–600) *[22.4in (1.6–23.6)]*,[601] ♂ 706mm (75–760) *[27.8in (3.0–29.9)]*.[601] **Weight:** ♀ 5.4kg *[11.9lb]*, ♂ 10.2kg *[22.5lb]*.[601] **Intermembral index:** 84.[234] **Adult brain weight:** NA.

Habitat Primary and secondary dry forest and *Acacia* woodlands.[234]

Diet Fruit and seeds, 73–78%; leaves, 11–14%; flowers, 1%; animal prey (including invertebrates, small vertebrates, and bird eggs), 1–3%.[865] These mangabeys use 61 species of plants for food, but a single palm tree species *(Phoenix reclinata)* provides 62% of the diet. Humans compete for this same wild food.[448]

Life History Weaning: NA. **Sexual maturity:** NA. **Estrus cycle:** 30d.[346] **Gestation:** 180d.[446] **Age 1st birth:** 78mo.[708] **Birth interval:** NA. **Life span:** 19y.[708] Females have a bright red perineal postconception swelling for 2 months.[446] A birth that was observed in the wild took 3–4 minutes. The infant moved its tail 2 minutes after delivery and within 20 minutes moved its head to search for a nipple. The group's females approached the new mother and infant. The males showed little interest.[446]

Locomotion Quadrupedal.

Social Structure Multimale-multifemale groups with twice as many females as males.[865] **Emigration:** Males leave; females remain in their natal group.[447] **Group size:** 17–36.[865] **Home range:** 17–100ha, depending on food abundance.[865] **Day range:** 1098m

(615–1480) *[3603ft (2018–4856)]*.[865]

Behavior Diurnal, arboreal, and terrestrial. Tana River mangabeys stay low in the forest, with 72.8% of their feeding on or within 2 meters *[7ft]* of the ground. They spend 30% of the day searching for invertebrates.[865] Intergroup interaction is more aggressive when food is not uniformly distributed and one source can be monopolized.[447] Males herd their receptive females away from neighboring groups.[447] Fights between groups consist of "all members of the opposing groups lining up on the ground, facing one another and lunging with their tails arched forward, heads down, eyelids lowered."[447] Physical contact, however, is rarely observed.[447] Groups rarely interact when fruit resources are in short supply.[447] When food is abundant, different groups have been seen to merge for hours and groom one another.[447]

Mating: Pregnant females have a postconception estrus, the purpose of which may be to confuse the males about who the father is.[446]

Kenya

White-collared Mangabey *Cercocebus torquatus*

Nigeria to Angola

white eyelids. The back is dark brown to black, and the dark tail has a white tip. The underparts are white.

Physical Characteristics Head and body length: ♂ 602mm (560–625) *[23.7in (22.0–24.6)].*[601] **Tail length:** 684mm (602–750) *[26.9in (23.7–29.5)].*[601] **Weight:** ♂ 10,750g *[23.7lb].*[601] **Intermembral index:** 83.[234] **Adult brain weight:** 109.6g *[3.9oz].*[346]

Habitat Primary and secondary, swamp, mangrove, and dry forest close to drinking water.[601]

Diet Fruits, leaves, animal prey, flowers. These mangabeys reportedly raid crops and cacao plantations.[898]

Life History Weaning: NA. **Sexual maturity:** 32mo.[346] **Estrus cycle:** 33d.[346] **Gestation:** 168d.[446] **Age 1st birth:** 54mo.[708] **Birth interval:** 13.[346] **Life span:** 27y.[708]

Locomotion Quadrupedal.

Social Structure Multimale-multifemale groups.[484] **Group size:** 14–60.[898] **Home range:** NA. **Day range:** NA.

Behavior Diurnal, arboreal, and terrestrial. White-collared mangabeys use understory trees.[234] They have been observed to travel on the ground in single file and feed on the forest floor and in emergent trees.[339] **Association:** This species associates with mona monkeys *(Cercopithecus mona),* white-throated guenons *(C. erythrogaster),*[631] western red colobuses *(Procolobus badius),* and black-and-white colobuses.[339] It occasionally associates with mustached guenons *(Cercopithecus cephus).*[566]

Vocalizations: A loud and distinct bark with 2 syllables.[864]

Taxonomy Disputed. Includes *C. t. atys* as a subspecies,[315] which some taxonomists believe is a full species. Because little is known about *C. torquatus* and a great deal of information on the behavior of *C. t. atys* is available from captive studies, *C. t. atys* has been given its own space in this book.

Distinguishing Characteristics The white-collared mangabey has a red cap, a white collar, and

Sooty Mangabey *Cercocebus torquatus atys*

S. McGRAW (NATURAL HABITAT)

Taxonomy Disputed. Some taxonomists recognize this as a valid species.[315]

Distinguishing Characteristics Sooty mangabeys are brownish gray, with white underparts.[601] Their faces are pink to gray.

Physical Characteristics Head and body length: NA. **Tail length:** NA. **Weight:** 8593g *[18.9lb].*[708]

Intermembral index: NA. **Adult brain weight:** NA.

Habitat Primary and secondary, flooded, dry, swamp, mangrove, and gallery forest.[601]

Diet Fruits, seeds, animal prey.[89]

Life History Infant: 12mo.[324] **Juvenile:** 12–60mo.[324] **Subadult:** 60–84mo.[324] **Sexual maturity:** ♀ 36.5mo.[324] **Estrus**

cycle: 34.5d.[708] **Gestation:** 167d.[166] **Age 1st birth:** 56.5mo.[303] **Birth interval:** 13mo.[166] **Life span:** 18y.[708] Females have a postconception estrus.[303]

Locomotion Quadrupedal.

Social Structure Multimale-multifemale groups. In captivity no matrilineal hierarchy has been observed. Females groom and interact more with other females of similar age than with their kin.[192] **Group size:** 95.[540] **Home range:** Large.[540] **Day range:** NA.

Behavior Diurnal, mostly terrestrial. They spend 70% of the day (and always travel) on the ground. They have powerful jaws used to crack hard nuts that other monkeys cannot.[540] In captivity, males carry infants to protect them rather than to use them as social buffers,[92] but infanticide has been recorded. One deposed alpha male carried and protected only the infants he might have sired.[192] These mangabeys are the only other primate besides humans to acquire leprosy from another of their own species.[304]

Association: These mangabys seek out arboreal monkeys in order to feed on dropped fruit.[540] **Mating:** Sexual behavior begins before age 1. Juvenile males mount sexually mature females 3 times more frequently than do sexually mature males.[324] In a captive experiment the alpha male could distinguish between a female's maximal fertile swelling and a postconception swelling.[325]

Sierra Leone to Ghana

ENDANGERED   USESA

ENDANGERED   USESA

Gray-cheeked Mangabey *Lophocebus albigena*

NATURAL HABITAT

6804–7711g *[15.0–17.0lb]*.[601] **Intermembral index:** 78.[625] **Adult brain weight:** NA.

Habitat Primary, tropical, semideciduous, and flooded forest.[601]

Diet Fruit, 59%; leaves, 5%; flowers, 3%; animal prey (including reptiles), 11%.[96] Of 63 plant species eaten, figs are preferred. These mangabeys reportedly raid crops.[898]

Life History Weaning: 7mo.[346] **Sexual maturity:** ♀ 36mo, ♂ 60–84mo.[551] **Estrus cycle:** 31.2d (30–38).[166] **Gestation:** 184–189d.[861] **Age 1st birth:** NA. **Birth interval:** 33mo (17.5–49.2).[601] **Life span:** 32.6y.[898] Birth season: May in Uganda. Pregnant females develop dark red nipples and perineum.[601]

Locomotion Quadrupedal.

Social Structure Variable. Multimale-multifemale,[601] 1 male–multifemale.[865] There was little range overlap at one study site;[601] ranges overlapped at another site.[383] **Group size:** 15[383] (6–28).[865] **Home range:** 13–26,[625] 41ha.[865] **Day range:** 1270m (580–2250) *[4167ft (1903–7382)]*.[865]

Behavior Diurnal and arboreal. Gray-cheeked mangabeys prefer the middle and upper canopy and only rarely come to the ground to drink.[865] An aggressive adult male can erect its shoulder hair to make itself look physically larger.[860] A slow head shake is a gesture given by a subordinate about to be supplanted by an approaching dominant.[860] **Association:** This species is frequently found with mustached guenons *(Cercopithecus cephus)*, mona monkeys *(C. mona)*, and putty-nosed guenons *(C. nictitans)* in Gabon.[601] It also associates with red-eared guenons *(C. erythrotis)*,[863] Wolf's guenons *(C. wolfi)*,[925] and, in Zaire, red-tailed guenons *(C. ascanius)*.[798] **Mating:** Males solicit copulation with a gesture known as a "head flag."[861] **Vocalizations:** 5.[347] These mangabeys produce a loud call that has been described as a whoop-gobble. It is audible for 1800m *[5906ft]*, or twice the diameter of the median home range.[85]

Nigeria to Kenya and Tanzania to Angola

Taxonomy Disputed. The genus was a subgenus of *Cercocebus* until 1979. The black mangabey *(L. aterrimus)* is considered by some to be a subspecies.[315]

Distinguishing Characteristics The gray-cheeked mangabey has short, whitish hair on the cheeks; a mantle of long, brownish hairs on the shoulders; and an occipital crest.[601] Both genders have

throat sacs, but the male's is larger.[625]

Physical Characteristics Head and body length: ♀ 522mm (497–559) *[20.6in (19.6–22.0)]*, ♂ 561mm (510–620) *[22.1in (20.1–24.4)]*.[601] **Tail length:** ♀ 740mm (669–770) *[29.1in (26.3–30.3)]*, ♂ 801mm (758–880) *[31.5in (29.8–34.6)]*.[601] **Weight:** ♀ 5671g (5443–5880) *[12.5lb (12.0–13.0)]*, ♂

Black Mangabey *Lophocebus aterrimus*

Taxonomy Disputed. Elevated from a subspecies of *L. albigena*.[317]

Distinguishing Characteristics: This mangabey is black,[509] with long, grayish hair on the cheeks and a pointed crest.[601]

Physical Characteristics Head and body length: ♀ 530mm *[20.9in]*.[601] **Tail length:** ♀ 750mm *[29.5in]*.[601] **Weight:** ♀ 15.3kg (13.0–18.0) *[33.7lb (28.7–39.7)]*, ♂ 21.0kg *[46.3lb]*.[601] **Intermembral index:** NA. **Adult brain weight:** NA.

Habitat Primary and secondary rain forest, swamp and gallery forest.[484]

Diet Fruits, seeds, young leaves, bark, flowers, animal prey. Black

mangabeys use 32 tree species.[382]

Life History NA.

Locomotion Quadrupedal; leaps to 5m *[16ft]*. Individuals occasionally suspend themselves by 1 arm or 2 legs when foraging.[382]

Social Structure Multimale-multifemale groups.[382] When foraging, a group may divide into subgroups.[382] A male may travel alone to visit an estrous female in another group but will return by nightfall to his group.[382] **Group size:** 9–10, to 16.[925] **Home range:** 48–70ha.[382] **Day range:** NA.

Behavior Diurnal and arboreal. Black mangabeys use all layers of the forest, but most feeding (73%) is in the middle and upper canopy[925] (12–30m *[39–98ft]*).[382] Home ranges of groups studied overlapped extensively (60–75%) and no territorial defense was observed.[382] In a 2-year study the researcher witnessed 31 attacks by the crowned hawk eagle

Black Mangabey *continued*

(Stephanoaetus coronatus). When this bird is sighted, black mangabeys alarm-bark and dive for dense cover, where they remain quiet for several hours.[382] They are hunted by humans for meat and do not live in any protected area.[484] **Association:** Black mangabeys commonly associate with Wolf's guenon *(Cercopithecus wolfi*—57%), red-tailed guenons (*C. ascanius*—20%), or both species (16%).[925] They also associate with blue monkeys *(C. mitis)*[280] and, in Zaire, Angolan black-and-white colobuses *(Colobus angolensis).*[539] They associate with and groom Pennant's red colobuses *(Procolobus pennantii).*[795] **Vocalizations:** The whoop-gobble loud call, which facilitates intergroup location and spacing, is usually given

(74%) in the morning (5–11 a.m.).[382] Black mangabeys reportedly vocalize loudly and frequently when they are active.[382] **Sleeping site:** The whole group sleeps in the same tree.[382]

Zaire, Angola, Zambia

Allen's Swamp Monkey *Allenopithecus nigroviridis*

Taxonomy Disputed. Placed in a monotypic genus in the 1970s.[315] This species is thought to be the closest guenon to the ancestral stock of *Cercocebus* and *Papio.*[279]

Distinguishing Characteristics Allen's swamp monkeys is stockily built and has webbing between the fingers and toes.[925] The coat is a greenish gray agouti, with white to orange underparts.[509] Males have a whitish blue scrotum; females have a prominent estrous swelling.[601]

Physical Characteristics Head and body length: 450–460mm *[17.7–18.1in].*[601] **Tail length:** 500mm *[19.7in].*[601] **Weight:** ♀ 3700g *[8.2lb]*, ♂ 5950g *[13.1lb].*[279] **Intermembral index:** 84.[234] **Adult brain weight:** 62.5g *[2.2oz].*[346] Diploid chromosome number: 48, the lowest number of all the cercopithecines.[601]

Habitat Primary lowland, swamp

Male.

forest.[601]

Diet Fruit, 81%; pith, 2%; roots, flowers, nectar, animal prey (insects, worms, fish), 17%.[280] These swamp monkeys lick the nectar of and thereby pollinate *Daniella pynaertii*, a flowering emergent tree in Zaire.[281]

Life History Weaning: NA. **Sexual maturity:** NA. **Estrus cycle:** NA. **Gestation:** NA. **Age 1st birth:** NA. **Birth interval:** NA. **Life span:** 28y.[708]

Locomotion Quadrupedal.

Social Structure Multimale-multifemale groups.[625] **Group size:** To 40.[625] **Home range:** NA. **Day range:** NA.

Behavior Diurnal, arboreal, and terrestrial. Allen's swamp monkeys use the lower strata of the forest (below 15m *[<49ft]*) and often forage on the ground.[925] They are good swimmers[925] and will dive into a river to avoid predators or flee on the ground.[280] They have been observed to scoop out small fish from muddy pools during the dry season.[925] They are hunted by humans from boats and are very shy.[925] **Association:** These monkeys occasionally associate with red-tailed guenons *(Cercopithecus ascanius)* and Wolf's guenons *(C. wolfi)*[925] when the latter 2 enter the swamp.[925] **Scent marking:** This species scent-marks with the sternum.[272] **Vocalizations:** These monkeys make a gobble call, high-pitched

Female.

chirps, and copulatory calls.[272] They reportedly make mooing sounds and have grunting contact calls.[925] **Sleeping site:** Trees by riverbanks.[625]

Zaire, Angola

Dwarf Guenon or Southern Talapoin Monkey *Miopithecus talapoin*

LOWER RISK

CITES II

Taxonomy Disputed. There may be 2 species of *M. talapoin,* 1 in the northern part of its range and 1 in the southern. The northern species had not been named as of 1995.[630]

Distinguishing Characteristics Dwarf guenons are the smallest Old World monkeys. They look superficially like squirrel monkeys *(Saimiri)* in that they have olive yellow agouti fur.[601] The underparts are light, and the males have a blue scrotum. Faces are gray with brown cheek stripes.[509]

Physical Characteristics Head and body length: ♂ 348mm [13.7in].[630] **Tail length:** ♂ 525mm [20.7in].[630] **Weight:** ♀ 745–820g [26.3–28.9oz], ♂ 1255–1280g [44.3–45.1oz].[601] **Intermembral index:** 83.[234] **Adult brain weight:** 37.7g [1.3oz].[346] Females have sexual swellings, which are rare for guenons.

Habitat Primary and secondary, gallery, mangrove, and swamp forest[601] up to 700m [2297ft].[898] Dwarf guenons rarely venture more than 450 meters [1476ft] from a river.[625]

Diet Fruit, 43%; leaves, flowers, and animal prey, including eggs, freshwater shrimp,[279] reptiles, and insects.[94] Mobile prey such as butterflies, moths, grasshoppers, caterpillars, beetles, and spiders are caught.[279] Dwarf guenons reportedly raid crops and steal manioc roots that villagers leave to soak.[96]

Life History Weaning: 6[346]–12mo.[96] **Subadult:** ♂ 36–84mo, ♀ 36–60mo.[601] **Sexual maturity:** ♀ 48mo, ♂ 114mo.[346] **Estrus cycle:** 35d.[625] **Gestation:** 165d.[551] **Age 1st birth:** 52mo.[708] **Birth interval:** 12mo.[346] **Life span:** 27.7y.[708] Birth season: Nov–Apr in Gabon.[279] Offspring: 1. Females have perineal swelling.[551] The infant's weight is 20% of the mother's.[601]

Locomotion Quadrupedal.

Social Structure Multimale-multifemale groups.[601] Dwarf guenons form single-sex subgroups, with

males interacting with males, and females with females.[714] Home ranges reportedly are well defined but not contiguous. **Group size:** 40–50,[601] 112.[903] Groups living near humans are double the size and triple the density of groups in natural habitats.[898] **Home range:** 122,[279] 400–500ha.[601] **Day range:** 2323m (1500–2950) [7622ft (4922–9679)].[551]

Behavior Diurnal and arboreal. Dwarf guenons are good swimmers.[601] They are known to use mobbing behavior on predators,[274] strange objects, and even other group members.[601] Grooming partners often entwine their tails.[625] Males stay higher in trees than females. "Talapoins [dwarf guenons] and the guenons of the genus *Cercopithecus* have fewer facial expressions than macaques and lack conciliatory or appeasement facial expressions."[897] Dwarf guenons rely more on vocalizations and posture to communicate. The open-mouth threat gesture is given quickly with the teeth exposed (like biting), which is different from the threat gestures of all other Old World monkeys.[897] Females are aggressive. **Vocalizations:** 17.[274] Contact calls

are used extensively. Every movement by an adult female in the wild is preceded and followed by a *coo* vocalization that is often answered by another female.[714] One notable call is a short explosive *k-sss!*[625] When entering a fruit tree, after an alarm, or during an aggressive interval, dwarf guenons may perform a chorus in which "all the calls of the repertoire are then uttered by more or less all troop members."[897] **Sleeping site:** Trees along a river edge.[625]

Angola, Zaire

Male.

148

Patas Monkey *Erythrocebus patas*

Male.

Taxonomy Disputed. 4 subspecies. The genus name was changed in 1970.[315]

Distinguishing Characteristics Patas monkeys are large and have long, slender limbs. The back and sides are reddish brown, and the underparts are grayish white. The face varies from black to light gray, with a white mustache.[601] Males have a bright blue scrotum.

Physical Characteristics Head and body length: ♀ 490mm *[19.3in]*, ♂ 600–875mm *[23.6–34.4in]*.[625] **Tail length:** ♀ 490mm *[19.3in]*, ♂ 621mm (430–724) *[24.4in (16.9–28.5)]*.[601] **Weight:** ♀ 4.0–7.0kg *[8.8–15.4lb]*, ♂ 7.0–13.0kg *[15.4–28.7lb]*.[625] **Intermembral index:** 92.[234] **Adult brain weight:** 106.6g *[3.8oz]*.[346] Females are half the size of males.[601]

Habitat *Acacia* woodland, savanna with a marked dry season.[601]

Diet Fruits, seeds, grass, and animal prey, including insects, lizards, and bird eggs. The highest percentage of the diet is from *Acacia* tree parts; the lowest is from grasses.[115] Patas monkeys will eat prickly pear cactus *(Opuntia),* which has been introduced to Kenya.[115] They need to drink daily.[115]

Life History Infant: 0–12mo.[115] **Weaning:** 12mo.[115] **Subadult:** ♂ 12–60mo, ♀ 12–30mo.[115] **Sexual maturity:** ♀ 36mo, ♂ 48mo.[133] **Estrus cycle:** 30d.[96] **Gestation:** 167d.[115] **Age 1st birth:** 36mo (25–46).[115] **Birth interval:** 11.8mo (7.3–23.1).[115] **Life span:** 21.6y.[708] Mating season: Variable (during the wet season). Birth season: Variable (during the driest months).[115] Females have no estrous swelling. Birth occurs during the day.[114] The infant's weight is 6% of the mother's weight. Infants are black until 2 months old. Females are first sexually receptive at 18 months but do not breed until age 3.[346]

Locomotion Quadrupedal. Patas monkeys are the fastest in the primate world[601] and have been clocked at 55km per hour *[34mi/h]*.[625] They walk on their fingers (digitigrade), not on the flat of their hands (palmigrade).

Social Structure 1 male–multifemale groups.[601] Females lead the group, and additional males may flood the group during breeding season.[625] Males are not territorial and rarely join in the females' boundary disputes.[714] A resident male will chase lone males and all-male groups.[714] Territorial ranges overlap 44–58%.[115] **Emigration:** Females remain in their natal group; males leave at age 3.[115] **Group size:** 5–34.[601] **Home range:** 3200–5200ha,[133] to 8000.[898] **Day range:** 4330m (700–11,800) *[14,207ft (2297–38,716)]*.[133]

Behavior Diurnal, mostly terrestrial, partially arboreal. The resident male patas monkey watches vigilantly for predators and other males.[601] "He reacts to any disturbance by bouncing noisily on bushes and trees and running away from the group, thereby diverting attention from them."[601] During dry seasons in Cameroon, several patas groups share the same water hole.[625] The female defense of the group involves aggression and long chases. The strategy for avoiding predators is vigilance, hiding, and rapid flight.[115] There is very little social grooming.[597] Female infants are allomothered by adult females significantly more than are male infants, which are

The male is twice the size of the female and is the fastest primate.

Patas Monkey *continued*

peripheralized and play only with other males.[505] **Mating:** Patas monkeys have single-mount matings.[503] "Sex is initiated by the female, who crouches in front of the male" and puffs out her cheeks.[115] **Vocalizations:** "Patas monkeys are very silent."[601] A male barks when males from other groups are encountered.[601] The alarm call and response for dogs and jackals are different from those for feline predators. The former call is a low chatter followed by flight into the woodland; the call for feline predators is a loud 2-phase bark in the predator's direction, and the troop moves up into trees.[115] Males and females have different alarm calls.[115] **Sleeping site:** During the day a troop may rest in the shade

of a tree. Each adult sleeps in a separate tree.[625] Patas monkeys do not sleep in the same trees 2 nights in a row.[115]

West Africa to Ethiopia, Kenya, and Tanzania

Female with a black infant that is less three months old.

Vervet, Grivet, or Green Monkey *Chlorocebus aethiops*

S. M. ROWE (NATURAL HABITAT)

Vervets (*C. a. pygerythrus* shown here) often forage on the ground.

Taxonomy Disputed. The genus name was changed in 1989 from *Cercopithecus*.[315] This species has 22 subspecies, of which 4 are believed by some taxonomists to be valid species—*C. a. aethiops*, *C. a. pygerythrus*, *C. a. sabaeus*, *C. a. tantalus*.[601]

Distinguishing Characteristics Vervet monkeys have a yellowish to olive agouti back and crown, which vary slightly with each subspecies. The underparts are white, the abdominal skin is blue, and the black facial skin is encircled with a white brow band and white cheeks.[509] The male has a red penis and a blue scrotum.[601]

Physical Characteristics Head and body length: ♀ 426mm (300–495) *[16.8in (11.8–19.5)]*, ♂ 490mm (420–600) *[19.3in (16.5–23.6)]*.[601] **Tail length:** ♀ 560mm (406–661) *[22.0in (16.0–26.0)]*, ♂ 630mm

(462–764) *[24.8in (18.2–30.1)]*.[601] **Weight:** ♀ 3257g (1500–4870) *[7.2lb (3.3–10.7)]*, ♂ 4582g (3100–6380) *[10.1lb (6.8–14.1)]*.[601] **Intermembral index:** 83.[234] **Adult brain weight:** 59.8g *[2.1oz]*.[346]

Habitat Savanna woodland, edge habitat near water from lowland swamp to the dry Sahel to montane forest up to 4500m *[14,765ft]*.[898] Vervets have been introduced to Caribbean islands (St. Kitts, Nevis, and Barbados). They are the most widespread of all African monkeys.[898]

Diet Fruit, seeds, leaves, and animal prey, including invertebrates, reptiles, birds, and mammals.[94] One study found that vervets ate mostly fiddler crabs in a mangrove forest.[213] They raid crops and take handouts from tourists.[898]

Vervet, Grivet, or Green Monkey *continued*

Life History Weaning: 8.5mo.[96] **Sexual maturity:**
♀ 4.5mo, ♂ 5mo.[551] **Estrus cycle:** NA. **Gestation:**
163d.[551] **Age 1st birth:** 60mo.[708] **Birth interval:** 16mo
(11–24).[551] **Life span:** 31y.[346] Mating season: Variable
(Mar–Jun).[885] Birth season: 1 per year, variable
(Oct–Jan).[885] Estrus is signaled by perineal changes
in color from white to pink.[213] Infants are weaned
during periods of greatest abundance.[903] They take
some solid food at 3 months.[96]

Locomotion Quadrupedal.

Social Structure Multimale-multifemale groups with 1.5
females to 1 male.[213] Males and females have a
dominance hierarchy.[714] **Emigration:** Females remain
in their natal group. Males emigrate often to neighboring groups with relatives or known individuals from
their natal group.[213] **Group size:** 5–76.[213] **Home range:**
41.8ha (18–96).[551] **Day range:** 950,[903] 2000m[885]
[3117, 6562ft].

Behavior Diurnal, semiterrestrial, semiarboreal. Vervets
adapt to seasonal changes in food by changing their
daily travel, diet, range, and intergroup relations.[213]
They have low levels of intragroup aggression.[213]
High-ranking females control access to limited
resources such as water and food and have been
found to be healthier than other members of the
troop. These females "conceived earlier, weaned
sooner and gave birth earlier the next year."[213] Vervet
groups that are provisioned near lodges are up to
twice as large as normal troops (28 vs. 16 average), a
result of more protection from predators, year-round
access to water, and a less seasonal birth rate.[483] In
a captive study, juvenile females that allomothered
more had better success raising their first infant.[212]
Baboons (*Papio*) and crowned hawk eagles are 2 of
the 16 potential predators upon vervets.[793] **Mating:**
Vervets have single-mount matings. Aliphatic acids
may be the basis of sexual attractiveness in the vaginal secretions of these primates.[885] **Vocalizations:**
21.[347] They have 3 different alarm calls and

NATURAL HABITAT

Vervet males have red, white, and blue displays.

responses for 3 predators. They
climb trees for carnivores, look up
for birds of prey, and look on the
ground for snakes. Field experiments using tape recordings have
proved that vervets recognize
individual calls from the members
of their own group and neighboring groups and can classify individuals into kin groups and their
place in the hierarchy.[113] They
are reported to use calls to
deceive other members of the
group.[113] **Sleeping site:** Fever
trees *(Acacia xanthophloea)* and,
to a lesser extent, umbrella trees
(A. tortilis) are preferred.[793]

Sub-Saharan Africa

J. M. LERNOULD

C. a. djamdjamensis is found in the mountains of Ethiopia.

C. a. tantalus.

151

White-throated guenon *Cercopithecus erythrogaster*

Guenons

Guenons   *Genus: Cercopithecus*

Cercopithecus, the genus of guenons, includes some of the most colorful monkeys. The name "guenon" comes from the French word for fright. When guenons are excited, they expose their teeth in a grimace. The variability of color occurs on the basic form of a monkey with a long tail, long hind limbs, a slender body, and a rounded head. Jonathan Kingdon has said that the distinctive facial patterns "look like experiments in signal geometry."[445] The facial patterns probably serve several purposes. They enhance species recognition and thus help maintain the reproductive isolation of each species. Within each species, an individual uses its patterned face to signal other troop members about where it is, where it is looking, and which way it may go.[445]

The majority of guenons are arboreal fruit eaters that disperse the seeds they eat. They help to regenerate the forest species on which they depend. Guenon groups are organized around a core of females (matrifocal) that live their lives in their birth group, defend their territory, and groom each other. One male is usually associated with a group of females, though there may be an influx of males during some seasons in some species. Young males generally leave the group before they mature and may travel alone or in a group of other males.

Many guenon species often associate with other primates, such as colobus monkeys and mangabeys, as well as other guenons. There are two explanations for these mixed-species associations. One hypothesis is that at least one species is improving its dietary intake. The other hypothesis is that species that travel together may be reducing the likelihood of predation by being better able to detect raptors, reptiles, and carnivorous mammals. Both of these explanations may be at work at the same time, because researchers have reported polyspecific groups to be using more open and more productive types of forest than one species alone.[280]

There is a great deal of disagreement among taxonomists about how many species or subspecies of guenons should be recognized. This is due in part to the group's recent evolution, which occurred as a result of climatic fluctuations over the last million years that fragmented the forests of Africa.[444] Isolated populations of guenons have had time to diverge into visually distinct forms but not enough time to become reproductively isolated species. Struhsaker has documented crosses and backcrosses between red-tailed guenons *(C. ascanius)* and blue monkeys *(C. mitis)* in the Kibale Forest in Uganda.[800] The hybrids are assimilated into their natal group, even though they may be larger than the majority of their close relatives and have different markings. It is thought that hybridization may be a successful reproductive strategy for some individuals. A female red-tailed guenon that mates with a male blue monkey will produce offspring with a larger body size than a purebred redtail. Thus the hybrid may be dominant in feeding disputes and may have food preferences different from those of either of its parents, which may decrease its competition for food. Guenons are in the middle of an evolutionary radiation that will be affected by the fragmentation of their forest habitats, caused by the recent unprecedented scale of human disturbances. Their future evolution will be determined by humans. If we destroy the forest habitat, only a few species will survive to adapt.

Red-tailed Guenon *Cercopithecus ascanius*

C. a. schmidti.

C. a. whitesidei.

E. THETFORD

Taxonomy 5 subspecies. In the *cephus* group.[601]

Distinguishing Characteristics
Red-tailed guenons are speckled yellow brown, with pale underparts. The lower end of the tail is chestnut brown. The face is black, with blue around the eyes, and has white cheek fur.[509] The nose color varies from white to yellow to black, depending on the subspecies.[601]

Physical Characteristics Head and body length: ♀ 425mm (320–463) *[16.7in (12.6–18.2)],* ♂ 494mm (405–629) *[19.4in (15.9–24.8)].*[601] **Tail length:** ♀ 651mm (530–779) *[25.6in (20.9–30.7)],* ♂ 772mm (627–892) *[30.4in (24.7–35.1)].*[601] **Weight:** ♀ 3300g (3175–3425) *[7.3lb (7.0–7.5)],* ♂ 4212g (3200–4835) *[9.3lb (7.1–10.7)].*[601] **Intermembral index:** 79.[234] **Adult brain weight:** 66.5g *[2.3oz].*[346]

Habitat Primary and secondary lowland rain forest, gallery, swamp, *Acacia* woodland, montane forest up to 2000m *[6562ft].*[898]

Diet Fruit, 62%; animal prey, 25%; leaves, 7.2%; gums, 2.8%.[133] Red-tailed guenons eat 104 species of plants, but 82% of the diet is from 20 species.[279] They eat more leaves when fruit is scarce.

Life History Infant: 18–24mo.[96] **Juvenile:** 24–48mo. **Sexual maturity:** ♀ 54mo, ♂ 72mo.[133] **Estrus cycle:** NA. **Gestation:** NA. **Age 1st birth:** 48–60mo.[134] **Birth interval:** 17.8mo[96] (18–52).[134] **Life span:** 22.5y.[708] Birth peak: Apr–Nov.[288] Of the guenons that have been studied, redtails are the only ones to give birth throughout the year, but only 3% of births are in June–August.[96]

Locomotion Quadrupedal.

Social Structure 1 male–multifemale groups.[898] A male's tenure in the same group is 11–72 months.[799] At some times of the year there are multimale influxes.[134] **Emigration:** Males emigrate; females remain and defend the territory.[799] **Group size:** 7–35.[898] **Home range:** 22–55ha.[133] **Day range:** 1543m *[5063ft].*[133]

Behavior Diurnal and arboreal. Red-tailed guenons use the lower strata 32% of the time (to forage for insects in the lianas);[134] the middle, 56.5%; and the upper strata, 11.5%.[280] A large group (50) has been observed to split (fission) into 2 groups (of 35 and 15), which took 4–6 months. "Both new groups traveled further each day than did the original group."[799] The rate of intergroup encounters and conflicts for both new groups increased "despite their presumed close kinship."[799] Intragroup aggression lessened in the smaller group. Furthermore, the "birth rate doubled in the small group but was unchanged in the large group."[799] After a new male took over a redtail troop, he killed and ate 2 newborns.[796] Redtails have hybridized with blue monkeys *(C. mitis)* at several sites.[800] The hybrid's coloration is intermediate but more like that of blue monkeys. Hybrids are larger, with less distinct color.[800] **Association:** Redtails initiate associations with red colobus monkeys *(Procolobus* spp.) 91.6% of the time, whereas red colobus monkeys form associations with redtails only 4.1%.[798] These 2 primate groups have the least dietary overlap at the Ugandan study site.[798] Redtails serve as sentinels in polyspecific groups by being the first to alarm-call.[925] Gray-cheeked mangabeys *(Lophocebus albigena)* are attracted to redtails 74.9%, and vice versa 24.1%.[798] Blue monkeys *(C. mitis)* form associations with redtails 61.5%; redtails join blues 30.8%.[798] Redtails occasionally mingle with Abyssinian black-and-white colobuses *(Colobus guereza).*[796] In general, associations are formed by less common species joining a more common one.[798] Smaller social groups join species with larger social groups. Monkeys with larger home ranges join species with smaller home ranges.[798] In Zaire, redtails associate with Wolf's guenons *(Cercopithecus wolfi),* black mangabeys *(Lophocebus aterrimus),* and Angolan black-and-white colobuses *(Colobus angolensis).*[804] When redtails enter the swamp, they

occasionally associate with Allen's swamp monkeys *(Allenopithecus nigroviridis).*
Mating: In a study, 11 long-term males performed 78.2% of all matings observed; the 19 short-term males, the other 21.8%.[799]

NATURAL HABITAT

Uganda, Zaire, Kenya, Zambia, Angola, Central African Republic

Campbell's Guenon *Cercopithecus campbelli*

Taxonomy 2 subspecies. In the *mona* group.[601]

Distinguishing Characteristics Campbell's guenon has a yellowish gray back and crown and a dark gray to black rump. The chin and underparts are white. It has pink lips, yellow whiskers, and a brow band that is white or yellow, depending on the subspecies.[601]

Physical Characteristics Head and body length: ♀ 400mm (360–430) *[15.7in (14.2–16.9)]*, ♂ 498mm (425–546) *[19.6in (16.7–21.5)]*.[601] **Tail length:** ♀ 638mm (580–680) *[25.1in (22.8–26.8)]*, ♂ 723mm (489–851) *[28.5in (19.3–33.5)]*.[601] **Weight:** ♀ 2.2kg *[77.6oz]*, ♂ 4.3kg (3.9–4.6) *[9.5lb (8.6–10.1)]*.[601] **Intermembral index:** NA. **Adult brain weight:** 65.8g *[2.3oz]*.[346]

Habitat Primary, secondary, and gallery forest;[280] abandoned farms.[231]

Diet Fruit, 78%; animal prey, 15.2%; leaves and flowers, 6.5%.[280] Campbell's guenons use 47 plant species.[77] They will raid unripe mango trees[340] and eat spider webs.[77]

Life History Weaning: 10–26mo.[458] **Sexual maturity:** ♀ 36mo, ♂ 54mo.[133] **Estrus cycle:** NA. **Gestation:** 180d.[77] **Age 1st birth:** 36–48mo.[77] **Birth interval:** 12mo.[133] **Life span:** NA. Mating season: Jun–Sep. Birth season: Nov–Jan (dry season in Ivory Coast).[77]

Locomotion Quadrupedal walking, climbing, and leaping.

Social Structure 1 male–multifemale groups.[280] The male's tenure in a group lasts on an average of 32.4mo.[134] **Emigration:** Young males emigrate at age 4–5 during the birth season.[77] **Group size:** 14.[231] **Home range:** 40ha.[231] **Day range:** NA.

Behavior Diurnal, arboreal, and partly terrestrial.[231] Campbell's guenons are the most terrestrial guenon; 20% of their locomotion is on the ground. They forage primarily in the understory and may drop to the ground and run long distances.[540] They will cross open ground.[77] They use the middle forest levels the most (54%), then the lower levels (32%) and the upper levels (14%).[280] The adult male is vigilant, often being higher in trees than others of its species and calling loudly at any sign of danger.[77] Head bobbing is the most common agonistic behavior. Chin thrusting is a mild threat. Subadults and juveniles are the most active and inquisitive and the first to approach unknown objects.[77] The alpha male is aggressive to maturing males, which eventually emigrate with same-age and peripheral females.[392] Infanticide has been

observed.[134] **Association:** Campbell's guenons form mixed-species groups with all available species of monkeys,[540] especially lesser spot-nosed guenons *(C. petaurista)* and Diana monkeys *(C. diana)*.[280] **Vocalizations:** Campbell's guenons make "noisy warning sneezes."[272] A male makes a croaking loud call and a booming cry.[339] Male loud calls are given between 8 a.m. and 3 p.m. or during and after thunderstorms[77] and at snakes, especially cobras.[77] The alarm call is a *crr crr*.[77]

Gambia to Ghana

Mustached Guenon *Cercopithecus cephus*

Taxonomy Disputed. In the *cephus* group. Formerly included *C. erythrotis* and *C. sclateri* as subspecies.[315]

Distinguishing Characteristics Mustached guenons have a black face with blue skin around the eyes, a white mustache bar, and white cheeks. The body is reddish brown agouti, and the lower part of the tail is red. The throat and underparts are white.[509]

Physical Characteristics Head and body length: 483–559mm *[19.0–22.0in]*.[509] **Tail length:** ♀ 693mm (670–720) *[27.3in (26.4–28.3)]*, ♂ 780mm *[30.7in]*.[601] **Weight:** ♀ 2.9kg *[102.3oz]*, ♂ 4.1kg *[144.6oz]*.[445] **Intermembral index:** 79.[234] **Adult brain weight:** 63.6g *[2.2oz]*.[346]

Habitat Primary and secondary rain forest, gallery, flooded forest.

Diet Fruit and seeds, 81%; animal prey, 12%; leaves, 6%.[279] The fruit of 57 species and other plant parts of 50 more are eaten, but 81% of the diet is from only

Mustached Guenon *continued*

20 species.[279] Oil palm fruit is a staple.[898] Prey includes slow-moving caterpillars and fast grasshoppers,[279] as well as bird eggs and fledglings.[94] "67% of the fruit sampled was orange or yellow in color and 86% possessed succulent pulp."[279]

Life History Weaning: NA. **Sexual maturity:** ♀ 48mo, ♂ 60mo. **Estrus cycle:** NA. **Gestation:** NA. **Age 1st birth:** 60mo.[708] **Birth interval:** 27.4mo.[279] **Life span:** 22y.[708]

Locomotion Quadrupedal.

Social Structure 1 male–multifemale groups.[279] These guenons spend 90% of their time in 44% of their home range.[279] **Group size:** 5–35.[898] **Home range:** 18–45ha.[898] **Day range:** 1298–1980m *[4259–6496ft]*.[133]

Behavior Diurnal and arboreal. Mustached guenons prefer the middle strata (61%) of dense second growth forest and use the upper strata (9%).[279] They feed on fruits and seeds in early morning and late afternoon and forage for prey during midday.[279] There is a sex-related difference in diet: males eat more fruit, and females eat more leaves and insects. Lactating females probably have a greater need for protein-rich food.[276] **Association:** These guenons associate with putty-nosed and crowned guenons (*C. nictitans, C. pogonias*) 5–45% of the time[280] and occasionally with white-collared and gray-cheeked mangabeys (*Cercocebus torquatus, Lophocebus albigena*).[566] All 5 species may form a mixed group.[566] When traveling with other species, mustached guenons range farther, use higher forest levels, retrace their steps less, and are the first to detect ground predators.[279] They have been observed to hybridize with red-eared guenons (*Cercopithecus erythrotis*) in the wild. **Vocalizations:** 8.[347]

Cameroon to Angola

Diana Monkey *Cercopithecus diana*

Taxonomy Disputed. In the *diana* group. Includes *C. roloway,* which some taxonomists consider to be a valid species, as 1 of 2 subspecies.[315]

Distinguishing Characteristics Diana monkeys are gray agouti, with a brownish red back, a black tail, and white underparts. They have a distinctive white stripe on the thigh and a red or cream-colored rump. The face is black, with a white beard.[509]

Physical Characteristics Head and body length: ♀ 445mm (440–450) *[17.5in (17.3–17.7)]*, ♂ 570mm (525–615) *[22.4in (20.7–24.2)]*.[601] **Tail length:** ♀ 713mm (700–725) *[28.1in (27.6–28.5)]*, ♂ 861mm (800–901) *[33.9in (31.5–35.5)]*.[601] **Weight:** ♀ 5.4kg *[11.9lb]*,[540] ♂ 5.0kg *[11.0lb]*.[708] **Intermembral index:** 79.[234] **Adult brain weight:** 77.3g *[2.7oz]*.[346]

Habitat Primary, gallery, and semi-deciduous mature forest; some secondary forest.[898]

Diet Fruit, seeds, leaves, flowers, exudates, insects.[641]

Life History Weaning: NA. **Sexual maturity:** NA. **Estrus cycle:** NA. **Gestation:** NA. **Age 1st birth:** 65mo.[708] **Birth interval:** 12mo.[708] **Life span:** 34.8y.[708]

Locomotion Quadrupedal.

Social Structure 1 male–multifemale[641] groups in which females are bonded and the male is peripheral.[360] **Group size:** 5[280]–50.[898] **Home range:** 37,[601] 93,[280] 250ha.[898] **Day range:** 1892m *[6208ft]*.[133]

Behavior Diurnal and arboreal. Diana monkeys use the lower forest strata 10%, the middle 57%, and the upper 33%.[280] A Diana group will respond more aggressively to nearby groups at its territorial boundaries and other areas of overlap.[360] **Association:**

Diana monkeys lead mixed-species groups of Campbell's guenon (*C. campbelli*) and lesser spot-nosed guenons (*C. petaurista*).[280] Olive colobuses (*Procolobus verus*) associate with Diana monkeys by following group movements and interspersing themselves within the group.[641] Both agonistic and affiliative interactions between the latter 2 species are rare, although juveniles may play together.[641] Western red colobuses (*P. badius*) move to be closer to a Diana group when the calls of chimpanzees (*Pan troglodytes verus*) are heard.[641] **Vocalizations:** Territorial encounters include loud calls and chases by males and agonistic calls by all members.[623] Female "chatter screams" are heard more frequently than male calls.[360] Diana monkeys have a whistle for an alarm call,[445] which is given in response to the alarm calls of squirrels, duikers, and birds.[360] They have a different loud call for eagles and leopards.

Sierra Leone to Ghana

Juvenile.

Dryas Guenon *Cercopithecus dryas*

S. NASH

Taxonomy Disputed. In the *diana* group. This species may be in the genus *Chlorocebus,* because of its similarity to *C. aethiops.*[315] The species *Cercopithecus salongo,* which was first described in 1977 from 1 juvenile specimen, was in 1991 documented to be not a new species but a member of this species.[128]

Distinguishing Characteristics Dryas guenons have a black face encircled with white. The ears have prominent white tufts.[366]

Physical Characteristics Head and body length: 350mm *[13.8in].*[366] **Tail length:** NA. **Weight:** ♀ 2220g *[4.9lb],* ♂ 3000g *[6.6lb].*[128] **Intermembral index:** NA. **Adult brain weight:** NA.

Habitat Tropical rain forest.[128]

Diet NA.

Life History NA.

Locomotion Quadrupedal.

Social Structure NA. **Group size:** NA. **Home range:** NA. **Day range:** NA.

Behavior Diurnal. Dryas guenons are believed to be semiterrestrial,[128] but little is known about them in their natural habitat.

Zaire

White-throated Guenon *Cercopithecus erythrogaster*

The gray-bellied form of the white-throated guenon is found in Nigeria.

white-throated guenons will raid crops.[898]

Life History NA. Infants have been observed in April.[631]

Locomotion: Quadrupedal walking and running on narrow supports.[631]

Social Structure 1 male–multifemale groups.[631] **Group size:** <30.[631] **Home range:** NA. **Day range:** NA.

Behavior: Diurnal and arboreal. White-throated guenons prefer the lower levels of forest in thick undergrowth below 15 meters *[49ft],*[631] where they move quietly. They hold their tail in a "vertical question mark," with the tip away from the head and body.[631] **Association:** These guenons associate with mona monkeys *(C. mona)* and white-collared mangabeys *(Cercocebus torquatus).* **Vocalizations:** The male has a "croaking loud call."[631]

Taxonomy Disputed. In the *cephus* group. Elevated from a subspecies of *C. cephus.* Some taxonomists include *C. sclateri* in this species as a subspecies.[631]

Distinguishing Characteristics White-throated guenons have a black face, a ruff of white hair under the lower jaw, a triangular crown of yellowish agouti fur, and a white nose.[631] The back is brown agouti.[509] The ventral surface of the tail is pale white. The belly varies from red in Benin to gray in Nigeria.[634]

Physical Characteristics Head and body length: 457mm *[18.0in].*[509] **Tail length:** NA. **Weight:** ♀ 2400g *[5.3lb],*[631] ♂ 2860g *[6.3lb].*[509] **Intermembral index:** NA. **Adult brain weight:** NA.

Habitat Moist evergreen, semideciduous, coastal, and secondary high forest.[634]

Diet Fruit, insects, leaves.[509] When restricted to remnant forests,

Nigeria, Benin

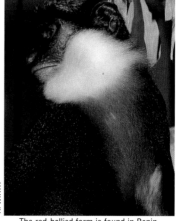

R. WIRTH

The red-bellied form is found in Benin.

Red-eared Guenon *Cercopithecus erythrotis*

Taxonomy Disputed. In the *cephus* group. Elevated from a subspecies of *C. cephus*.[315]

Distinguishing Characteristics Red-eared guenons have a brown agouti back and gray black limbs. The facial skin is blue, and the nose and ear tips are red. The fur on the cheeks is yellow. The tail is brown agouti at the base, gradually becoming all red toward the tip.[509]

Physical Characteristics Head and body length: ♂ 356–400mm *[14.0–15.7in]*.[509] **Tail length:** NA. **Weight:** ♂ 4082–4990g *[9.0–11.0lb]*.[509] **Intermembral index:** NA. **Adult brain weight:** 65.4g *[2.3oz]*.[346]

Habitat Younger (secondary) forest is preferred over mature forest.[898]

Diet Fruit, leaves, animal prey (including insects), and shoots.[509] Red-eared guenons reportedly raid crops.[898]

Life History Weaning: 6mo.[346] **Sexual maturity:** NA. **Estrus cycle:** NA. **Gestation:** NA. **Age 1st birth:** NA. **Birth interval:** NA. **Life span:** NA.

Locomotion Quadrupedal.

Social Structure Probably 1 male–multifemale groups. **Group size:** 4–29.[898] **Home range:** NA. **Day range:** NA.

Behavior Diurnal and arboreal.[509] Red-eared guenons have not been well studied in their natural habitat. In a captive study the gut passage rate was about 26.7 hours—much longer than the rates for frugivorous New World monkeys like spider monkeys (8h) and more like those for the folivorous howler monkeys (18–30h). African guenons' diets are believed to be more flexible and can include leaves and fruit.[517] **Association:** Red-eared guenons associate with gray-cheeked mangabeys *(Lophocebus albigena)* 41% of the time[863] and have been observed to hybridize with mustached guenons *(C. cephus)* in their natural habitat.[584]

Nigeria, Cameroon, Bioko

E. GADSBY - PANDRILLUS

Owl-faced Monkey *Cercopithecus hamlyni*

Taxonomy Monotypic.[898] In the *hamlyni* group.

Distinguishing Characteristics Owl-faced monkeys are dark gray, with a contrasting white vertical stripe down the nose. Males have a bright blue scrotum.[601] Neonates have a light yellowish brown coat that changes to adult gray with age.[601]

Physical Characteristics Head and body length: 559mm *[22.0in]*.[509] **Tail length:** NA. **Weight:** ♀ 3680g (2880–4480) *[8.1lb (6.3–9.9)]*,[288] ♂ 5486g *[12.1lb]*.[127] **Intermembral index:** NA. **Adult brain weight:** NA.

Habitat Bamboo, rain forest at 900–4554m *[2953–14,942ft]* elevation.[898]

Diet Fruit, leaves, animal prey.[509]

Life History NA. Birth season: May–Oct in Zaire.[288]

Locomotion Quadrupedal.

Social Structure 1 male–multifemale groups. **Group size:** 10.[898] **Home range:** NA. **Day range:** NA.

Behavior Diurnal and arboreal. Owl-faced monkeys prefer the lower forest levels, a preference that is believed to be a primitive trait among guenons.[215] As of 1995, this species had never been the subject of a long-term study. **Scent marking:** In captivity, owl-faced monkeys scent-mark with the sternal glands of their chest.[215] **Vocalizations:** The most notable is a loud "boom" call.[272]

Zaire, Rwanda

L'Hoest's Monkey *Cercopithecus lhoesti*

Taxonomy Disputed. In the *lhoesti* group. Used to include *C. preussi* as a subspecies.[315]

Distinguishing Characteristics L'Hoest's monkeys have a dark face, with a muzzle framed by a white bib. They are mostly gray agouti, with a reddish saddle on the back. Males have a white perineal patch.[601] The tail tip is hook shaped.[509]

Physical Characteristics Head and body length: 457–559mm [18.0–22.0in].[509] **Tail length:** NA. **Weight:** ♀ 3500g (2970–4030) [7.7lb (6.5–8.9)],[288] ♂ 5966g [13.1lb].[127] **Intermembral index:** NA. **Adult brain weight:** 76g [2.7oz].[346]

Habitat Lowland rain forest, savanna, and montane forest.[898]

Diet Fruit, leaves, animal prey.[509] L'Hoest's monkeys occasionally raid crops.[898]

Life History NA. Birth peak: Dec–Feb.[133] Birth season: Apr–Nov.[288]

Locomotion Quadrupedal.

Social Structure 1 male–multifemale groups.[509] **Group size:** 5–17.[898] **Home range:** 700–1000ha.[898] **Day range:** NA.

Behavior Diurnal, arboreal, and terrestrial.[509] L'Hoest's monkeys travel long distances on the ground and are reported to forage chiefly in the lower part of the forest.[280] **Association:** These guenons are rarely found in association with other guenon species.[280] **Scent marking:** This species does not appear to use

L'Hoest's monkeys travel long distances on the ground.

scent marks.[498] **Vocalizations:** L'Hoest's guenons have a low-pitched cohesion call but no high-pitched contact calls.[272] They produce several chirps and "hack" vocalizations, which are said to be similar to those of mustached guenons *(C. cephus)*. Males have a distinctive loud call.[794]

Zaire, Rwanda, Uganda, Burundi

Female with nursing infant.

Blue Monkey *Cercopithecus mitis*

Taxonomy Disputed. In the *mitis* group.[134] 22 subspecies in 2 groups—*C. m. mitis* and *C. m. albogularis*.[601] The latter is considered by some taxonomists to be a separate species.[315]

Distinguishing Characteristics The blue monkey's coloration varies with the subspecies. Members of the *C. m. mitis* group have a dark gray to light brown, orange, or olive back; a contrasting black crown; a gray face; and gray underparts. Members of the *C. m. albogularis* group are yellowish gray or olive overall, lack the contrasting crown, and have a white throat and white ear tufts.[601]

Physical Characteristics Head and body length: ♀ 468mm (390–587) [18.4in (15.4–23.1)], ♂ 565mm (462–710) [22.2in (18.2–28.0)].[601] **Tail length:** ♀ 686mm (541–880) [27.0in (21.3–34.6)], ♂ 755mm (601–953) [29.7in (23.7–37.5)].[601] **Weight:** ♀ 4231g (2700–5443) [9.3lb (6.0–12.0)], ♂ 7350g (5897–8955) [16.2lb (13.0–19.7)].[601] **Intermembral index:** 82.[234] **Adult brain weight:** 75g [2.6oz].[346]

Habitat Evergreen, semideciduous, bamboo, and dry scrub from lowland to montane up to 3300m [10,827ft]. Water and trees must be nearby.[898]

Female *(C. m. stuhlmani)*.

159

Blue Monkey *continued*

Diet Fruit, 54.6%; leaves, 18.9%; animal prey, 16.8%; gums, 1.9%.[133] They do not hunt for reptiles, birds,[131] and bush babies[95] but have been observed to capture them when the opportunity presented itself. Golden mitis monkeys *(C. mitis kandti)* reportedly eat bamboo.[293]

Life History Infant: 16–30mo.[134] **Weaning:** 30mo.[134] **Sexual maturity:** NA. **Estrus cycle:** NA. **Gestation:** 140d.[346] **Age 1st birth:** 71mo.[708] **Birth interval:** 24–54mo.[134] **Life span:** 20y. Mating and birth seasons: Variable.

Locomotion Quadrupedal.

Social Structure 1 male–multifemale groups. When several females are in estrus, several males (3–6, up to 11) may temporarily join the group.[134] A resi-

Female and infant *(C. m. albogularis).*

dent male's tenure lasts 14–94 months.[134] **Emigration:** Males emigrate. **Group size:** 12–70.[898] **Home range:** 23–50ha.[133] **Day range:** 1140–1300m *[3740–4265ft].*[903]

Behavior Diurnal and arboreal. The resident male in a troop of blue monkeys "acts as a reference point for his group."[714] Only the females defend territory, while the resident male chases away strange males.[714] At low densities, groups are more unstable and there is more male–male competition for females, more non-resident male intrusions, shorter male tenure, and more infanticide.[97] **Association:** Blue monkeys associate with red-tailed guenons *(C. ascanius),* red colobuses *(Procolobus* spp.), gray-cheeked mangabeys *(Lophocebus albigena),*[798] and, in Tanzania, Angolan black-and-white colobuses *(Colobus angolensis).*[925] When traveling with redtails, blue monkeys travel more and cover a larger area to find food.[280] Fertile F_2 hybrids have been observed between redtails and blue monkeys at several sites.[800] **Mating:** Females mate with several partners if they are available. Of the matings observed, 5–20% occurred without the resident male's knowledge.[134] Blue monkeys have multiple-mount matings.[169]

Vocalizations: The most notable call is the male's loud *pyow.* When it is given, group members move toward the resident male.[714]

Ethiopia to South Africa, Zaire, Angola

Juvenile *(C. m. albogularis).*

Mona Monkey *Cercopithecus mona*

Taxonomy Disputed. Monotypic. In the *mona* group.[601]

Distinguishing Characteristics
Mona monkeys have a brown agouti back with a white rump. The face is bluish gray, with a pink muzzle and a dark stripe from each eye to each ear. The cheek hair is yellowish, and the forehead is a yellowish white. The underparts and the inside of the leg are white. The outside of the leg is black, as is the tail.[509]

Physical Characteristics Head and body length: ♀ 419mm (375–457) *[16.5in (14.8–18.0)]*, ♂ 545mm (410–631) *[21.5in (16.1–24.8)]*.[601] **Tail length:** ♀ 587mm (533–655) *[23.1in (21.0–25.8)]*, ♂ 767mm (635–879) *[30.2in (25.0–34.6)]*.[601] **Weight:** ♂ 2733g (2400–5300) *[6.0lb (5.3–11.7)]*.[601] **Intermembral index:** 86.[234] **Adult brain weight:** 66g *[2.3oz]*.[346]

Habitat Primary and secondary, mangrove, and riparian forest up to 2000m *[6562ft]*.[898]

Diet Fruit, leaves, shoots, insects. Mona monkeys reportedly raid maize and cacao crops.[898]

Life History Weaning: NA. **Sexual maturity:** NA. **Estrus cycle:** NA. **Gestation:** NA. **Age 1st birth:** NA. **Birth interval:** NA. **Life span:** 22y.[708]

Locomotion Quadrupedal.

Social Structure 1 male–multife-male. **Group size:** 8–35.[898] **Home range:** 3ha.[898] **Day range:** NA.

Behavior Diurnal and arboreal. Mona monkeys prefer the lower levels of second-growth forest. An introduced population that is slightly different lives on the island of Grenada in the Caribbean. **Association:** Mona monkeys associate with white-throated guenons *(C. erythro-gaster)*[631] and white-collared mangabeys *(Cercocebus torqua-tus)*[339] and are frequently found with gray-cheeked mangabeys *(Lophocebus albigena)*.[602] They have been observed to hybridize with crowned guenons *(Cercopithecus pogonias)* in the wild.[800] **Vocalizations:** Mona monkeys have a "noisy warning sneeze." The loud call is a "boom."[272]

Ghana to Cameroon

Juvenile.

De Brazza's Monkey *Cercopithecus neglectus*

Taxonomy Monotypic.[601] In the *neglectus* group.

Distinguishing Characteristics De Brazza's monkeys are gray agouti with a reddish brown back and black limbs and tail. They have a white rump, a white thigh stripe, an orange crescent-shaped forehead, and white eyelids, muzzle, and beard. Males have a blue scrotum. The neonatal coat is light yellowish brown.[509]

Physical Characteristics Head and body length: ♀ 464mm (394–543) *[18.3in (15.5–21.4)]*, ♂ 560mm (484–595) *[22.0in (19.1–23.4)]*.[601] **Tail length:** ♀ 530mm (470–574) *[20.9in (18.5–22.6)]*, ♂ 691mm (591–784) *[27.2in (23.3–30.9)]*.[601] **Weight:** ♀ 4458g (4100–4816) *[9.8lb (9.0–10.6)]*; ♂ 7050, 8045g *[15.5, 17.7lb]*.[601] **Intermembral index:** 82.[234] **Adult brain weight:** 70.8g *[2.5oz]*.[346]

Habitat Primary and secondary, swamp, bamboo, and dry montane forest up to 2100m *[6890ft]*.[898]

Diet Fruit and seeds, 74%; leaves, 9%; arthropods, 5%; flowers, 3%; mushrooms, other animal prey. De Brazza's monkeys eat reptiles, caterpillars, and ants.[279] They occasionally raid millet crops.[898]

Life History Weaning: 12mo. **Subadult:** 56mo.[708] **Sexual maturity:** ♀ 42–48mo, ♂ 60–72mo.[133] **Estrus cycle:** NA. **Gestation:** 168d.[346] **Age 1st birth:** 53.5mo.[346] **Birth interval:** 27.4mo.[133] **Life span:** 22y.[458] Birth season: Variable. De Brazza's infants mature earlier than most other guenon infants. They are able to forage more than nurse by 5 months.[458]

Locomotion Quadrupedal.

Social Structure Variable. Monogamous 1 male–1 female groups and 1 male–multifemale groups.[714] This is the only cercopithecine monkey reported to be monogamous. Groups have overlapping territories with no known territorial defense.[898] **Group size:** 2–10, up to 35.[898] **Home range:** 6.6,[714] 13ha.[898] **Day range:** 530m *[1739ft]*.[280]

Behavior Diurnal, arboreal, and terrestrial. De Brazza's monkeys are known for their cryptic behavior. They will remain in one place without moving for up to 5 hours or flee silently.[280] They stay low in the forest (under 5m *[16ft]*) 70% of the time and are on the ground 20%.[280] They are said to use their food resources intensively, because they were observed to visit the same fruit tree 4 times a day for 2 weeks and not leave any half-eaten fruit.[280] De Brazza's monkeys have been observed to dig for invertebrates in leaf litter and mud.[925] When a predator is detected, the male shakes branches and barks to divert attention from his group. He may even attack the predator. The white beard may accentuate a silent yawn that males perform more than females.[272] De

Juvenile.

Brazza's monkeys can swim.[925] **Association:** This species does not associate with other primate species.[714] **Scent marking:** Marks are made with the sternal gland.[215] Males perform more scent-marking behavior than females.[498] **Vocalizations:** 6 discrete (not graded).[274] De Brazza's monkeys have a "boom" for a loud call, which is produced by inflating the vocal sac. They do not use a low-pitched cohesion call to avoid detection by predators.[272]

Cameroon to Uganda and Angola, Kenya, Ethiopia, and Sudan

The white beard may accentuate a silent yawn, which males perform more than females.

Putty-nosed or Greater Spot-nosed Guenon *Cercopithecus nictitans*

Taxonomy 3 subspecies.[898] In the *mitis* group.

Distinguishing Characteristics Putty-nosed guenons are dark olive agouti and have a dark face and a white spot on the nose. The underparts and tail are black, as are the hands and feet.[509]

Physical Characteristics Head and body length: 432–660mm *[17.0–26.0in]*.[509] **Tail length:** NA. **Weight:** ♀ 4082g *[9.0lb]*, ♂ 6350g *[14.0lb]*.[509] **Intermembral index:** 82.[234] **Adult brain weight:** 78.6g *[2.8oz]*.[346]

Habitat Primary and older secondary flooded forest, rain forest.[898]

Diet Fruit, up to 90%; leaves, animal prey. In some seasons these guenons eat more leaves when fruit is scarce. They prey on caterpillars and ants and reportedly steal chickens from human settlements.[898]

Life History Weaning: NA. **Sexual maturity:** ♀ 48mo, ♂ 60–72mo.[133] **Estrus cycle:** 28d.[346] **Gestation:** NA. **Age 1st birth:** 60mo.[96] **Birth interval:** 24mo.[96] **Life span:** NA. Birth peak: Dec–Jan.[96]

Locomotion Quadrupedal.

Social Structure 1 male–multifemale groups.[276] **Group size:** 20[903] (7–60).[898] **Home range:** 37–75ha,[898] to 174.[133] **Day range:** 1825m *[5988ft]*.[133]

Behavior Diurnal and arboreal. Putty-nosed guenons alternately rest and forage for insects in undergrowth during the heat of midday.[280] During the rest of the day they prefer the higher forest levels, where they are found at 27.2 meters *[89ft]* on average.[282] Males, which are responsible for identifying predators and alarming the troop, are found at higher levels than females. Males spend less time feeding than females, probably because of their less taxing physiological requirements (no lactation) and their higher dominance, which allows them priority access to the highest-quality food. They also feed faster

than females.[276] **Association:** In Gabon, putty-nosed guenons associate with crowned guenons *(C. pogonias)* in 97% of all sightings.[282] They associate occasionally with gray-cheeked mangabeys *(Lophocebus albigena)*[566] and frequently with mustached guenons *(C. cephus)*.[280] When associating with the latter, the puttynose watches for aerial predators, particularly the crowned hawk eagle *(Stephanoaetus coronatus),* while the mustached guenon watches below for ground predators.[282] Another advantage of mixed species' foraging together is a more diversified diet and a more efficient search for food.[282]

Liberia to Ivory Coast, Nigeria to Zaire to Angola

Juvenile.

Lesser Spot-nosed Guenon *Cercopithecus petaurista*

Taxonomy 2 subspecies. In the *cephus* group.[601]

Distinguishing Characteristics Lesser spot-nosed guenons have a greenish gray back and a black face. The nose, throat, ear tufts, and underside of the tail are white.[509]

Physical Characteristics Head and

Lesser spot-nosed monkeys are observed to associate with all of the other diurnal monkeys in Ivory Coast.

body length: ♀ 411mm (290–493) *[16.2in (11.4–19.4)]*, ♂ 492mm (440–527) *[19.4in (17.3–20.7)]*.[601] **Tail length:** ♀ 649mm (556–690) *[25.6in (21.9–27.2)]*, ♂ 712mm (523–787) *[28.0in (20.6–31.0)]*.[601] **Weight:** ♀ 3020g (2600–3800) *[6.7lb (5.7–8.4)]*, ♂ 3820g (3400–4500) *[8.4lb (7.5–9.9)]*.[601] **Intermembral index:** NA. **Adult brain weight:** NA.

Habitat Primary and secondary rain forest and coastal, swamp, and scrub forest near rivers.[601]

Diet Fruit, especially figs; flowers;[280] animal prey, including insects.[540] These guenons will raid maize and cacao crops.[898]

Life History NA.

Locomotion Quadrupedal.

Social Structure 1 male–multifemale groups. **Group size:** 15–20.[898] **Home range:** 700m *[2297ft]*.[540] **Day range:** NA.

Behavior Diurnal and arboreal. Lesser spot-nosed guenons are reported to "move in a quiet skulking fashion in thick growth."[641] **Association:** Found in mixed-species groups that include Campbell's guenon *(C. campbelli)* and Diana monkeys *(C. diana);* the latter lead the group.[280] In Sierra Leone, they associate with olive colobuses *(Procolobus verus)* and *C. campbelli*.[641] In Ivory Coast they associate with all the other monkey species to some degree.[540]

Vocalizations: This species has a loud call.[540] When disturbed by a human observer, it has a purring alarm call that is usually given by the male, whose whole head bobs at the intruder. The rest of the troop retreats while the male displays.[339]

Gambia to Benin

Crowned Guenon *Cercopithecus pogonias*

Taxonomy Disputed. 4 subspecies. In the *mona* group.[601]

Distinguishing Characteristics Crowned guenons have 3 black stripes on the forehead, with white or yellow in between. The sagittal crest is black, the muzzle is pink, the face is blue gray, and the back, hands, feet, and lower part of the tail are brown agouti.[509]

Physical Characteristics Head and body length: ♀ 462mm (440, 483) *[18.2in (17.3, 19.0)]*, ♂ 536mm (525, 547) *[21.1in (20.7, 21.5)]*.[601] **Tail length:** ♀ 737mm *[29.0in]*, ♂ 810mm (750, 870) *[31.9in (29.5, 34.3)]*.[601] **Weight:** ♀ 3181g *[7.0lb]*, ♂ 4545g *[10.0lb]*.[509] **Intermembral index:** NA. **Adult brain weight:** 71.1g *[2.5oz]*.[346]

Habitat Primary and secondary tropical rain forest.

Diet Fruit, leaves, shoots,[509] and animal prey, especially grasshoppers. Crowned guenons eat more

C. p. pogonias.

animal prey when fruit is scarce.[280]

Life History Weaning: NA. **Sexual maturity:** ♀ 48mo, ♂ 60–72mo.[133] **Estrus cycle:** NA. **Gestation:** 165d.[89]

Age 1st birth: 60mo.[708] **Birth interval:** 24mo.[134] **Life span:** 20y.[708] Birth season: Dec–Apr in Cameroon.[601]

Locomotion Quadrupedal.

Social Structure 1 male–multifemale groups.[280] **Group size:** 9–19.[276] **Home range:** 55–100ha,[280] to 174.[133] **Day range:** 825–1980m *[2707–6496ft]*.[133]

Behavior Diurnal and arboreal. Crowned guenon females eat more leaves and insects, and the males eat more fruit.[276] Males are more vigilant in detecting aerial predators and territorial intruders, and females are more concerned with social interactions with their offspring.[274] They all flee from predators through the trees, not on the ground. They use the lower strata of the forest 11% of the day, the middle strata 67%, and the upper strata 22%.[280] **Association:** When traveling in mixed-species groups with

Crowned Guenon *continued*

putty-nosed guenons (*C. nictitans*) and mustached guenons *(C. cephus),* crowned guenons vocalize the most[280] and are first to alarm-call at aerial predators. Each species in a mixed-species group has a different dietary emphasis. In the wild, crowned guenons will hybridize with mona monkeys *(C. mona).*[800]
Vocalizations: 17 graded calls.[274] The trill is the most frequent call, along with grunts, barks, and chirps. Gecker, grunt, trill, isolation trill, and boom are less frequent. Only the male gives the alarm bark and the morning loud

C. p. grayi.

call.[274] Young infants may vocalize up to 10.25 times per minute.[274] The amount of vocalizations given by infants decreases as they mature. The captive hybrid offspring of a male crowned guenon and a female red-tailed guenon (*C. ascanius*) made vocalizations different from those of either parent species.[274]

Nigeria, Cameroon, Gabon, Congo, Zaire, Central African Republic

Preuss's Monkey *Cercopithecus preussi*

Taxonomy Disputed. 2 subspecies.[601] In the *lhoesti* group. Some taxonomists think this species is a subspecies of L'Hoest's monkey *(C. lhoesti).*[509] Its karyotype is reported to be identical with the latter.[186]

Distinguishing Characteristics Preuss's monkeys are gray agouti, with a reddish saddle on the back. They have only a small amount of white on the chin.[601]

Physical Characteristics Head and body length: ♂ 574mm (550–605) *[22.6in (21.7–23.8)].*[601] **Tail length:** ♂ 646mm *[25.4in].*[601] **Weight:** NA. **Intermembral index:**

81.5. **Adult brain weight:** NA.
Habitat Primary and secondary moist forest above 1000m[898] *[3281ft].*
Diet Fruit, leaves, insects.[509]
Life History NA.
Locomotion Quadrupedal.
Social Structure 1 male–multifemale groups. **Group size:** 2–8.[280] **Home range:** NA. **Day range:** NA.
Behavior Diurnal and arboreal. Preuss's guenons use the midcanopy 15% of the day and the lower forest parts 85%. They flee on the ground from threats.[280]
Vocalizations: Preuss's guenons rarely use an antipredator call.[280]

Cameroon, Bioko

E. GADSBY - PANDRILLUS

Sclater's Guenon *Cercopithecus sclateri*

Taxonomy Disputed. In the *cephus* group. Elevated from a subspecies of *C. erythrotis* in 1980.[315]

Distinguishing Characteristics Sclater's guenons have a white nose bridge, white ear tufts, and a black band on the head. The black-tipped tail is red on the underside and dark gray on the upper side.[637]

Physical Characteristics Head and body length: NA. **Tail length:** NA. **Weight:** ♀ 2.0–3.0kg *[4.4–6.6lb]*, ♂ 4.0kg *[8.8lb]*.[637] **Intermembral index:** NA. **Adult brain weight:** NA.

Infant.

Habitat Swamp, riverine forest, small remnant sacred tree groves.[637]

Diet Fruit, leaves, animal prey. When restricted to remnant forests near villages, Sclater's guenons will raid crops, including oil palm nuts.[637]

Life History NA.

Locomotion Quadrupedal.

Social Structure Multimale-multifemale groups. **Group size:** 15–30.[637] **Home range:** NA. **Day range:** NA.

Behavior Diurnal and arboreal. Sclater's guenons were first described from a captive animal in 1904, but the first scientific field observations were not made until 1987. No long-term ecological study had been undertaken as of 1995.[638] This species is the most endangered guenon in Africa.[638]

Nigeria

A young Sclater's guenon threatens the photographer.

After a mother guenon is shot, the infant is often kept as a pet.

Female and infant.

Sun-tailed Guenon *Cercopithecus solatus*

Taxonomy Disputed. In the *lhoesti* group. Discovered in 1984. Some taxonomists think this species is a subspecies of *C. lhoesti,* but *C. solatus* has 3 chromosomal rearrangements when compared with *C. lhoesti.*[186]

Distinguishing Characteristics Sun-tailed guenons have a dark gray back with a chestnut orange saddle. The throat is white, and the tail toward the tip is yellow orange. The male has a blue scrotum.[343]

Physical Characteristics Head and body length: NA. **Tail length:** NA. **Weight:** ♀ 3.5kg *[7.7lb],* ♂ 6.0kg *[13.2lb].*[280] **Intermembral index:** NA. **Adult brain weight:** NA.

Habitat Lowland tropical rain forest. The entire range is limited to 10,360 sq km *[4000 sq mi].*[275]

Diet NA.

Life History NA.
Locomotion Quadrupedal.
Social Structure 1 male–multife-male groups. **Group size:** 5–15.[280] **Home range:** NA. **Day range:** NA.

Behavior Diurnal, arboreal, and terrestrial. Sun-tailed guenons are said to be secretive.[273] They forage on the ground[275] and flee from predators on the ground.[186] They are hunted with snares because of their terrestrial travel.[186] **Vocalizations:** No loud call has been reported for sun-tailed guenons. The alarm bark is reported to be similar to that of L'Hoest's monkeys *(C. lhoesti).*[275]
Sleeping site: Trees.[275]

Gabon

The sun-tailed guenon was first discovered in 1984.

Wolf's Guenon *Cercopithecus wolfi*

Taxonomy Disputed. 3 subspecies. In the *mona* group. Some taxonomists consider *C. w. denti* to be a full species.[315] Wolf's guenons differ from crowned guenons *(C. pogonias grayi)* by 1 inversion in the karyotype.[591]

Distinguishing Characteristics Wolf's guenons are dark gray with a reddish saddle on the back. The underparts are white or yellow, depending on the subspecies. The arms are black, the legs are brownish red, and the tip half of the tail is black. The head has a pale brow and a broad band of black from the eye to the ear. The ear tufts are red or white.[601]

Physical Characteristics Head and body length: ♂ 485mm (445–511) *[19.1in (17.5–20.1)].*[601] **Tail length:** ♂ 779mm (695–822) *[30.7in (27.4–32.4)].*[601] **Weight:** ♀ 2760g (2380–3140) *[6.1lb (5.2–6.9)],*[288] ♂ 3800–4200g *[8.4–9.3lb].*[127] **Intermembral index:** NA. **Adult brain weight:** NA.

Habitat Primary and secondary lowland rain forest, swamp forest.[925]

Diet Fruit, leaves, flowers, nectar, insects. In Zaire, Wolf's guenons feed on nectar and pollinate *Daniellia pynaertii,* an emergent leguminous tree.[281]

Life History NA. Birth season: Jun–Dec.[288]
Locomotion Quadrupedal.
Social Structure NA. **Group size:** 1–12.[925] **Home range:** NA. **Day range:** NA.

Behavior Diurnal and arboreal. Wolf's guenons split into smaller groups to forage for insects.[925] They pre-fer to forage 15 meters *[49 ft]* off the ground. They are preyed upon by the crowned hawk eagle *(Stephanoaetus coronatus).* When an eagle is seen, they sound an alarm and plunge from the branches.[519] **Association:** Wolf's guenons associate with black mangabeys *(Lophocebus aterrimus)* 80% of the time.[288] They also associate with red-tailed guenons *(C. ascanius)* and occasionally with Angolan black-and-white colobuses *(Colobus angolensis).*[539] When Wolf's guenons enter the swamp, they occasionally associate with Allen's swamp monkeys *(Allenopithecus nigroviridis).*[925] **Vocalizations:** 7.[590] Wolf's guenons have 2 contact calls, 2 travel calls, and 3 alarm calls.[590] When foraging, they make a grunting call to stay in vocal contact.[925] More contact calls are given when a group is more dispersed or when it is in thick secondary growth and visibility is poor.[590] They call more when hunting insects than when eating fruits.[590]

Zaire, Uganda

Red-shanked douc langur *Pygathrix nemaeus*

Leaf-eating Monkeys

Leaf-eating Monkeys   *Subfamily:* Colobinae

The Colobinae, which is the other large subfamily of Old World monkeys in the family Cercopithecidae (besides Cercopithecinae), are known as the leaf-eating monkeys. They evolved on two continents—the colobus monkeys in Africa and the langurs and leaf monkeys in Asia. All colobines have a specially adapted stomach, which is sacculated and supports bacterial colonies. These bacteria make it possible for colobines to digest the cellulose in their diet of leaves, unripe fruit, and seeds. The molar teeth have high, pointed cusps to break down leaves. Colobines do not have cheek pouches. Generally colobines are arboreal and have longer hind legs and tails than other Old World monkeys. Most colobines have short thumbs, and some colobus species have no thumbs at all.[234]

Most species of African and Asian leaf-eating monkeys live in a social group with one male and several females. The females remain in their natal group and maintain the group's territory. The male often has a relatively short tenure in a female group and is forced out by another male in as little as two years. In Hanuman langurs this short residency has led to a distinctive male reproductive strategy: a male that practices infanticide (the killing of unweaned infants) will have more offspring than males that don't practice infanticide, because females come into estrus (ovulate) either after their infant is weaned at 12–18 months or soon after they have lost an infant. The infants of many species of this subfamily are often born with a coat color that differs from that of the mother. The strikingly bright orange natal coat of the silvered langur *(Trachypithecus cristatus)* changes to adult gray in the first three months. The adaptive significance of the striking natal coat in species in which infanticide is common has yet to be fully explained. Allomothering has been documented in many colobines. This behavior is the care of infants by troop members other than the mother, usually by juvenile females that may or may not be aunts of the infant. It is hypothesized that this baby-sitting may help the mother feed more efficiently and may provide the other caretakers with practice for future infant care.

The colobus monkeys of Africa are divided into two genera. Each genus has a distinct coloration, as well as distinct anatomical features.

The members of the genus *Colobus* have various patterns of long black and white fur. They have a three-chambered stomach, a large larynx, and a subhyoid sac, and the females have no sexual swelling. The red colobus monkeys *(Procolobus)* have variable amounts of red fur. The genus *Procolobus* also includes the olive colobus, which is the smallest African leaf-eating monkey. *Procolobus* species have a four-chambered stomach, a small larynx, and no subhyoid sac, and females have sexual swellings.

The Asian leaf monkeys are more diverse, with five currently recognized genera. Members of the genus *Presbytis* are found on the Malay Peninsula and in Indonesia. The natal coat in this genus is usually whitish, with a dark-colored stripe down the back. The most studied leaf monkey is the common gray Hanuman langur, of the genus *Semnopithecus.* It is more terrestrial and lives in a wider range of habitats than any other colobine. Its infants have black fur at birth. The brow-ridged langurs are placed in the genus *Trachypithecus* and are named for their cranial feature, which resembles raised eyebrows. The natal coat of most of the leaf monkeys in this genus is bright yellow or orange. The final two colobine genera, *Pygathrix* and *Nasalis,* have been termed the odd-nosed monkeys. Three of the four snub-nosed monkeys of the genus *Pygathrix* and subgenus *Rhinopithecus* live in China and range in deciduous forests that are snow covered for several months each year. The most colorful leaf monkeys are the two douc langurs (subgenus *Pygathrix),* of which little is known. They live only in Southeast Asia and are highly endangered because of the numerous human conflicts that have destroyed their habitat. Their attractiveness makes them prime targets for hunters that supply an illegal pet trade. Tragically most douc langurs quickly die in captivity because it is so difficult to replicate their natural diet. The pig-tailed langur *(Nasalis concolor)* is found only on the Mentawai Islands of Indonesia and has the smallest tail of any langur. The male proboscis monkey *(Nasalis larvatus)* from Borneo is the most memorable leaf monkey, with its elongated, tubular nose. All of the odd-nosed monkeys are endangered. The Tonkin snub-nosed monkey *(Pygathrix avunculus)* is critically endangered. Like many primates it is threatened by habitat destruction and human hunting.

169

Angolan Black-and-white Colobus *Colobus angolensis*

Female and juvenile.

Taxonomy 6 subspecies.[82]

Distinguishing Characteristics Angolan black-and-white colobuses are black with a white brow band, cheeks, and throat. They have long-haired white epaulettes on the shoulders.[80] The lower half of the tail is white.[635]

Physical Characteristics Head and body length: ♀ 529mm (483–593) *[20.8in (19.0–23.3)]*, ♂ 590mm (553–660) *[23.2in (21.8–26.0)]*.[602] **Tail length:** ♀ 706mm (629–763) *[27.8in (24.8–30.0)]*, ♂ 829mm (762–920) *[32.6in (30.0–36.2)]*.[602] **Weight:** ♀ 7403g (6804–8895) *[16.3lb (15.0–19.6)]*, ♂ 9670g (9072–10285) *[21.3lb (20.0–22.7)]*.[602] **Intermembral index:** NA. **Adult brain weight:** 73.5g *[2.6oz]*.[346] Infants are born white and change to gray and then black in the first 6 months.[311]

Habitat Primary and secondary lowland to montane forest up to 3000m *[9843ft]*.[835]

Diet Seeds, young leaves, unripe fruit.[635] Of a total of 46 plant species eaten, 47% of the diet comes from just 5.[518] Seeds of leguminous trees, which grow in poor soils, provide a protein-rich food that makes up 39% of the diet. These colobuses also eat unripe figs and "trunk-encrusting lichen."[635] Laboratory analysis of the diet from one study site identified the nutritional value of each food type. Young leaves have 14.8% crude protein, 6.1% condensed tannins, and 3.8% lipids. Mature leaves have 10.9% crude protein, 3.7% condensed tannins, 9.0% lipids, and 55.6% water. Seeds have 12.7% crude protein, 5.2% condensed tannins, 9.05% lipids, and 68.9% water. Unripe fruits have 13.5% crude protein, 5.1% condensed tannins, no lipids, and 85.6% water.[518]

Life History NA. Birth season: May–Dec in Zaire.[835]

Locomotion Quadrupedal; leaping.[311]

Social Structure Variable. 1 male–multifemale,[311] multimale-multifemale groups with up to 5 males.[62] Groups forming large aggregations of up to 300 individuals have been observed in Nyungwe Forest, Rwanda.[635] **Group size:** 2–16, up to 50.[898] **Home range:** >400ha.[62] **Day range:** 500–2000m *[1641–6562ft]*.[62]

Behavior Diurnal and arboreal. Angolan black-and-white colobus monkeys occasionally come to the ground near streams[539] to eat herbaceous vegetation.[635] They prefer the highest forest levels, including emergent trees. When a troop is threatened by a predator, the male jumps and roars until the rest of the troop has fled.[311] Very little grooming or interaction has been observed within groups.[311] **Association:** This species associates with blue monkeys *(Cercopithecus mitis)* in Tanzania.[925] In Zaire it is found in mixed-species groups with red-tailed guenons *(C. ascanius)*, Wolf's guenons *(C. wolfi)*, and black mangabeys *(Lophocebus aterrimus)*.[539]

Angola, Zaire, Rwanda, Burundi, Zambia, Kenya, Tanzania

Abyssinian, Guereza, or Eastern Black-and-white Colobus *Colobus guereza*

At 3 months, an infant's hands and feet turn black.

Taxonomy This species was named *C. abyssinicus* until 1956.[706] It is the most derived black-and-white colobus species.[640]

Distinguishing Characteristics This colobus has a U-shaped mantle of long white fur that descends from its shoulders and rounds its back.[80] It has white hair surrounding a black face. The tail is 5–95% white, depending on the subspecies.[602]

Physical Characteristics Head and body length: ♀ 576mm (521–673) *[22.7in (20.5–26.5)]*, ♂ 615mm (543–649) *[24.2in (21.4–25.6)]*.[602] **Tail length:** ♂ 687mm (528–797) *[27.0in (20.8–31.4)]*, ♀ 667mm (521–826) *[26.3in (20.5–32.5)]*.[602] **Weight:** ♂ 13.5kg *[29.8lb]*, ♀ 7.9–9.2kg *[17.4–20.3lb]*.[639] **Intermembral index:** 79.[234] **Adult brain weight:** 82.3g.[346]

Habitat Primary and secondary forest, riverine forest, wooded grassland.[80] These colobuses are often abundant in young forests with a few tree species and a prolonged dry season.[805] They are found at higher

By age 1, the juvenile has adult coloration but still has a pink face.

170

Abyssinian, Guereza, or Eastern Black-and-white Colobus *continued*

densities in logged forest than in primary forest.[417]

Diet Leaves, fruit. The diet is sometimes monotonous. In Kibale, Uganda, 68.5% of the diet is from 1 species *(Celtis durandii).*[805]

Life History Weaning: 13mo.[346] **Sexual maturity:** NA. **Estrus cycle:** NA. **Gestation:** NA. **Age 1st birth:**

58mo.[708] **Birth interval:** 18–24mo.[185] **Life span:** 22.2y.[708] Births: Year-round.[804]

Locomotion Quadrupedal; leaping.[706]

Social Structure Variable. 1 male–multifemale,[805] multimale-multifemale.[772] These colobus monkeys are territorial and will aggressively chase other groups from their core area.[628] **Emigration:** Maturing males leave the group.[805] Solitary subadults have been observed. **Group size:** 9 (7–11).[855] **Home range:** 15–16ha.[805] **Day range:** 535m (288–1004) *[1755ft (945–3294)].*[805]

Behavior Diurnal and arboreal. Abyssinian black-and-white colobus monkeys rest more than half the day.[772] They sunbathe in early morning high in the trees and later feed in the lower levels.[628] They will travel on the ground to feed on aquatic plants, including duckweed *(Lemna minor).*[628] In Ethiopia, "groups fission when their size results in fewer than 10 trees per individual within their territory. Daughter groups emigrate to suboptimal areas."[185] Social behavior is uncommon between individuals except mother and infant.[628]

Newborn infants are carried and handled by other group members.[805] This species is endangered because it was hunted for its pelts, which were made into wall decorations for tourists.[898] **Association:** This species occasionally mingles with red-tailed guenons *(Cercopithecus ascanius)* and rarely with Pennant's red colobuses *(Procolobus pennantii).* **Mating:** This species is said to have multiple-mount matings.[169] **Vocalizations:** These colobus monkeys produce a low-pitched, roaring chorus.[541] They roar often at night and around dawn.[634]

Nigeria to Ethiopia, Kenya, Uganda, Tanzania

Female with a 1-month-old infant.

King or Western Black-and-white Colobus *Colobus polykomos*

Adult male.

Taxonomy Formerly included *C. vellerosus* as a subspecies.[310]

Distinguishing Characteristics The king black-and-white colobus is black, with white or grayish white around the face and on its shoulders. The tail is white the entire length and is not bushy.[80] Neonates have pink skin and white fur.[635]

Physical Characteristics Head and body length: ♀ 600mm (570–648) *[23.6in (22.4–25.5)],* ♂ 633mm (584–680) *[24.9in (23.0–26.8)].*[602] **Tail length:** ♀ 886mm (806–959) *[34.9in (31.7–37.8)],* ♂ 874mm (718–1000) *[34.4in (28.3–39.4)].*[602] **Weight:** ♀ 8.3kg *[18.3lb],* ♂ 9.9kg *[21.8lb].*[639] **Intermembral index:** 78.[234] **Adult brain weight:** 76.7g *[2.7oz].*[346]

Habitat Moist forest.[898]

Diet Seeds, leaves. In Sierra Leone, seeds are eaten September–February, young leaves March–July, and mature leaves July–September.[150] Over 63% of the leaves eaten are from liana vines rather than trees.[150] Most fruit is eaten unripe, and the seeds are extracted from the fruit.[150]

Life History Weaning: NA. **Sexual maturity:** NA. **Estrus cycle:** NA. **Gestation:** 170d.[346] **Age 1st birth:** 102mo.[346] **Birth interval:** 12.6mo.[346] **Life span:** 30.5y.[708] Birth season: Dec–Feb.[635]

Locomotion Quadrupedal; leaping.

NATURAL HABITAT

LOWER RISK

CITES II

King or Western Black-and-white Colobus *continued*

Social Structure Multimale-multifemale groups. **Group size:** 5–21.[898] **Home range:** 18–22ha.[898] **Day range:** NA.

Behavior Diurnal and arboreal. The king black-and-white colobus is often seen in tall emergent trees. During times when high-energy foods are unavailable, it adapts, becoming less active and thus conserving energy.[149] Allomothering has been reported in captivity.[531] **Association:** This species associates with the olive colobus *(Procolobus verus)* and at Tiwai Island, Sierra Leone, with the western red colobus *(P. badius)*. Red colobus adults have been observed handling black-and-white colobus infants. **Vocalizations:** Several males will roar as a group.[634]

A colobus leaps down to another tree.

Gambia to Ivory Coast

Black Colobus *Colobus satanas*

Taxonomy 2 subspecies.[82]

Distinguishing Characteristics The black colobus is entirely black. The infant's coat is brown, not white as in other colobuses.[635]

Physical Characteristics Head and body length: ♀ 635mm *[25.0in]*, ♂ 670mm *[26.4in]*.[602] **Tail length:** 800mm *[31.5in]*.[602] **Weight:** 10.9kg *[24.0lb]*.[541] **Intermembral index:** NA. **Adult brain weight:** 80.2g *[2.8oz]*.[346]

Habitat Primary and secondary forests, swamps, up to 1000m *[3281ft]*.[898]

Diet Seeds, young leaves, flowers, buds, and, rarely, mature leaves.[345] The diet changes every 3 months with the season.[541] This species feeds heavily on seeds, which are high in nutrients and are found in large clumps, allowing a whole group to feed without going far.[541] The leaves eaten are mostly from rare plants rather than common plants. Like other colobines, these colobuses prefer foods rich in minerals and nitrogen and low in lignin and tannin, which inhibit digestion.[542] They eat soil that is rich in sodium.[635]

Life History Weaning: 16mo.[346] **Sexual maturity:** NA. **Estrus cycle:** NA. **Gestation:** 195d.[346] **Age 1st birth:** NA. **Birth interval:** NA. **Life span:** NA. During estrus the female's clitoris protrudes.[640]

Locomotion Quadrupedal; leaping.

Social Structure Multimale-multifemale groups[639] with up to 3 males in a group. **Group size:** 15[541] (2–30).[898] **Home range:** 59.5ha.[541] **Day range:** 459m *[1506ft]*.[804]

Behavior Diurnal and arboreal. Black colobuses spend more than 60% of the day resting and only 22% feeding. They travel most in the afternoon to new feeding sites. **Association:** This species associates with guenons, which during one association were observed to deter an attack by a crowned hawk eagle *(Stephanoaetus coronatus)*.[283] Enhanced predator detection and avoidance are probably one of the main reasons that different species travel together. **Vocalizations:** The black colobus does not have a daily territorial vocalization.[541] Its loud call is higher pitched and pulses at a faster rate than that of other species.[640]

Equatorial Guinea, Bioko, Cameroon

Geoffroy's or White-thighed Black-and-white Colobus *Colobus vellerosus*

Taxonomy Disputed. Elevated from a subspecies of *C. polykomos* in 1983.[310] *C. polykomos dollmani* is a hybrid of *C. vellerosus* and *C. polykomos* that formerly occurred in an area between the Sassandra and Bandama rivers. That area is now inundated by a dam's reservoir, and the population may be extinct.[310]

Distinguishing Characteristics Geoffroy's black-and-white colobus is mostly black, with a bushy white facial fringe, a white stripe on the thigh, and a white tail. It has a black crown and no epaulettes.[602]

Physical Characteristics Head and body length: ♀ 610–660mm *[24–26in]*, ♂ 610–641 *[24–25.2in]*.[317] **Tail length:** 754–816mm *[29.7–32.1in]*.[640] **Weight:** ♀ 8.3kg *[18.3lb]*, ♂ 9.9kg *[21.8lb]*.[639] **Intermembral index:** NA. **Adult brain weight:** NA. This species has the largest larynx of all *Colobus* monkeys.[640]

Habitat Lowland moist, deciduous gallery forest and savanna.[635]

Diet Young leaves, mature leaves, fruit pulp, seeds. In the dry season, fruit and seeds are eaten in a smaller part of the yearly range.[635]

Life History NA.

Locomotion Quadrupedal; leaping.

Social Structure Multimale-multifemale groups with up to 4 males and an average of 6.5 females.[635] **Group size:** 16.[635] **Home range:** 48.[635] **Day range:** 307m (75–752) *[1007ft (246–2467)]*.[635]

Behavior Diurnal and arboreal. Geoffroy's black-and-white colobus travels on the ground between forest patches when it lives in a savanna habitat.[635]

Vocalizations: The male loud call is given to intimidate predators when they are sighted and as a response to the loud calls of other males.[640] It is not used during territorial confrontations.[635] The loud call is said to be most similar to that of the Abyssinian black-and-white colobus *(C. guereza)*.[640]

J. OATES (NATURAL HABITAT)

Ivory Coast to Nigeria

Western Red Colobus *Procolobus (Piliocolobus) badius*

Taxonomy Disputed. 3 subspecies. *P. b. temminckii* is included with this species.[635] All 14 subspecies of red colobus used to be included in *Colobus badius*.[602] Groves recognizes 4 separate species.[82] Other taxonomists use a different genus name, *Piliocolobus,* for the red colobuses.[602]

Distinguishing Characteristics The western red colobus is orange red to whitish orange except for its head, back, and tail, which are gray to blackish. The pubic area is white.[80]

Physical Characteristics Head and body length: ♀ 528mm (470–584) *[20.8in (18.5–23.0)]*, ♂ 569mm (470–690) *[22.4in (18.5–27.2)]*.[602] **Tail length:** ♀ 665mm (515–750) *[26.2in (20.3–29.5)]*, ♂ 665mm (540–800) *[26.2in (21.3–31.5)]*.[602] **Weight:** ♀ 8.2kg *[18.1lb]*, ♂ 8.3kg *[18.3lb]*.[639] **Intermembral index:**

87.[234] **Adult brain weight:** 73.8g.[346]

Habitat Primary and secondary rain forest,[898] dry deciduous forest, savanna woodland, and gallery forest. Rain forest habitats that provide young growth throughout the year are preferred.[805]

Diet Leaves, shoots, flowers, fruit. Red colobuses are selective feeders.[124] When feeding on mature leaves, they eat only the leaf stem or tip, depending on the tree species.[124] They eat parts of 54 plant species.[124]

Life History Weaning: NA. **Sexual maturity:** NA. **Estrus cycle:** NA. **Gestation:** 198d.[635] **Age 1st birth:** NA. **Birth interval:** 17.3mo.[346] **Life span:** NA. **Birth season:** Dry season in Gambia.[635]

Locomotion Quadrupedal; leaping.

Social Structure Multimale-multifemale groups with a ratio of

NATURAL HABITAT

Female and infant.

Western Red Colobus *continued*

NATURAL HABITAT

1 male to 1.5–3 females.[778] **Emigration:** Females leave their natal troop.[778] **Group size:** 19–80.[805] **Home range:** 35ha.[805] **Day range:** 649m (223–1185) [2129ft (732–3888)].[805]

Behavior Diurnal and arboreal. The western red colobus in Gambia lives in groups in which the dominant male is alpha for only 1 breeding season.[778] Females groom more with females, and favorite grooming partners transfer from their natal troop to new groups.[778] In Gambia, females chase away alien males. Two females and a male were observed killing an alien male after an infant disappeared. Infanticide was suspected.[778] Activity budget: Feeding, 44.5%; rest, 34.8%; travel, 9.2%.[805]

Association: The red colobus associates with white-collared mangabeys (*Cercocebus torqua-* *tus*)[339] and, in Sierra Leone, with king black-and-white colobuses (*Colobus polykomos*). In the Taï National Park in Ivory Coast, *P. badius* associates with Diana monkeys (*Cercopithecus diana*) more during the season when chimpanzees hunt. These associations are believed to help prevent predation by detecting predators more quickly. Diana monkeys are more alert than red colobuses, and there is very little dietary competition between the 2 species.[623] **Vocalizations:** Females give a quavering call during mating.[635]

Gambia to Ghana

Pennant's Red Colobus *Procolobus (Piliocolobus) pennantii*

Taxonomy Disputed. 9 subspecies. Includes *P. p. gordonorum, P. p. kirkii, P. p. tephrosceles,* and *P. p. tholloni*. Some recognize *P. p. kirkii* and *P. p. tholloni* as species.

Distinguishing Characteristics The coloration of Pennant's red colobuses varies with the subspecies. Generally they have a blackish red crown, with dark brown back, feet, and tail. The rump, forearms, and legs are a light fawn color.[602]

Physical Characteristics Head and body length: ♀ 530mm *[20.9in]*, ♂ 630mm *[24.8in]*.[795] **Tail length:** ♀ 605mm *[23.8in]*, ♂ 700mm *[27.6in]*.[795] **Weight:** ♀ 7.0kg *[15.4lb]*, ♂ 10.5kg *[23.1lb]*.[639] **Intermembral index:** NA. **Adult brain**

weight: NA.

Habitat Primary and secondary rain forest, *Acacia* woodlands, and swamp, scrub, gallery, and montane forest up to 2000m *[6562ft]*.[898]

Diet Young leaves, 51%; mature leaves, 21%; flowers, 12%; fruit, 6%.[124] Nutritional analysis has shown that the young leaves of one of the chief foods, *Celtis durandii,* have twice as much protein as the mature leaves.[795]

Life History Weaning: NA. **Sexual maturity:** ♀ 38–46mo, ♂ 46.5mo.[804] **Estrus cycle:** NA. **Gestation:** NA. **Age 1st birth:** NA. **Birth interval:** 25.5mo.[804] **Life span:** NA. **Births:** Year-round.[804] **Birth peaks:** Apr–Jun, Nov.[804] Females have a perineal swelling during

T. STRUHSAKER (NATURAL HABITAT)

Kirk's red colobus *(P. p. kirkii)* is considered a valid species by some taxonomists.

estrus.[804]

Locomotion Quadrupedal. This species reportedly makes "suicidal leaps."[795]

Social Structure Multimale-multifemale. Groups are patrilineal, with twice as many females (6–21) as males (3–10).[582] **Emigration:** Females emigrate.[795] **Group size:** 20[903] (5–57).[804] **Home range:** 35.3ha (177.7–53).[795] **Day range:** 557m[804] (398–958)[795] *[1828ft (1306–3141)]*.

NATURAL HABITAT

A leaping red colobus *(P. p. tephrosceles)*.

Pennant's Red Colobus *continued*

Behavior Diurnal and arboreal. This red colobus feeds in the canopy and emergent trees.[635] Females invest more parental care in sons, which remain in the troop and cooperate in group defense, than in daughters, which emigrate.[806] A male red colobus was seen killing an infant that was born within his natal group.[803] Crowned hawk eagles *(Stephanoaetus coronatus)* prey on infants, juveniles, and adults.[801] Activity budget: Feeding, 45%; rest, 35%; travel, 9%; grooming, 5%; play, 3%; other, 4%.[772] **Association:** These colobuses associate with blue monkeys *(Cercopithecus mitis)* 40% of the time and rarely with

These Kirk's red colobus monkeys are eating charcoal from a burned stump. It may help them detoxify the secondary compounds found in their leafy diet.

T. STRUHSAKER (NATURAL HABITAT)

NATURAL HABITAT

Red colobuses *(P. p. tephrosceles)* have learned to eat parts of eucalyptus trees, which have been imported to Africa from Australia.

Abyssinian black-and-white colobuses *(Colobus guereza).*[802] They also associate and groom with red-tailed guenons *(Cercopithecus ascanius)* and gray-cheeked mangabeys *(Lophocebus albigena).* **Mating:** Matings are multiple-mount.[169] **Vocalizations:** 12.[347] The most notable calls are the group spacing call, described as *nyow,* and an alarm call described as *chist.* A rapid quavering call is given by a mating pair that is being harassed.[635] **Sleeping site:** Tall emergent trees.[635]

Congo Republic, Zaire, Uganda, Tanzania, Zanzibar (Tanzania)

Preuss's Red Colobus *Procolobus (Piliocolobus) preussi*

Taxonomy Disputed. Monotypic.[80] Elevated from a subspecies of *P. badius.* [315] A population that was discovered in the Niger Delta in 1993 may be described as a new subspecies.[674]

Distinguishing Characteristics Preuss's red colobus has a black head and back. The back is ticked with orange; the tail, limbs, and cheeks are bright red;[602] and the underparts are a whitish orange.[80] The Niger Delta subspecies has white on its arms, shoulders, and cheeks.[711]

Physical Characteristics Head and body length: NA. **Tail length:** NA. **Weight:** NA. **Intermembral index:** NA. **Adult brain weight:** NA.

Habitat Tall primary forests near water; not in second growth forests.[898]

Diet Young leaves, mature leaves, fruit, buds.[80]

NATURAL HABITAT

Life History NA. Birth season: May–Jun.[602] Females have the largest sexual swelling of all red colobuses. It is pink and appears to be one-fourth of the female's body volume.[602]

Locomotion Quadrupedal; leaping.

Social Structure NA. **Group size:** 47 (24–80).[80] **Home range:** NA. **Day range:** NA.

Behavior: Diurnal and arboreal. Preuss's red colobus has not been studied thoroughly in the field. **Mating:** Females have a quavering copulation call that is given before, during, and after mating.[635]

Cameroon

ENDANGERED

USESA

Tana River Red Colobus *Procolobus (Piliocolobus) rufomitratus*

ENDANGERED

USESA

Taxonomy Disputed. Monotypic. Elevated from a subspecies of *P. badius*.[315] The cranium is very distinct.[82]

Distinguishing Characteristics
The Tana River red colobus has a dark gray back and tail. The cheeks and limbs are lighter gray, and the underparts are yellowish white.[80] The crown of the head is light red with a black-tipped crest.[161]

Physical Characteristics Head and body length: NA. **Tail length:** NA. **Weight:** NA. **Intermembral index:** NA. **Adult brain weight:** NA.

Habitat Evergreen gallery forest.[161]

Diet Young leaves, 52%; fruit, 25%; mature leaves, 11%; flowers, 6%. There is not as much diversity in the tree species of the Tana River forest. Of the 22 species used, 92% of the diet comes from 10.[524]

Life History NA.

Locomotion Quadrupedal; leaping.

Social Structure Variable. 1 male–multifemale, [635] multimale-multifemale.[161] **Group**

L. LELAND (NATURAL HABITAT)

size: 18–25.[804] **Home range:** 9.5–10.25ha.[804] **Day range:** 603m *[1978ft]*.[804]

Behavior Diurnal and arboreal. The Tana River red colobus adapts in several ways to the food supply reduction caused by drought and habitat fragmentation. It eats less preferred foods such as mature leaves and rotates food choices, using no more than

10 species at any one time. It moves less and visits the lesser-used parts of its range.[161] This red colobus uses only the closed canopy of the interior forest. This habitat is becoming increasingly restricted by forest fragmentation, which opens the canopy near the forest edge.[161]

Kenya

Olive Colobus *Procolobus (Procolobus) verus*

RARE

IUCN

Taxonomy Monotypic.[602]

Distinguishing Characteristics
The olive colobus, the smallest of the colobines, has an olive gray brown back and white underparts.[641] Infants are darker than adults.[635]

Physical Characteristics Head and body length: ♀ 466mm (380–546) *[18.3in (15.0–21.5)]*, ♂ 480mm (455–510) *[18.9in (17.9–20.1)]*.[602] **Tail length:** ♀ 573mm (430–710) *[22.6in (16.9–28.0)]*, ♂ 559mm (500–610) *[22.0in (19.7–24.0)]*.[602] **Weight:** ♀ 4.2kg *[9.3lb]*, ♂ 4.7kg *[10.4lb]*.[639] **Intermembral index:** 80.1.[425] **Adult brain weight:** 57.8g *[2.0oz]*.[346]

Habitat Swamps, rain forest, and dry semideciduous forest.[898]

Diet Young leaves, 63%; seeds, 14%; mature leaves, 10%; flowers, 7%; fruit, 3%.[469] At one study site this colobus uses more than 50 plant species but does not feed on the 5 most common trees, which form almost half of the forest biomass. The high fiber content of those leaves makes them undesirable.[632]

Life History Weaning: NA. **Sexual maturity:** NA. **Estrus cycle:** NA. **Gestation:** 150–180d.[804] **Age 1st birth:** NA. **Birth interval:** NA. **Life span:** NA. **Mating season:** Apr–Aug.[635] **Birth season:** Nov–Feb.[804] Females have perineal swellings during estrus.[804]

Locomotion Quadrupedal; leaping.

Social Structure Variable. 1 male–multifemale, 2 male–multifemale.[602] **Emigration:** Females may emigrate.[635] **Group size:** 8 (5[641]–20[898]). **Home range:** 28.5ha.[641] **Day range:** NA.

Behavior Diurnal and arboreal. The olive colobus is a cryptic species that prefers feeding sites in thick low growth and tangles below 15m *[49ft]*.[641] Its small body size makes it potentially more vulnerable to predators, including leopards, chimpanzees, and crowned hawk eagles. Its drab color provides camouflage, which, along with its cryptic behavior, helps protect it.[632] This is the only anthropoid monkey to carry infants by mouth.[641] **Association:** The olive colobus is almost

S. McGRAW (NATURAL HABITAT)

The female is carrying her infant in her mouth. This is the only Old World monkey known to do so.

always within 50 meters *[164ft]* of another primate species, including the king black-and-white colobus *(Colobus polykomos)*.[641] It associates especially with large dispersed groups of Diana monkeys *(Cercopithecus diana)*, which do not compete with it for food but stay in the upper canopy, where they are likely to spot predators and give an alarm call. This warning system enables the olives to forage over a larger area.[641] One olive colobus group maintained an association with Diana monkeys 80% of the time for 3 years.[641] When traveling with lesser spot-nosed

Olive Colobus *continued*

NATURAL HABITAT

Olive colobuses are the smallest colobine monkeys.

guenons *(C. petaurista),* which are smaller and have smaller groups, the olives are quiet and more cryptic and prefer tangles.[641] **Vocalizations:** The olive colobus makes relatively few vocalizations and is often inactive, making it difficult to locate in the forest.[898] During territorial interactions, adult males make loud "laughing" calls.

Sierra Leone to
Togo, Nigeria

Grizzled Leaf Monkey *Presbytis comata*

ENDANGERED

IUCN

E. THETFORD (DIGITAL ILLUSTRATION)

Taxonomy Disputed. 2 subspecies. The name was changed from *P. aygula* in 1983.[315] *P. c. fredericae* has been proposed as a separate species.[82]

Distinguishing Characteristics Grizzled leaf monkeys have a gray back, white underparts, and a black head and crest.[602]

Physical Characteristics Head and body length: ♀ 535mm (475–570) *[21.1in (18.7–22.4)]*, ♂ 505mm (430–595) *[19.9in (16.9–23.4)]*.[602] **Tail length:** ♀ 659mm (590–720) *[25.9in (23.2–28.3)]*, ♂ 655mm (560–724) *[25.8in (22.0–28.5)]*.[602] **Weight:** ♀ 6671g *[14.7lb]*, ♂ 6396g *[14.1lb]*.[234] **Intermembral index:** 76.[234] **Adult brain weight:** 80.3g *[2.8oz]*.[346]

Habitat Montane and submontane forest at 1200–1800m *[3937–5906ft]* elevation.[719]

Diet Young leaves, 59%; fruit, 14%; flowers, 7%; mature leaves, 6%; seeds, 1%;[773] branch tips, fungi, pseudobulbs,[46] soil.[719] These monkeys use 74 plant species, including 28 for leaves[677] and 36 for fruit.[719]

Life History Infant: 12mo.[719] **Juvenile:** 18mo.[719] **Sexual maturity:** NA. **Estrus cycle:** NA. **Gestation:** NA. **Age 1st birth:** NA. **Birth interval:** NA. **Life span:** NA.

Locomotion Quadrupedal; leaping.

Social Structure 1 male–multifemale, but monogamous pairs have been observed at 1 site.[46] Adult males are dominant. Groups are territorial and very cohesive.[719] **Group size:** 7 (3–12).[46] **Home range:** 12–22ha,[46] to 40.[719] **Day range:** 400–600m *[1312–1969ft]*.[319]

Behavior Diurnal and arboreal. Grizzled leaf monkeys use the middle and upper layers of the canopy.[677] They occasionally come to the ground to eat reddish soil.[719] Intertroop encounters can last up to 50 minutes, during which males display by calling, running, and leaping. They have not been observed to physically fight.[719] Male juveniles play more often with one another than female juveniles play with one another.[719] Activity budget: Rest, 60%; feeding, 30%; travel, 5%.[719] Allomothering has been observed. Fewer than 3000 of these langurs remain.

Vocalizations: 11. The most notable is the male loud call *kik*.[719] **Sleeping site:** Trees at least 20 meters *[66ft]* above the ground.[719]

Java (Indonesia)

R. & J. SEITRE (NATURAL HABITAT)

P. c. fredericae.

Banded Leaf Monkey *Presbytis femoralis*

P. f. robinsoni.

R. WIRTH

P. f. natunae.

Taxonomy Disputed. 11 subspecies were separated from the 17 subspecies of *P. melalophos* in 1977.[602] It has been proposed that this species be divided into *P. femoralis* with 5 subspecies and *P. siamensis* with 6.[82]

Distinguishing Characteristics Banded leaf monkeys have a dark brown back and gray, brown, or white underparts. The color pattern on the head varies, with differing amount of gray, black, and white.[543]

Physical Characteristics Head and body length: 432–610mm *[17.0–24.0in].*[80] **Tail length:** 610–838mm *[24.0–33.0in].*[80] **Weight:** 5897–8165g *[13.0–18.0lb].*[80] **Intermembral index:** NA. **Adult brain weight:** NA.

Habitat Primary and secondary rain forest, swamp and mangrove forest,[80] rubber plantations.[898]

Diet Fruit, 49% (but not ripe figs in Malaysia).[152] These monkeys are reported to raid crops occasionally.[898]

Life History NA.

Locomotion Quadrupedal; leaping.

Social Structure NA. **Group size:** 10 (6–20).[898] **Home range:** 9–21ha.[898] **Day range:** NA.

Behavior Diurnal and arboreal. Banded leaf monkeys have not been well studied in their natural habitat. The Natuna Island subspecies is threatened by offshore gas production and defense facilities.[883] **Vocalizations:** The adult male's loud call has 2 phrases:[6] "a harsh dull trill or purring croak followed by a loud and distinctive *chak, chak, chak, chak.*[543] The alarm call has been compared to the sound of a machine gun.[506]

Malay Peninsula,
Singapore, Sumatra,
Natuna Islands
(Indonesia), Borneo

179

White-fronted Leaf Monkey *Presbytis frontata*

Taxonomy Disputed. Possibly 2 subspecies.[82]

Distinguishing Characteristics White-fronted leaf monkeys have a dark gray brown body. The head is black with a crest.[602] The common name comes from a white patch of naked skin on the forehead.[543]

Physical Characteristics Head and body length: ♀ 545mm (525–570) *[21.5in (20.7–22.4)]*, ♂ 600mm *[23.6in]*.[602] **Tail length:** ♀ 726mm *[28.6in]*, ♂ 740mm *[29.1in]*.[602] **Weight:** ♀ 5660g *[12.5lb]*, ♂ 5600g *[12.3lb]*.[490] **Intermembral index:** 76.[234] **Adult brain weight:** NA.

Habitat Lowland dipterocarp forest[654] below 300m *[984ft]*.[543]

Diet NA.

Life History NA.

Locomotion Quadrupedal; leaping.

Social Structure NA. **Group size:** NA. **Home range:** NA. **Day range:** NA.

Behavior Diurnal and arboreal. The white-fronted leaf monkey reportedly is "rarely seen and occurs at low population densities or is nomadic."[654] It has never been studied in its natural habitat. In Sarawak, Malaysia, this species, along with other leaf monkeys, is hunted for its gallstones. Known as bezoar stones, they are sought for magical and medicinal purposes and as a pigment. The monkeys must be killed and dissected to find the gallstones, which not all individuals have.[898]

S. NASH

Borneo

Hose's Leaf Monkey *Presbytis hosei*

P. RODMAN (NATURAL HABITAT)

Taxonomy Disputed. 3 subspecies. Separated from *P. aygula* (now called *P. comata*) in 1958.[82] Some taxonomists still include this species as a subspecies of *P. comata*.[301]

Distinguishing Characteristics Hose's leaf monkey has a gray back, white underparts, and blackish hands and feet.[654] The color of the head varies with subspecies. Females of one subspecies have a darker head than males. Infants are white, with a darker spinal stripe.[543]

Physical Characteristics Head and body length: 480–557mm *[18.9–21.9in]*.[654] **Tail length:** 646–840mm *[25.4–33.1in]*.[654] **Weight:** ♂ 6200g *[13.7lb]*,[639] ♀ 5570g *[12.3lb]*.[234] **Intermembral index:** 75.[234] **Adult brain weight:** NA.

Habitat Lowland and mountainous tropical rain forest[654] up to 4000m *[13,124ft]*.[543]

Diet Leaves, 65%; seeds and unripe fruit, 30%; flowers, 2%;[568] animal prey, including bird eggs and nestlings.[301] These monkeys eat 30–47 plant species.

Life History NA.

Locomotion Quadrupedal; leaping.

Social Structure 1 male–multifemale groups. Lone individuals have also been observed.[46] **Emigration:** Males emigrate.[46] **Group size:** 7 (6–12).[654] **Home range:** 34ha.[568] **Day range:** NA.

Behavior Diurnal and arboreal. Female Hose's leaf monkeys groom other females more than they groom the male, which has been observed to be separated from the group.[568] This species is reportedly more likely to hide rather than flee from human hunters.[543] Like several other leaf monkeys, it is hunted by humans for its gallstones (bezoar stones). **Association:** Sympatric with and reported to associate with maroon leaf monkeys *(P. rubicunda)*.[543] It is unusual for Asian colobine monkeys to share the same habitat. **Vocalizations:** Males give a "gurgle growl" call at dawn.[568] The alarm call is said to be "grunt-like."[301]

Borneo

Mitered Leaf Monkey *Presbytis melalophos*

Alarm calling *(P. m. melalophos).*

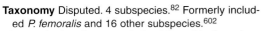

Sumatra

Mitered leaf monkey and infant *(P. m. mitrata).*

LOWER RISK

CITES II

E. THETFORD

Taxonomy Disputed. 4 subspecies.[82] Formerly included *P. femoralis* and 16 other subspecies.[602]

Distinguishing Characteristics Mitered leaf monkeys
have a pointed crest on the crown of the head. The
coat color varies with the subspecies, from white to
gray to reddish orange.[602] Neonates are white, with a
dark stripe down the back and across the
shoulders.[602]

Physical Characteristics Head and body length:
♀ 494mm (420–555) *[19.4in (16.5–21.9)]*, ♂ 492mm
(423–565) *[19.4in (16.7–22.2)]*.[602] **Tail length:**
♀ 711mm (636–821) *[28.0in (25.0–32.3)]*, ♂ 711mm
(618–824) *[28.0in (24.3–32.4)]*.[602] **Weight:** ♀ 5.8kg
[12.8lb], ♂ 5.9kg *[13.0lb]*.[639] **Intermembral index:**
78.[234] **Adult brain weight:** 80g *[2.8oz]*.[346]

Habitat Primary lowland rain forest, selectively logged
forest, plantations.[152]

Diet Young leaves, 28%; seeds, 25%; fruit, 24%; flowers, 12%; mature leaves, 8%.[469] These monkeys
avoid sweet succulent fruits like figs, preferring dry
unripe bitter fruit. Ripe succulent fruit can acidify the
neutral pH of the forestomach, destroying some of the
beneficial bacteria that help leaf monkeys digest food.
They eat parts of 55 plant species.[152]

Life History NA.

Locomotion Leaping, 42.5%; hopping, 25.0%;
quadrupedal walking and running, 20.7%; climbing,
8.4%; arm swinging, 3.4%. Mitered leaf monkeys use
smaller supports than other leaf monkeys.[233]

Social Structure Variable. 1 male–multifemale,
multimale-multifemale.[46] Territories overlap, but no
territorial defense has been observed. Males defend

their mates, not the resources.[46] **Emigration:** Males
and some females emigrate.[46] **Group size:** 15
(12–18).[46] **Home range:** 14–29.5ha.[46] **Day range:**
300–1360m *[984–4462ft]*.[45]

Behavior Diurnal and arboreal. Mitered leaf monkeys
feed at all forest levels[46] but more often in
understory.[233] When feeding on high-nutrient fruits,
seeds, or flowers rather than leaves, they must travel
farther in a day and are more likely to encounter
other troops looking for the same resources.
Allomothering is rare.[46] The response of these monkeys to selective logging has been
extensively studied. They have
adapted by changing their home
range size and position.[417] They are
forced to use lower canopy levels
because the large trees were cut.
They split into small groups, eat
more leaves, travel less, and rest
more.[417] Logging caused high levels of infant mortality at the study
site while it lasted.[417] **Vocalizations:**
Adult males give a distinctive low
kakakakaka call around dusk and
dawn and at intervals through the
night. The territorial call is usually
given at 5–6 a.m. and
6–9 p.m.[387]

P. m. nobilis.

E. THETFORD

Mentawai Island Leaf Monkey *Presbytis potenziani*

Taxonomy 2 subspecies.[82]

Distinguishing Characteristics Mentawai Island leaf monkeys have a black back and a black tail. The cheeks, brow band, chin, and throat are white.[543] The underparts are light orange, reddish orange, or brown.[80] The genital region is yellowish white, and males have a white scrotum.[840] Neonates are whitish gray, with a dark stripe down the back.[840]

Physical Characteristics Head and body length: ♂ 500mm *[19.7in]*.[772] **Tail length:** ♂ 580mm *[22.8in]*.[772] **Weight:** ♀ 6.4kg *[14.1lb]*, ♂ 6.5kg *[14.3lb]*.[639] **Intermembral index:** 81.[309] **Adult brain weight:** NA.

Habitat Primary and secondary evergreen rain forest; swamp, logged, and mangrove forest.[898]

Diet Leaves, 55%; fruit and seeds, 32%; other, 14%. These monkeys are one of the few primates that can eat the leaves of the dipterocarp tree family.[46]

Life History Infant: 0–12mo.[973] **Juvenile:** 12–36mo.[973] **Subadult:** 36–60mo.[973] **Sexual maturity:** NA. **Estrus cycle:** NA. **Gestation:** NA. **Age 1st birth:** NA. **Birth interval:** NA. **Life span:** NA.

Locomotion Quadrupedal; leaping.

Social Structure Monogamous 1 male–1 female groups.[867] This species is the only monogamous colobine monkey.[254] It lives in highly cohesive groups and is territorial,[46] though some groups have a 30% range overlap.[975] **Emigration:** Males emigrate; females may leave their troop at adulthood.[973] **Group size:** 2–6.[46] **Home range:** 11.5–40ha.[975] **Day range:** 540m (60–1120) *[1772ft (197–3675)]*.[975]

Behavior Mentawai Island leaf monkeys have a low frequency of social interaction.[973] Groups usually maintain their distance with spacing calls. During intertroop encounters there are female–female agonistic threat displays. On the island of Siberut when potential predators are seen, the

M. OLSON (DIGITALLY COMPOSITED)

male performs a distraction display that includes leaping and branch bouncing while the female and offspring escape. **Association:** These leaf monkeys often associate with pig-tailed langurs *(Nasalis concolor)* and pig-tailed macaques *(Macaca nemestrina pagensis)*. They are displaced from food sources by Kloss's gibbons *(Hylobates klossii)*. They avoid feeding competition with gibbons by leaving their sleeping tree before sunrise to arrive at food trees earlier.[602] **Vocalizations:** A territorial call is described as a

loud repetitive rapid *bagok*.[840] The female's alarm calls are high-pitched shrill meows.[973] **Sleeping site:** Tall emergent trees with many liana vines. Human hunters on Siberut climb the vines at night to shoot these monkeys.[823]

Mentawai Islands
(Indonesia)

Maroon Leaf Monkey *Presbytis rubicunda*

C. MARSH (NATURAL HABITAT)

Taxonomy 5 subspecies.[82]

Distinguishing Characteristics The coloration of maroon leaf monkeys varies with the subspecies. In general they are dark maroon[898] or blackish red to red orange.[80] They have variations of brown or black on the hands and feet.[80]

Physical Characteristics Head and body length: ♀ 495mm (410–500) *[19.5in (16.1–19.7)]*, ♂ 570mm *[22.4in]*.[602] **Tail length:** ♀ 702mm (675–725) *[27.6in (26.6–28.5)]*, ♂ 730mm *[28.7in]*.[602] **Weight:** ♀ 5700g *[12.6lb]*, ♂ 6211g *[13.7lb]*.[639] **Intermembral index:** 76.[234] **Adult brain weight:** 92.7g *[3.3oz]*.[346]

Habitat Tropical rain forest.[639]

Diet Young leaves, 36%; seeds, 30%; fruits, 19%; flowers, 11%; mature leaves and petioles, 1%; occasionally termites. Maroon

leaf monkeys are highly selective feeders. Only 6.5% of the diet is from the most common trees (82%) in the forest. These monkeys prefer to eat young leaves and seeds from large fleshy fruits without bright colors.[151] They are seed predators on all but the tiniest seeds.[151] They use 103 plant species in total and eat soil from termite mounds. Their foregut fermentation is able to digest cellulose and carbohydrates, as well as to detoxify unpalatable alkaloid chemicals.[152]

Life History NA.

Locomotion Quadrupedal.

Social Structure 1 male–multifemale groups. The male defends his females, not the resources.[853] Home ranges overlap 10%.[151] **Emigration:** Both genders emigrate.[46] **Group size:** 7 (3–10).[772] **Home range:** 37–

Maroon Leaf Monkey *continued*

84ha.[151] **Day range:** 746–846m *[2448–2776ft]*.[853]
Behavior Diurnal and arboreal. Maroon leaf monkey groups split into subgroups to forage.[46] When a new male invaded a 1-male group, the group was observed to divide into 2 separate groups. The 2 females that had infants stayed with the resident male. The 2 females without infants went with the new

male.[46] This observation supports the hypothesis that the threat of infanticide leads males and females to remain close.[853] **Association:** Sympatric and reported to associate with Hose's leaf monkeys *(P. hosei)*.[543]

Borneo, Karimata Islands (Indonesia)

Thomas's Leaf Monkey *Presbytis thomasi*

Taxonomy Disputed. 2 subspecies. Elevated from a subspecies of *P. comata* in 1942.[82]

Distinguishing Characteristics Thomas's leaf monkeys have a gray back, black hands and feet, white underparts, and a black crest with 2 white stripes above the eyes.[602] The tail is gray above and whitish below.[602]

Physical Characteristics Head and body length: ♀ 550mm *[21.7in]*, ♂ 550mm *[21.7in]*.[602] **Tail length:** ♀ 760mm *[29.9in]*, ♂ 690mm *[27.2in]*.[602] **Weight:** NA. **Intermembral index:** NA. **Adult brain weight:** NA.

Habitat Primary and secondary rain forest, rubber and fruit plantations.[318]

Diet Fruit and seeds, 58%; leaves, 32%; flowers, 8%.[318] These monkeys eat 20 species of fruit and seeds, 24 species of leaves, and 10 species of flowers.[469]

Life History NA. Births: Year-round.[318]

Locomotion Quadrupedal.

Social Structure 1 male–multifemale groups.[781] The male defends his females, not the resources.[853] **Emigration:** Males emigrate.[46] All-male groups (up to 10 members) have a small range (only 1.7ha).[318] **Group size:** 6 (3–21).[46] **Home range:** 12.3–15.7,[318] 37.7ha.[853] **Day range:** 640m (150–1300) *[2100ft (492–4265)]*.[781]

Behavior Diurnal and arboreal. All-male groups of Thomas's leaf monkeys have a dominance hierarchy that is maintained by

NATURAL HABITAT

embracing, mounting and grooming.[319] Researchers have observed that when an alpha male disappeared from his group, the females stayed in their home range. It took 4 months for a new male to be accepted by the females.[781] Intertroop encounters happen where territories of both troops overlap. Although these agonistic interactions are mostly vocal, physical fights do occur.[318] A resident male was killed during a fight with another male that subsequently took over the troop.[318] Infanticide was observed during another aggressive encounter in the wild.[318] Activity budget: Rest, >50%; feeding, 25–40%; travel, <10%.[318]

Vocalizations: 13.[719] Alpha males use their loud call during intertroop encounters. The loud call is described as "loud wheezing cackles followed by 4–8 deep resonant nasal sounds, 'ngkung' and a boom cough sound."[318]

Sleeping site: Each troop has 6–9 sleeping sites.[318]

Sumatra

LOWER RISK

CITES II

Hanuman Langur *Semnopithecus entellus*

Adult male.

NATURAL HABITAT

Taxonomy Disputed. 15 subspecies. The genus *Semnopithecus* includes only 1 species. Some taxonomists consider this genus to be a subgenus of *Presbytis*.[315] Others include all species of the genus *Trachypithecus* in *Semnopithecus*.[82]

Distinguishing Characteristics The Hanuman langur's coloration varies across subspecies, from gray to dark brown to golden, with varying amounts of black.[602] The subspecies from the northern part of the range have larger bodies than those from the south. The subspecies from Sri Lanka is the smallest.[602]

Physical Characteristics Head and body length: ♀ 406–680mm *[16.0–26.8in]*, ♂ 510–780mm *[20.1–30.7in]*.[602] **Tail length:** ♀ 693–1016mm *[27.3–40.0in]*, ♂ 761–978mm *[30.0–38.5in]*.[602] **Weight:** ♀ 11.2kg (6.7–15.6) *[24.7lb (14.8–34.4)]*, ♂ 18.3kg (10.6–19.8) *[40.3lb (23.4–43.6)]*.[639] **Intermembral index:** 83.[234] **Adult brain weight:** 135.2g *[4.8oz]*.[346]

Habitat Tropical, subtropical, dry thorn scrub, pine, semievergreen, and alpine forest and urban areas; not evergreen forests.[46] These langurs live from sea level to 4267 meters *[14,000ft]* in altitude, the widest range for any primate other than humans.[898]

Diet Mature leaves, 35%; fruit, 24%; buds, 11%; flowers, 9.5%; young leaves, 4%; animal prey, 3%; exudates, bark, soil, herbs.[46] Hanuman langurs eat 53 plant species and more insects than any other colobine.[46] The Himalayan subspecies rely on pinecones, bark, and twigs during snowy winter months.[46] Troops living near temples rely on offerings

from humans. These langurs will raid gardens and crops.[898]

Life History Infant: 0–15mo.[683] **Weaning:** 13–20mo.[389] **Juvenile:** 15–48mo.[683] **Subadult:** 48–98mo.[683] **Sexual maturity:** 46.5–47mo.[804] **Estrus cycle:** 24.1d.[683] **Gestation:** 168[346]–200d.[683] **Age 1st birth:** 51mo.[346] **Birth interval:** 16.7mo.[683] **Life span:** 20y.[346] Births: Year-round except in extremely seasonal habitats.[804]

Locomotion Quadrupedal.

Social Structure Variable. 1 male–multifemale, multimale-multifemale.[683] Larger groups may break into subgroups in some seasons.[46] The male defends his mates; females defend their resources.[853]

Emigration: At age 4, males emigrate,[683] forming all-male bands. **Group size:** Range of means 11–64 for 262 groups;[855] entire range 5–100.[772] **Home range:** 50–130,[683] 200–1200ha.[46] **Day range:** 360m (60–1300) *[1181ft (197–4265)]*.

Behavior Diurnal, terrestrial, and arboreal. Hanuman langurs are the most terrestrial of any colobine. They spend up to 80% of the day on the ground, and almost all feeding is within 5 meters *[16ft]* of the ground.[46] In 1-male groups the alpha male's tenure is usually less than 2 years. When a new male takes over a troop, he systematically kills infants sired by the previous alpha male. This reproductive strategy was first documented in this species.[389] Juvenile males usually leave when a new male takes over the group.[537] In one study, 66% of the juvenile males were chased out and the rest simply left with the ousted male, which was probably their father.[683] Fewer than 50% of these juveniles survive after 2 years in an all-male band. Allomothering was first documented in this langur.[389] This is the sacred monkey of India,

184

Hanuman Langur *continued*

named after the Hindu monkey-god Hanuman.
Association: Sympatric with purple-faced leaf monkeys
(Trachypithecus vetulus).[46] Hanuman langurs may
associate with rhesus macaques *(Macaca mulatta);*
the two compete for human offerings around temples.
Hanuman langurs have hybridized with Nilgiri langurs
(Trachypithecus johnii).[898] **Vocalizations:** 16,[347] 6 of
which are made only by males.[375] The most notable
is the whoop call, which is a group spacing call.[389] In
the Himalayas these langurs have an "Au" contact
call. Infants have a specific play vocalization.[531]
Sleeping site: Trees and high places, including hotel
ledges.[711]

India, Pakistan,
Bangladesh,
Sri Lanka, Burma

This langur species, which is named after the Hindu god Hanuman, is often fed at temples.

Ebony Langur *Trachypithecus (Trachypithecus) auratus*

Taxonomy Disputed. 3 subspecies.[82] Elevated from a subspecies of *T. cristatus* in 1985.[309] Formerly in the genus *Presbytis*. Some taxonomists include this species in *Semnopithecus*.[82]

Distinguishing Characteristics Ebony langurs are black except for 1 population that is reddish brown. They have a slight crest and prominent cheek tufts.[130] The facial skin is bluish. Infants are orange.

Physical Characteristics Head and body length: 460–750mm *[18.1–29.5in]*.[130] **Tail length:** 610–820mm *[24.0–32.3in]*.[130] **Weight:** 7.1kg *[15.6lb]*.[467] **Intermembral index:** NA. **Adult brain weight:** NA.

Habitat Remnant primary and secondary forest; teak plantations.[178]

Diet Young leaves, 46%; ripe fruit, 27%; unripe fruit, 8%; flowers, 7%; mature leaves, 1%; insects, 1%.[469] The top 15 species of a total of 88 contribute 70% of the diet. In a study of 2 ebony langur groups, 66 plant species were eaten exclusively by 1 group but not the other. All groups prefer leaves that are lower in fiber and thus easier to digest.[467] When preferred foods are scarce, the diet is diversified to include more abundant but less nutritious foods.[469]

Life History NA.

Locomotion Quadrupedal.

Social Structure 1 male–multifemale groups. There is little overlap of territories.[46] **Emigration:** Males emigrate and travel alone or in an all-male band.[46] **Group size:** 6–23.[46] **Home range:** 5ha (2.5–8).[46] **Day range:** NA.

Behavior Diurnal and arboreal. Ebony langur infants are commonly allomothered by the group's females and juveniles.[46] These langurs have learned to eat the young leaves and the midribs of mature leaves of teak trees *(Tectona grandis)* when other food is unavailable.[469]

Java, Bali, Lombok
(Indonesia)

E. THETFORD

One population of ebony langurs has reddish brown as well as black individuals.

E. THETFORD

Silvered Langur *Trachypithecus (Trachypithecus) cristatus*

A juvenile may still suckle occasionally.

Taxonomy Disputed. 4 subspecies.[82] Formerly in the genus *Presbytis*. Some taxonomists include this species in *Semnopithecus*.[82]

Distinguishing Characteristics Silvered langurs are brownish gray to black, with grayish or yellowish hair tips that give them a silvered appearance. The groin and the underside of the tail are yellowish.[602] The crest on the crown of the head varies; some subspecies have more of a crest than others. Infants are born orange and gradually change to gray in 3 months.[654]

Physical Characteristics Head and body length:
♀ 489mm (460–514) *[19.3in (18.1–20.2)]*, ♂ 544mm (503–580) *[21.4in (19.8–22.8)]*.[602] **Tail length:**
♀ 715mm (678–751) *[28.1in (26.7–29.6)]*, ♂ 699mm

(671–750) *[27.5in (26.4–29.5)]*.[602]
Weight: ♀ 5.7kg *[12.6lb]*, ♂ 6.6kg *[14.5lb]*.[639] **Intermembral index:** 82.[234] **Adult brain weight:** 64g *[2.3oz]*.[346]

Habitat Primary and secondary, coastal, mangrove, and riverine forest; plantations. These langurs prefer lowlands below 458m *[1503ft]* but reportedly occur as high as 1737m *[5699ft]*.[898]

Diet Leaves, 80%; shoots, 10%; fruit (especially figs), 10%; soil. A total of 94 plant species are eaten. Silvered langurs have not been observed to eat any animal prey.[84]

Life History Infant: 18mo.[346]
Juvenile: 18–48mo.[346] **Subadult:** 36–48mo.[346] **Sexual maturity:** NA.
Estrus cycle: NA. **Gestation:** NA.
Age 1st birth: NA. **Birth interval:** 18mo.[346] **Life span:** NA.

Locomotion Quadrupedal; hopping in high grass.[51]

Social Structure 1 male–multifemale groups. The male defends his females, not the resources.[853] **Emigration:** All males and some females emigrate.[772] **Group size:** 15–28.[853]
Home range: 4.7[84]–20ha.[853] **Day range:** 350–530m *[1148–1739ft]*.[853]

Behavior Diurnal and arboreal.[654] The silvered langur male maintains his troop's unity and is vigilant for predators and other groups. In a captive study, the bright orange infants under 6 weeks old often spent more time away from their mothers than with them because they were being allomothered by females and juveniles in the troop. The adult male paid little attention to the infant.[711] He spends only 24% of

the day feeding; females spend 30.4%.[84] Infanticide was reported at a Malaysian study site.[895]
Mating: The female solicits mating by shaking her head at the male.[978] **Vocalizations:** Several, including a *snick* alarm call and a *kwah* travel call.[51] Infants have specific play vocalizations.[531]

Burma and Indochina to Borneo

Infants are born orange.

Occasionally infants are forcibly taken from their mothers to be allomothered by more-dominant females.

Infants gradually change to gray.

Delacour's Langur *Trachypithecus (Trachypithecus) delacouri*

Taxonomy Disputed. Elevated from a subspecies of *T. francoisi*.[82;317]

Distinguishing Characteristics Delacour's langurs are glossy black, with a white to cream-colored area on the outside of the thighs and across the lower back. The face is black, with white from the corners of the mouth to the ears. The hair on the tail is especially long on the middle part of the tail.[80] Infants are brownish and have more white on their head.[595]

Physical Characteristics Head and body length: 559–838mm *[22.0–33.0in]*.[80] **Tail length:** 813–864mm *[32.0–34.0in]*.[80] **Weight:** NA. **Intermembral index:** NA. **Adult brain weight:** NA.

Habitat Karst forest.[873]

Diet In captivity Delacour's langurs will eat leaves from 40 tree species. They prefer the leaves of trees that are associated with karst.[595]

Life History NA.

Locomotion Quadrupedal.

Social Structure NA. **Group size:** 3–6.[595] **Home range:** NA. **Day range:** NA.

Behavior Diurnal and arboreal. Delacour's langurs have been difficult to study in their natural habitat because they are rare and heavily hunted. They are critically endangered, with their population estimated to be fewer than 200 individuals. They are believed to exist in only 1 national park in Viet Nam. **Sleeping site:** Natural caves of limestone (karst) cliffs.[595]

Viet Nam

T. NADLER

Francois's Langur *Trachypithecus (Trachypithecus) francoisi*

Taxonomy Disputed. 5 subspecies. Formerly in the genus *Presbytis.* Some taxonomists include this species in *Semnopithecus*.[80] Four subspecies (*T. f. hatinhensis*, *T. f. laotum*, *T. f. leucocephalus*, and *T. f. poliocephalus*) may be recognized as full species after further study.[315]

Distinguishing Characteristics Francois's langurs are glossy black, with varying amounts of white or yellow on the head, body, and tail, depending on the subspecies.[602] They all have a pointed crest.[80]

Physical Characteristics Head and body length: ♀ 570mm (550–590) *[22.4in (21.7–23.2)]*, ♂ 548mm (485–635) *[21.6in (19.1–25.0)]*.[602] **Tail length:** ♀ 852mm (830–887) *[33.5in (32.7–34.9)]*, ♂ 849mm (820–872) *[33.4in (32.3–34.3)]*.[602] **Weight:** 5896g *[13.0lb]*.[80] **Intermembral index:** NA. **Adult brain weight:** NA.

Habitat Forests that grow on karst (limestone) formations.[976]

Diet NA.

Life History NA.

Locomotion Quadrupedal.

Social Structure 1 male–multifemale.[976] **Group size:** NA. **Home range:** NA. **Day range:** NA.

Francois's langurs are usually associated with limestone formations and often sleep in caves.

Francois's Langur *continued*

The white-capped black langur *(T. f. leucocephalus)* is found only in China.

An infant stripe-headed black langur *(T. f. laotum)*.

Behavior Diurnal, arboreal, and terrestrial.[976]
Francois's langurs are reported to travel mainly
between 7 and 10 a.m. and between 3 and 4 p.m.[976]
Allomothering has been reported in captivity.[531]
Activity budget: Rest, 69.1%; feeding, 13.8%; travel,
11.7%.[976] **Vocalizations:** These langurs have a loud
call for territorial spacing.[976] Infants have a specific
play vocalization.[531] **Sleeping site:** Caves in karst
cliffs. One group in China has 4 caves that it uses,
depending on weather, food, and perhaps other
groups.[997]

Viet Nam, Laos,
China

Infant Francois's langurs *(T. f. francoisi)*
gradually change to the black
coloration of adults.

Francois's langurs *(T. f. francoisi)* grooming.

189

Golden Langur *Trachypithecus (Trachypithecus) geei*

Taxonomy Disputed. Monotypic. Formerly in the genus *Presbytis*. Some taxonomists include this species in *Semnopithecus*,[82] others include it as a subspecies of *T. pileatus*.[315]

Distinguishing Characteristics Golden langurs are golden orange white, with semierect crown hairs. The coat color is reported to change seasonally from cream to chestnut.[772] Neonates are orange brown or gray.[46]

Physical Characteristics Head and body length: ♀ 490mm *[19.3in]*.[602] **Tail length:** ♀ 713mm *[28.1in]*.[602] **Weight:** ♀ 9.5kg *[20.9lb]*, ♂ 10.9kg *[24.0lb]*.[639] **Intermembral index:** NA. **Adult brain weight:** 81.3g *[2.9oz]*.[346]

Habitat Moist evergreen, dipterocarp, riverine, savanna, and deciduous forest up to 2400m *[7874ft]*.[898]

Diet Fruit, leaves, flowers. This species reportedly raids cardamom plantations.[898]

Life History NA. Mating season: Jan–Feb. Birth season: Jul–Aug.[588]

Locomotion Quadrupedal.

Social Structure 1 male–multifemale; occasionally 2 male–multifemale.[588] **Emigration:** Males emigrate. Lone males and all-male groups have been reported.[588] **Group size:** 16.7 (6–40).[898] **Home range:** 150–600ha.[898] **Day range:** NA.

Behavior Diurnal and arboreal. Golden langurs prefer the upper part of trees.[588] In the winter they have been observed to come to the ground to drink and to eat wet sand. They rest during midday in winter.[588]

Association: Golden langurs are displaced from feeding sites by rhesus macaques *(Macaca mulatta)* and Assamese macaques *(M. assamensis)*.[588]

Vocalizations: The most notable vocalization is a high-pitched *aeke-ke-aeke* alarm call. Unlike most other langur males, golden langur males do not have a loud territorial whoop call.[588]

India, Bhutan

G. CURBIT - BRUCE COLEMAN LTD. (NATURAL HABITAT)

Nilgiri Langur *Trachypithecus (Kasi) johnii*

J. OATES (NATURAL HABITAT)

India

Taxonomy Disputed. 2 subspecies.[82] Formerly in the genus *Presbytis*. Some taxonomists include this species in *Semnopithecus*. It has also been placed in the genus *Kasi*.[602]

Distinguishing Characteristics Nilgiri langurs are black, with white ticking that is more visible on the thighs. The crown, nape, and whiskers form a light brown hood.[602] A brown variety may be a hybrid with *Semnopithecus entellus*.[372]

Physical Characteristics Head and body length: ♂ 570mm (508–645) *[22.4in (20.0–25.4)]*.[602] **Tail length:** ♂ 860mm (756–965) *[33.9in (29.8–38.0)]*.[602] **Weight:** ♀ 10,896g *[24.0lb]*, ♂ 12,712g (11,804–13,620) *[28.0lb (26.0–30.0)]*.[602] **Intermembral index:** 80.[234] **Adult brain weight:** 84.6g *[3.0oz]*.[377]

Habitat Mostly evergreen, deciduous, and riverine forests; teak plantations;[34] deciduous woodlands at 150–2400m *[5.9–9.4in]*.[898]

Diet Fruit, 25%; leaves, 58%; flowers, 9%.[46] These langurs eat the shoots and flowers of teak trees[34] and 114 other plant species. They will raid potato, cauliflower, and cardamom crops.[898]

Life History NA.

Locomotion Quadrupedal.

Social Structure 1 male–multifemale groups with 1–5 females that have a well-defined dominance order.[151] **Emigration:** Males always emigrate and travel alone or form all-male bands. Females occasionally leave their natal troop.[582] **Group size:** 17[772] (3–25).[898] **Home range:** NA. **Day range:** NA.

Behavior Diurnal and arboreal. Nilgiri langur groups have been observed to split into 2 groups when the beta male matured and was able to give a loud call. Although males are antagonistic, no physical fights have been seen.[374] Female–female interactions are often augmented "by loud squealing and screeching vocalization by the subordinate." The females reconcile when the dominant gives a specific vocalization and embraces the other, which often leads to the subordinate's grooming of the dominant.[668] Threat behaviors of these langurs include staring, head bobbing, a grinning open mouth, and branch shaking.[668] A subordinate's presentation (turning its hindquarters) to the more dominant group member prevents an attack, signals recognition of the dominant's status, and allows the subordinate to gain proximity to the dominant. The dominant often mounts or touches the presenting animal.[668] Allomothering has been reported in a captive study.[531] Nilgiri langurs are hunted for "black monkey medicine," which local people use for liver and lung ailments and for presumed tonic effects.[898]

Association: Sympatric with lion-tailed macaques *(Macaca silenus)*.[704] Nilgiri langurs have hybridized with Hanuman langurs *(Semnopithecus entellus)*.[898]

Vocalizations: 16,[347] 5 of which are produced only by adult males. These include barks at predators and loud calls with whoops given by the group leader.[378] The most frequent vocalizations are the whistle (18%), the hiccup (16.5%), and a warble (13%). A grunt vocalization is made by the dominant in an agonistic interaction.[668] The hybrid with *S. entellus* has a call that combines elements of both species' calls.[372]

Dusky or Spectacled Leaf Monkey *Trachypithecus (Trachypithecus) obscurus*

Taxonomy Disputed. 10 sub-species.[315] *T. phayrei* has been proposed as a subspecies.[82]

Distinguishing Characteristics Spectacled langurs are named for the wide interrupted white rings around each eye on their dark face. The body is dark gray to blackish brown, and there is a white patch of skin on the upper and lower lips.[543] The underparts are yellowish brown to blackish gray.[80] Infants are born orange yellow[543] and change to adult coloration within 9 months.[385]

Physical Characteristics Head and body length: ♀ 425–595mm *[16.7–23.4in]*, ♂ 420–675mm *[16.5–26.6in]*.[602] **Tail length:** ♀ 723mm (635–813) *[28.5in (25.0–32.0)]*, ♂ 734mm (570–790) *[28.9in (22.4–31.1)]*.[602] **Weight:** ♀ 6602g (4994–8626) *[14.6lb (11.0–19.0)]*, ♂ 7319g (6129–9080) *[16.1lb (13.5–20.0)]*.[602] **Intermembral index:** 83.[234] **Adult brain weight:** 67.6g *[2.4oz]*.[346]

Habitat Primary, secondary, lowland, submontane, and montane forest to 1828m *[5998ft]*; scrub, rubber plantations,[469] gardens.[82]

Diet Fruit and seeds, 35%; leaves, 58%; flowers, 7%.[145] Dusky leaf monkeys eat ripe figs regularly and spend 13–80% of their feeding time on figs in some months.[145]

Life History Weaning: NA. **Sexual maturity:** NA. **Estrus cycle:** NA. **Gestation:** 150d.[346] **Age 1st birth:** NA. **Birth interval:** NA. **Life span:** NA.

Locomotion Quadrupedal walking and running, 50%; leaping, 40%; climbing, 9.2%. When these mon-

keys are feeding, 68% of their locomotion is quadrupedal walking and running, 15% is leaping, and 6% is climbing.[233] They use large branches as pathways within the canopy.[417]

Social Structure Variable. 1 male–multifemale,[853] multimale-multifemale.[46] The average group has 2.5 males.[853] The male defends his females, not the resources.[853] An average group of 17 splits into small subgroups (2.6) to forage in the afternoon.[145] No all-male groups have been observed.[145] **Group size:** 10.3.[903] **Home range:** 33ha. **Day range:** 950m *[3117ft]*.[903]

Behavior Diurnal and arboreal. Dusky leaf monkeys feed and travel in the upper canopy[46] (25–50m *[82–164ft]*). Because of their diet of more mature leaves and unripe fruit, they are reportedly more sedentary than mitered leaf monkeys *(Presbytis melalophos)*.[233] Allomothering has been observed.

Association: These monkeys associate with siamangs *(Hylobates syndactylus)*, but the 2 species' overlap of tree species used in the diet is only 16%.[145]

Vocalizations: The most notable vocalization is a loud snortlike hoot composed of 4 syllables.[543] Loud sharp calls made in the afternoon are believed to be alarm barks rather than long calls.[387]

Thailand, Malay Peninsula

Phayre's Leaf Monkey *Trachypithecus (Trachypithecus) phayrei*

Taxonomy Disputed. This species has been proposed to become a subspecies of *T. obscurus* or be given the name *Semnopithecus barbei*.[82]

Distinguishing Characteristics
Phayre's leaf monkeys are gray to blackish brown. The brow, hands, and feet are black, and the upper arms, legs, and tail are silvery gray.[80] They have a white patch around the eyes and on both lips.[704] Neonates are "straw colored."[704]

Physical Characteristics Head and body length: ♀ 516mm (442–570) *[20.3in (17.4–22.4)]*, ♂ 542mm (420–600) *[21.3in (16.5–23.6)]*.[602] **Tail length:** ♀ 764mm (719–795) *[30.1in (28.3–31.3)]*, ♂ 770mm (745–858) *[30.3in (29.3–33.8)]*.[602] **Weight:** ♀ 6946g (6356–7490) *[15.3lb (14.0–16.5)]*, ♂ 7931g (5675–9080) *[17.5lb (12.5–20.0)]*.[602] **Intermembral index:** NA. **Adult brain weight:** NA.

Habitat Primary and secondary, dense high forest;[898] bamboo forest along hillsides and streams.[704]

Diet Fruit, 24.4%; leaves, 58.4%; petioles, 9.7%.[771]

Life History NA.
Locomotion Quadrupedal.
Social Structure 1 male–multife-male groups.[704] **Group size:** 8.8 (3–30).[704] **Home range:** 75ha.[771] **Day range:** 1000m *[3281ft]*.[771]

Behavior Diurnal and arboreal. Phayre's leaf monkeys have not been well studied in their natural environment. They commonly have gallstones, presumably because they live in a limestone-rich environment. Local people hunt them for these stones, known as bezoar stones, which are an ingredient in folk medicine.[898] **Vocalizations:** The most notable vocalizations are a "high pitched roar" that is the loud call,[772] a 4-note alarm bark, and an agitated "purr."[315]

Laos, Burma,
Viet Nam, Thailand,
China

193

Capped Leaf Monkey *Trachypithecus (Trachypithecus) pileatus*

Taxonomy 5 subspecies.[82]

Distinguishing Characteristics Capped langurs are cream-colored or gray tinged with orange. They have a black face, "gray-whitish to orange" underparts, and dark gray or black hands, feet, and tail.[80] Infants up to 5 months old are creamy white, with a pink face; at 5–14 months they acquire adult coloration.[772]

Physical Characteristics Head and body length: ♀ 543mm (490–660) *[21.4in (19.3–26.0)]*, ♂ 618mm (533–710) *[24.3in (21.0–28.0)]*.[602] **Tail length:** ♀ 946mm (859–1025) *[37.2in (33.8–40.4)]*, ♂ 867mm (830–955) *[34.1in (32.7–37.6)]*.[602] **Weight:** ♀ 10.0kg *[22.0lb]*, ♂ 12.8kg *[28.2lb]*.[639] **Intermembral index:** 82.[234] **Adult brain weight:** NA.

Habitat Subtropical, broadleaf, evergreen, deciduous, and bamboo forest up to 1000m *[3281ft]*; teak plantations.[119]

Diet Fruit, 24%; young leaves, 22%; mature leaves, 20%; seeds, 9%; flowers, 7%; animal prey, 1.6%. These monkeys eat 35 plant species and 1 caterpillar species.[772] They spend almost 30% of their feeding time on 2 species of trees,[773] but only 10% of the diet is from the 3 most abundant trees.[772] During the rainy season when fruit is abundant, 50% of their feeding is on fruit, particularly figs. In the dry season they survive on mature leaves and some seeds.[773] Occasionally they eat gum and termite soil trails.[772] They drink water from tree cavities.[772]

Life History Infant: 14mo.[772] **Juvenile:** 15–24mo.[772] **Subadult:** 24–48mo.[772] **Sexual maturity:** NA. **Estrus cycle:** NA. **Gestation:** NA. **Age 1st birth:** NA. **Birth interval:** NA. **Life span:** NA. Mating season: Oct–Dec. Birth season: April.[772]

Locomotion Quadrupedal.

Social Structure 1 male–multifemale groups that are not territorial.[774] A male's tenure in a multifemale group is 13–26 months.[774] **Emigration:** Both genders emigrate. All-male bands and lone males have been reported.[772] **Group size:** 6.4 (2–13).[772] **Home range:** 21.6ha (16–24).[772] **Day range:** 325m (50–700) *[1066ft (164–2297)]*.[772]

Behavior Diurnal and arboreal. Capped langurs occasionally feed and play on the ground,[119] where they are at greater risk of predation. A golden jackal was observed killing a female as she traveled between trees.[772] These monkeys feed early and late in the day, resting during midday. Adult females lead the group 90% of the time.[772] Intergroup encounters are less aggressive than those of Hanuman langurs *(Semnopithecus entellus)* and purple-faced leaf monkeys *(T. vetulus)*.[46] The male herds his group and is aggressive to females that move in the direction of another male.[774] One male was observed to form a new group with 1 female at a time rather than attempting to take over the whole group. Females do

C. STANFORD (NATURAL HABITAT)

The capped leaf monkey is the largest species in this genus.

not defend their territory or take part in encounters with other troops. Allomothering takes place among adult females, increasing the mother's available feeding time. There is a low rate of social grooming and no rigid dominance hierarchy among females.[772] Females groom males, but the male does not reciprocate.[396] Activity budget: Rest, 40%; feeding, 35%; moving, 18%; play, grooming, and other, 6.8%.

Vocalizations: Capped leaf monkeys have no loud call or contact call. The alarm bark is *ah-ah,* and they have an agitation grunt.[772]

India, Burma, Bangladesh, China

Purple-faced Leaf Monkey *Trachypithecus (Kasi) vetulus*

Taxonomy Disputed. Formerly *Presbytis senex*, the species was reclassified in 1980.[315] Other taxonomists recognize *Kasi* as a valid genus rather than a subgenus.[639]

Distinguishing Characteristics Purple-faced leaf monkeys have a grayish brown to black back and a crown of light brown. The throat and whiskers are white to yellow. The tail is brown to blackish, with a light brown or yellow tip.[82] Infants are gray, with a brownish tinge; by 6 months they look like adults.[718]

Physical Characteristics Head and body length: ♀ 484–540mm *[19.1–21.3in]*, ♂ 495–608mm *[19.5–23.9in]*.[602] **Tail length:** ♀ 666–820mm *[26.2–32.3in]*, ♂ 618–853mm *[24.3–33.6in]*.[602] **Weight:** ♀ 5108–7491g *[11.3–16.5lb]*, ♂ 5675–9431g *[12.5–20.8lb]*.[602] **Intermembral index:** 76.[602] **Adult brain weight:** 64.9g *[2.3oz]*.[346]

Habitat Primary, secondary, semi-deciduous dry, riverine, coastal, scrub, and montane cloud forest up to 2195m *[7202ft]*.[898]

Diet Mature leaves, 40%; fruit, 28%; young leaves, 20%; flowers, 12%. Only 3 tree species make up 70% of the diet, and 12 species 90%.[369] Purple-faced leaf monkeys use 28 species of 61 present in the forest in which they were studied.[46] They raid potato and cauliflower crops.[898]

Life History Weaning: 7–8mo.[346] **Sexual maturity:** NA. **Estrus cycle:** NA. **Gestation:** 195–210d.[346] **Age 1st birth:** NA. **Birth interval:** 16–24mo.[718] **Life span:** NA. Mating season: Oct–Jan[718] (rainy season). Birth season: May–Aug. Birth peak: Jun (40% of births).[718] The birth peak, like that of some other *Trachypithecus* species, seems related to food availability and abundance in seasonal habitats.[717]

Locomotion Quadrupedal.

Social Structure 1 male–multifemale; occasionally 2 male–multifemale.[717] Groups are very territorial.[46] **Emigration:** Males emigrate. Some females leave their natal troop to join another group or form a new one.[582] **Group size:** 6–20.[898] **Home range:** 0.9–14.9ha.[898] **Day range:** NA.

Behavior Diurnal and arboreal. Purple-faced leaf monkey mothers have been observed to abandon their infants after a new male, who was not the father, took over the troop. The new male then attacked and injured the infants. By abandoning their infants, the females were less likely to sustain personal injury.[717] The diet is low in protein (11.5%

J. OATES (NATURAL HABITAT)

Male.

Purple-faced Leaf Monkey *continued*

Mother and infant.

by dry weight), and to compensate, these leaf monkeys minimize the energy they expend by having a small home range and long rest periods.[369] Allomothering has been observed. **Association:** Sympatric with Hanuman langurs *(Semnopithecus entellus).*[46] **Vocalizations:** Males have a long-distance territorial vocalization.[369]

Sri Lanka

Red-shanked Douc Langur *Pygathrix (Pygathrix) nemaeus*

Taxonomy Disputed. Formerly included *P. nigripes* as a subspecies.[82]

Distinguishing Characteristics Red-shanked douc langurs have a gray back and gray underparts. The upper part of the legs is black; the knee and below are orange red. The black hands, feet, brow, and shoulders contrast with the white cheeks and the white throat, which is bordered along the chest by reddish brown. The face is pale terra-cotta, with a white muzzle.[602] The tail, fore-arms, and genital region are white, as is the triangular patch at the tail base. The penis is red; the scrotum, white.[80] Infants are gray, with a black face and 2 pale stripes beneath the eyes.[602]

Physical Characteristics Head and body length: ♀ 597mm *[23.5in]*, ♂ 586mm (550–630) *[23.1in (21.7–24.8)]*.[602] **Tail length:** ♀ 597mm *[23.5in]*, ♂ 680mm (600–735) *[26.8in (23.6–28.9)]*.[602] **Weight:** ♀ 8.2kg *[18.1lb]*, ♂ 10.9kg *[24.0lb]*.[234] **Intermembral index:** 89,[602] 94.[234] **Adult brain weight:** 108.5g *[3.8oz]*.[346]

Female with a 3-week-old infant.

Habitat Primary and secondary tropical rain forest at 300–2000m *[984–6562ft]* elevation.[898]

Diet Leaves, buds, fruit, seeds, flowers. These langurs eat 50 plant species but no animal prey.[1009]

Life History Infant: 0–12mo.[492] **Weaning:** 9–13mo.[492] **Juvenile:** 12–36mo.[492] **Subadult:** 36–60mo.[492] **Sexual maturity:** ♀ 48mo, ♂ 60mo.[602] **Estrus cycle:** 28–30d.[625] **Gestation:** 165–190d.[625] **Age 1st birth:** 48mo.[492] **Birth interval:** 16.5mo.[346] **Life span:** >30y.[169] Young infants have been observed in January and May.[495]

Locomotion Quadrupedal.

Social Structure Multimale-multifemale groups with 2 females to 1 male. Males and females have their own hierarchies, and males are dominant to females.[625] **Emigration:** Both males and females emigrate.[46] **Group size:** 4–25, up to 40.[495] **Home range:** NA. **Day range:** NA.

Behavior Diurnal and arboreal. Red-shanked douc langurs use the high canopy.[493] They have a specific "play face" in which the eyes are closed, exposing very pale blue eyelids.[439] Their fixed stare is a threat display. A grimace with the mouth open and the teeth exposed is a submissive gesture

given in response to a stare. It is also used to initiate grooming or play.[439] Allomothering is common in captivity. A male was observed to care for an orphaned infant that was fed by 2 females in the group.[439] In a captive study, an infant rode ventrally on all members of the group.[439] These langurs are hunted most around the Vietnamese New Year (Tet).[1008] **Association:** Partially sympatric with black-shanked douc langurs *(P. nigripes)* in Viet Nam's central highlands.[495] **Mating:** Before mating, both genders give a sexual signal with the jaw forward, eyebrows raised and then lowered, and a head shake.[439] Single-mount[439] and multiple-mount matings[169] have been reported. **Vocalizations:** These langurs have a low-pitched growl that is given as a threat, and a short, harsh distress squeal.[439]

Male.

Viet Nam, Laos, Cambodia

Black-shanked Douc Langur *Pygathrix (Pygathrix) nigripes*

ENDANGERED USESA

Taxonomy Disputed. Elevated from a subspecies of *P. nemaeus*.[405] The taxonomy is currently unclear because a new form has been documented that may be a hybrid or subspecies.[496]

Distinguishing Characteristics Black-shanked douc langurs are similar to red-shanked douc langurs, except that the former have a bluish face[317] and black legs.[602] The scrotum and inside of the thighs are blue.[317] The new form has a pale face, gray legs and forearms, and a white belly and genital region.[496]

Physical Characteristics Head and body length: NA. **Tail length:** NA. **Weight:** NA. **Intermembral index:** NA. **Adult brain weight:** NA.

Habitat Primary and secondary tropical rain forest.[495]

Diet Primarily leaves; some fruit, especially figs.[495]

Life History NA.

Locomotion Quadrupedal.

Social Structure Multimale-multifemale groups with 2 females to 1 male. **Group size:** NA. **Home range:** NA. **Day range:** NA.

Behavior Diurnal and arboreal. Black-shanked douc langurs have not been observed to drink standing water or come to the ground.[495] The first study of their behavior began in 1996. They are listed as endangered under *P. nemaeus*. **Association:** Partially sympatric with red-shanked douc langurs (*P. nemaeus*) in Viet Nam's central highlands.[495]

Viet Nam, Cambodia

This previously undescribed form of black-shanked douc langur was confiscated from a market in Viet Nam in 1995.

Tonkin Snub-nosed Monkey *Pygathrix (Rhinopithecus) avunculus*

R. BOONRATANA (NATURAL HABITAT)

Tonkin snub-nosed monkeys are the most critically endangered primate species in Asia.

ernment has taken some modest but important steps to attempt to ensure the short-term survival of this species. **Vocalizations:** These monkeys have a noisy contact call and an alarm call that sounds like a hiccup, or *huu chhhk*.[75] The long-distance call is a doglike bark. **Sleeping site:** Low in trees that grow on steep mountainsides, away from the wind.[75]

Viet Nam

Taxonomy Disputed. Monotypic.[898] The subgenus name *Rhinopithecus* was formerly the genus name.[315]

Distinguishing Characteristics Tonkin snub-nosed monkeys have a brown to black back. The underparts are yellowish white to orange, and the elbows have a patch of the same color. The crown and ears are white to yellowish, the facial skin is blue, and the lips are pink. The brown tail has a contrasting whitish yellow or orange gray tip. The male's penis is black, and the scrotum is white.[80] Infants are born whitish gray and gradually darken.[75]

Physical Characteristics Head and body length: ♀ 540mm *[21.3in]*, ♂ 650mm *[25.6in]*.[75] **Tail length:** ♀ 650mm *[25.6in]*, ♂ 850mm *[33.5in]*.[75] **Weight:** ♀ 8.5kg *[18.7lb]*, ♂ 14.0kg *[30.9lb]*.[75] **Intermembral index:** NA. **Adult brain weight:** NA.

Habitat Primary forest on limestone hills (karst) at 200–1200m *[656–3937ft]*.[75] This species' known habitat is now restricted to 2 small forest reserves near the town of Na Hang in Viet Nam.

Diet Dec–Apr: fruit, 47%; young leaves, 38%; seeds, 15%.[75] These monkeys use 52 species of plants.[983]

Life History NA.

Locomotion Quadrupedal walking, climbing, leaping; occasional brachiation.[75]

Social Structure 1 male–multifemale groups with overlapping ranges.[75] One-male groups frequently form temporary large groups. Males emigrate to form all-male groups.[75] **Group size:** 25.[75] **Home range:** NA. **Day range:** NA.

Behavior Diurnal and arboreal. As of 1995, Tonkin snub-nosed monkeys had been the subject of only 1 field study. Because they have been heavily hunted, they are very shy of humans. Roughly 150 individuals remain. If no poaching occurs, only 80 individuals are needed in the founding population to maintain a 98% survival rate for 100 years.[967] The Vietnamese gov-

T. NADLER

Juvenile.

Black or Yunnan Snub-nosed Monkey *Pygathrix (Rhinopithecus) bieti*

ENDANGERED USESA

ZANG TAI (NATURAL HABITAT)

China

Black snub-nosed monkeys live in temperate montane forest that is snow covered for 6 months of the year.

Taxonomy Disputed. Monotypic.[315] Elevated from a subspecies of *P. roxellana* in 1988. The subgenus name *Rhinopithecus* was formerly the genus name.[130]

Distinguishing Characteristics The chest, limbs, and upper back of Yunnan snub-nosed monkeys are blackish gray; the hips, thighs, and lower back are whitish; and the belly, cheeks, throat, and ears are yellowish white. These monkeys have pink facial skin, a black brow, and a forward-drooping crest.[80] The black tail has a tuft at the end, like a cow's tail. The Chinese call this species the cow-tailed monkey.

Physical Characteristics Head and body length: ♀ 740–830mm *[29.1–32.7in]*,[130] ♂ 830mm *[32.7in]*.[459] **Tail length:** 518–747mm *[20.4–29.4in]*.[130] **Weight:** ♀ 9.2kg *[20.3lb]*, ♂ 15.0kg *[33.1lb]*.[414] **Intermembral index:** NA. **Adult brain weight:** NA.

Habitat Broadleaf deciduous, evergreen, conifer, montane, and temperate forest[46] up to 4500m *[14,765ft]*.[460]

Diet Lichen, 50%; grasses, leaves, fruit. The lichen is from the genus *Usnea* and grows on evergreen trees.[459]

Life History Infant: 12mo.[459] **Sexual maturity:** NA. **Estrus cycle:** NA. **Gestation:** 170d.[459] **Age 1st birth:** NA. **Birth interval:** NA. **Life span:** NA. Mating season: Aug–Sep. Birth season: Mar–May.[414]

Locomotion Quadrupedal.

Social Structure Multimale-multifemale bands made up of 1 male–multifemale and all-male groups. **Group size:** 23–200.[46] **Home range:** 2500ha (1000–4000).[459] **Day range:** NA.

Behavior Diurnal, arboreal, and terrestrial. Yunnan snub-nosed monkeys spend 10–35% of their time on the ground.[460] Adults are more terrestrial than juveniles.[460] These monkeys use the upper part of the forest but will cross treeless ridges. They spend 22% of the day on the ground, 68% in fir trees, and 10% in other trees.[499] They are adapted to cold mountain temperatures. In June the average temperature is 7°C *[45°F]*; in October, –0.3°C *[31°F]*. They do not move to lower elevations in particular seasons but will remain at lower altitudes for a few days during heavy snows.[460] Activity budget: Feeding, 39%; rest, 34%; social activities, 16%; travel, 11%.[499] Hunting and habitat destruction are the major threats to this species.[459] **Sleeping site:** Trees in valleys.[930]

C. KIRKPATRICK

Males weigh half again as much as females.

Guizhou Snub-nosed Monkey *Pygathrix (Rhinopithecus) brelichi*

The white patch between the shoulders indicates that this is a male.

Subadult males stay together and are peripheral to the main group.

Taxonomy Disputed. Monotypic.[82] Elevated from a subspecies of *P. roxellana* in 1970.[315] The subgenus name *Rhinopithecus* was formerly the genus name.[82]

Distinguishing Characteristics Guizhou snub-nosed monkeys have a grayish brown back, a whitish gray belly, and pale gray on the thighs. The upper arms, the inside of the shoulders, and the brow are yellowish orange. The crown of the head is brown, and the top of the ears is white. The facial skin is blue, and the lower lip is pink. Males have a white patch of fur on the back between the shoulders; their scrotum and nipples are white.[80] Infants are gray.

Physical Characteristics Head and body length: 660mm (660–762) [26.0in (26.0–30.0)].[80] **Tail length:** 559–771mm [22.0–30.4in].[80] **Weight:** NA. **Intermembral index:** NA. **Adult brain weight:** NA. The

male is 1.5–2 times the size of the female.[58]

Habitat Subtropical, evergreen or deciduous, broadleaf or coniferous forest at 1000–2300m [3281–7546ft].[57] These monkeys live only in 1 reserve in the province of Guizhou in China.[57]

Diet Winter—buds, 91%; spring—young leaves, 95%; summer—fruit, 35%; fall—magnolia buds, 27%; some arthropods.[58] The amount of leaves in the diet varies from 7% in January–March to 93% in April–June.

Life History NA.

Locomotion Quadrupedal; climbing, leaping, and semibrachiation.[46]

Social Structure 1 male–multifemale groups with 1–3 females. Several 1-male groups travel together in a larger band with all-male groups on the periphery. Larger bands may split and re-form.[58] The band may be

as large as 430.[46] **Emigration:** Young males join an all-male group.[46] **Group size:** 6 (3–10).[58] **Home range:** 3500ha.[57] **Day range:** NA.

Behavior Diurnal and arboreal. Guizhou snub-nosed monkeys are able to form large bands because the trees on which they feed occur in large patches. They will travel on the ground across deforested land to get to another area of forest.[58] Chinese attempts to breed this species in captivity have been unsuccessful. Fewer than 1000 individuals are believed to remain.[57]

Vocalizations: These monkeys have numerous graded vocalizations and contact calls. The most notable are the *waa-gek* alarm call and the male aggression call.[58] **Sleeping site:** Broadleaf evergreen trees with dense foliage.[57]

China

Infants are a lighter color than adults.

Sichuan Golden Snub-nosed Monkey *Pygathrix (Rhinopithecus) roxellana*

G. SCHALLER (NATURAL HABITAT)

Golden snub-nosed monkeys live in some of the same forests as the giant panda.

Taxonomy Disputed. 2 subspecies.[862] The subgenus name *Rhinopithecus* was formerly the genus name.[315]

Distinguishing Characteristics Golden snub-nosed monkeys have a dark brown back and tail, with longer (100mm *[4in]*), whitish orange hairs on the shoulders and tail tip. The chest, underparts, and legs are a whitish orange to bright orange, as are the throat and area around the face. The skin around the eyes is pale blue, and the muzzle is white. Males have granulomatous flanges on the sides of the mouth.[80]

Physical Characteristics Head and body length: 680–760mm *[26.8–29.9in]*.[309] **Tail length:** 646–722mm *[25.4–28.4in]*.[309] **Weight:** ♀ 6.5–10.0kg *[14.3–22.0lb]*, ♂ 15.0–39.0kg *[33.1–86.0lb]*.[234] **Intermembral index:** 95.[602] **Adult brain weight:** 121.5g *[4.3oz]*.[346]

Habitat Mixed bamboo, conifer, and deciduous forest up to 3150m *[10,335ft]*. These monkeys are adapted to the longest winter and the coldest temperatures that any primate withstands besides humans.[46] They prefer altitudes of 2000–2800 meters[234] *[6562–9187ft]* with winter temperatures of 0–9°C *[32–48°F]*. The record low in their range is −30°C *[−22°F]*.[808]

Diet Leaves, buds, lichen, fruit.[46]

Life History NA. Mating season: Oct–Dec. Birth season: Apr–Jun.[932]

Locomotion Quadrupedal; climbing, leaping, some suspensory.[808] These monkeys walk quadrupedally with the tail over the back.

Social Structure 1 male–multifemale groups are the basic reproductive unit, but several 1-male and all-male groups[46] may form a band of up to 200 individuals. Several bands may join together for short periods with a total of 600 individuals.[46] Bands are fission-fusion and are largest in the mating season. **Group size:** 20[234]–200, up to 600.[46] **Home range:** 2300ha (1000–3000).[46] **Day range:** 1000–2000m *[3281–6562ft]*.[808]

Behavior Diurnal, arboreal, and terrestrial. Sichuan golden snub-nosed monkeys travel on the ground.[808] The male protects his group and will carry juveniles. During travel, bachelor males both lead the band and bring up the rear.[932] In summer these monkeys move to above 3000 meters *[9843ft]* in altitude. In winter they stay lower (2000–2800m *[6562–9187ft]*).[46] Allomothering has been observed.[932] **Sleeping site:** Pairs sleep in trees.[932]

China

R. A. MITTERMEIER

A male threat display.

Proboscis Monkey *Nasalis (Nasalis) larvatus*

Taxonomy Monotypic.[80]

Distinguishing Characteristics Proboscis monkeys have a reddish orange crown and back. The shoulders, cheeks, throat, and nape are pale orange. The legs, belly, rump patch, and tail are whitish gray.[80] The face is pink in adults and dark blue in infants.[654] The penis is red; the scrotum, black.[80] Males have an elongated pendulous nose—hence the common name. The feet are partially webbed. Juveniles have a blue nose until age 3.

Physical Characteristics Head and body length: ♀ 620mm (610–640) *[24.4in (24.0–25.2)]*, ♂ 745mm (730–760) *[29.3in (28.7–29.9)]*.[602] **Tail length:** ♀ 573mm (550–620) *[22.6in (21.7–24.4)]*, ♂ 665mm (660–670) *[26.2in (26.0–26.4)]*.[602] **Weight:** ♀ 10.0kg *[22.0lb]*, ♂ 21.2kg *[46.7lb]*.[639] **Intermembral index:** 94.[234] **Adult brain weight:** 94.2g *[3.3oz]*.[346] Canine length: ♀ 6.1mm *[0.2in]*, ♂ 8.2mm *[0.3in]*.[490]

Habitat Coastal nipa palm, mangrove, lowland, riverine, and peat swamp forest below 245m *[804ft]* elevation.[898]

Diet Leaves, 44%; seeds, 20%; fruit, 17%; flowers, 3%; animal prey, 1%.[469] Proboscis monkeys obtain 60% of their diet from 4 of at least 55 plant species they ingest. They are seed predators on 3 of the most abundant tree species. The fruit they eat is usually unripe. They are frugivorous January–May and folivorous June–December.[924]

Life History Infant: 12mo. **Juvenile:** 12–36mo. **Sexual maturity:** ♀ 36–60mo, ♂ 60–84mo. **Estrus cycle:** NA. **Gestation:** 166d.[602] **Age 1st birth:** 55mo.[710] **Birth interval:** 12–24mo. **Life span:** 13.5y.[710]

Locomotion Quadrupedal; climbing and leaping.

Social Structure Variable. 1 male–multifemale groups, all-male bachelor troops. The 1-male groups may join other groups temporarily.[710] Groups are not territorial;[46] home ranges overlap up to 95%.[924] **Emigration:** Males and some females emigrate.[772]

Males display by staring, showing their canines, and blowing through their nose.

Proboscis Monkey *continued*

Group size: 9 (4–20).[46] **Home range:** 900ha.[47] **Day range:** 706m *[2316ft].*[47]

Behavior Diurnal and arboreal. Proboscis monkeys are good swimmers; to escape danger, the entire troop will dive into water. If startled while crossing a river, they can swim underwater for 20 meters *[66ft].*[47] They may have become restricted to riverine and coastal forests because the poor soils in the upland dipterocarp forests of Borneo lack minerals.[924] Allomothering has been observed.[46] **Association:** In south Borneo, proboscis monkeys compete with and are displaced by long-tailed macaques *(Macaca fascicularis)* and Borneo orangutans *(Pongo pygmaeus).*[924] **Mating:** Matings are multiple-mount.[169] **Vocalizations:** Males vocalize through the nose with a *kee honk* sound. Other vocalizations are "honks, groans, squeals, and loud roars."[654] **Sleeping site:** Trees over water.

Borneo

Infants are born with a dark blue face. The nose of this juvenile has not yet changed to the adult coloration.

Female and infant.

NATURAL HABITAT

Pig-tailed Langur *Nasalis (Simias) concolor*

R. TENAZA

The pig-tailed langur lives only on the Mentawai Islands
and is endangered.

before humans arrived 2000 years ago. Pig-tailed langurs are hypothesized to have adapted to human predation by reducing group size and becoming more cryptic.[838] **Association:** Pig-tailed langurs often associate with Mentawai Island leaf monkeys *(Presbytis potenziani).*[973] **Mating:** The male's loud call is emitted during courtship. The female replies with shrill squeals.[838] **Vocalizations:** Pig-tailed langurs are usually very quiet, with no contact call.[838] Their most notable vocalization is a loud call, exchanged by males, that can be heard 500 meters *[1641ft]* away.[639] The call is donkeylike,[886] with 2–25 nasal barks.[824] Loud calls, which may be heard in morning and afternoon, are emitted spontaneously 51% of the time, in response to other males 33%, in response to falling trees 13%, in response to thunder 4%, and during intergroup male fights 1%.[838] **Sleeping site:** Emergent trees with dense foliage.[838]

Mentawai Islands
(Indonesia)

Taxonomy Disputed. 2 subspecies.[82] The subgenus name *Simias* was the genus name until 1975.[639]

Distinguishing Characteristics Pig-tailed langurs are the only colobine with a short tail. They have 2 color phases. One-sixteenth to one-third of the population is cream buff with brown.[838] The rest are blackish brown, with a white penal tuft and a black face bordered with whitish hairs.[80]

Physical Characteristics Head and body length: ♀ 500mm *[19.7in]*, ♂ 515mm *[20.3in]*.[602] **Tail length:** ♀ 140mm *[5.5in]*, ♂ 155mm *[6.1in]*.[602] **Weight:** ♀ 7100g *[15.6lb]*,[639] ♂ 8750g *[19.3lb]*.[234] **Intermembral index:** 98.[602] **Adult brain weight:** NA.

Habitat Primary hilly tropical rain forest;[602] swamp forest; rarely secondary growth.[898]

Diet Leaves, 60%; fruit, 25–30%.[961]

Life History NA. Birth season: Variable; Jun–Jul in Siberut.[602]

Locomotion Quadrupedal.[838]

Social Structure Variable. 1 male–1 female, 1 male–multifemale, multimale-multifemale.[867] **Group size:** 4.6 (2–20).[602] **Home range:** 2.5–20ha,[867] to 30.[838] **Day range:** NA.

Behavior Diurnal and arboreal. Pig-tailed langur males approach each other to within 20 meters *[66ft]* during intertroop encounters that may last up to 30 minutes.[639] The Mentawai Islands had no large predators

R. TENAZA

Pig-tailed langurs are the only colobine with a short tail.

White-handed gibbon *Hylobates lar*

Apes

We humans are included in one of the three families of apes in the superfamily Hominoidea. Except for our species, all of the other members of this superfamily have restricted ranges in either Southeast Asia or central Africa. Apes differ from monkeys in several ways. Apes have no tail and generally have a larger body size and weight than most other primates. They have a more upright body posture and a broad chest. Because of their size, apes tend to hang below branches rather than balance on top of them as do most other primates. This specialization has led to numerous changes in ape skeletons and muscles. Apes rely more on vision than on smell and have a short broad nose rather than a snout, as Old World monkeys do. Apes have a larger brain relative to their body size than other primates do. The neurological complexity of apes requires them to have longer gestations and a lengthy maturation.

Paleontologists who study hominoid fossils have found a diverse radiation of apes in the Miocene fossil record, which began 23 million years ago. The gibbons split from the other apes between 20 million and 12 million years ago. All of the living apes have the following dental formula:[234]

$$\frac{2.1.2.3}{2.1.2.3} \times 2 = 32$$

Lesser Apes (Gibbons)   *Family:* Hylobatidae

The lesser apes are the gibbons of the family Hylobatidae, which all live in Southeast Asia. About 1000 years ago they were reported from Chinese literature to live as far north as the Yellow River. They diverged within the last million years or so by being separated and isolated during the periods of glaciation, and 11 species are currently recognized. *Hylobates,* the single genus of Hylobatidae, is divided into 4 subgenera, each with a different number of chromosomes and tending to have different coloration. The hoolock gibbon (subgenus *Bunopithecus*) has 38 chromosomes and is sexually dichromatic—the males and females are different colors. Members of the *lar* group (subgenus *Hylobates*) have 44 chromosomes and are variable in color. Members of the *concolor* group (subgenus *Nomascus*) have 52 chromosomes; the males are black, and the females are buff. The siamang (subgenus *Symphalangus*) has 52 chromosomes, and both males and females are black.[118]

Most gibbons inhabit only primary forest, and a majority of their diet is ripe fruit. The siamang *(Hylobates syndactylus)* is the largest of the family and eats more leaves. All gibbons have long forearms with hooklike fingers specialized to swing from branch to branch (brachiation—see Figure 4 on page 6). In captive experiments, gibbons do not pass the standard self-recognition mirror test.[1017] They have ischial callosities (two pads on the rump), sleep on tree branches, and do not build nests.[599]

Gibbons live in monogamous family groups. At adulthood (about age 10), the offspring are often forced out of the group. Males and females of the same species are about the same size, and both sexes have canine teeth of equal length. All the gibbons are territorial, and in most species both genders practice vocal dueting as a territorial spacing display. The great call of a pair of gibbons is one of the wonders of the primate world. The call of each gibbon species distinguishes it. The sonogram provided here for some species is a poor substitute for the eerie sirenlike crescendo that is heard in the forest.

Hoolock or White-browed Gibbon *Hylobates (Bunopithecus) hoolock*

Female.

PERTH ZOO - WESTERN AUSTRALIA

and social interaction take up less than 15% of the day.[12] Juveniles will come to the ground to cross gaps of more than 15 meters *[49ft]* that they cannot leap.[12] Intertroop encounters consist of singing and countersinging, with males chasing each other for 10–20 meters *[33–66ft]*. This is followed by growls, whoops, and more singing and countersinging as the intergroup distance increases as they leave to forage.[292] **Mating:** Mating has been observed only before 10:30 a.m., after a bout of grooming. The male initiates the mating. Postconception matings have been observed.[11] **Vocalizations:** The pair duets when another group is nearby.[589] A 15-to-18-minute duet is usually given in the morning and started by the male, with the female joining later.[12] **Sleeping site:** The female sleeps with her infant. The male and other off-spring sleep separately.[12]

India, Burma, China

Taxonomy 2 subspecies.[286]

Distinguishing Characteristics

Male hoolock gibbons are black, with white eyebrows that turn up at the end. Females are copper tan, with a whitish eyebrow band. Whitish hair encircles the face and contrasts with darker brown cheeks.[984] The neonatal coat is milky white; by 9 months the coat changes to gray, then dark gray. Juveniles of both genders are black, with white eyebrows. Female subadults change from black to gray to the tan color of adult females.[12]

Physical Characteristics Head and body length: ♀ 483mm *[19.0in]*.[985] **Weight:** ♀ 6.1kg *[13.4lb]*, ♂ 6.9kg *[15.2lb]*.[490] **Intermembral index:** 129.[234] **Adult brain weight:** 108.5g *[3.8oz]*.[346]

Habitat Primary evergreen and semideciduous[291] forest at 152–1370m *[499–4495ft]* elevation.[898]

Diet Fruit, 65%; leaves, 13%; buds, 12%; flowers, 5%; animal prey (including insects and bird

eggs[589]), 5%.[12]

Life History Infant: 24mo.[12] **Weaning:** 23mo.[346] **Juvenile:** 24–48mo.[12] **Subadult:** 48–84mo.[12] **Sexual maturity:** 84mo.[346] **Estrus cycle:** 28d.[346] **Gestation:** NA. **Age 1st birth:** NA. **Birth interval:** 36mo.[11] **Life span:** 42y.[985] Mating season: Rainy season.[120] Birth season: Nov–Feb.[292]

Locomotion Brachiation, leaping, bipedal walking.[12]

Social Structure Monogamous 1 male–1 female with offspring in a cohesive family group. Single individuals have been reported.[120] **Group size:** 3.5 (3–6).[120] **Home range:** 15–30,[12] 300–400ha.[589] **Day range:** 600m (300–1000) *[1969ft (984–3281)]*.[589]

Behavior Diurnal and arboreal. Hoolock gibbons feed as a group for 50% of the day in trees at a height of 18–22 meters *[59–72ft]*.[589] They spend less time feeding in the rainy season.[589] The female leads troop movements.[12] Grooming, play,

PERTH ZOO - WESTERN AUSTRALIA

Male.

kilohertz

0

5   10   15   20

seconds

Dark-handed or Agile Gibbon *Hylobates (Hylobates) agilis*

Female.

travel in the upper canopy, feed on fruit hanging from branches of small trees in the middle canopy, and rest sitting on large branches.[289] The monogamous social unit is stable because the female's territoriality prevents the male from having more than 1 female and because the male is aggressive to other males to ensure his paternity.[565] A resident female will approach and aggressively chorus at an intruding female; only the male approaches an intruding male to evict him.[562]

Association: Partially sympatric with white-handed gibbons *(H. lar),* partially sympatric and occasionally hybridizing with Mueller's Bornean gray gibbons *(H. muelleri),*[83] and sympatric with siamangs *(H. syndactylus).* Siamangs do not compete extensively with dark-handed gibbons for the same food.[682]

Vocalizations: 6.[347] The great call is sung from the highest trees in order to increase the distance it travels.[289] The female loud call is "a series of high pitched rising and falling notes."[654] The singing of mated and unmated males is similar; songs are probably not used by unmated males to attract a mate.[563] A more complex song elicits a faster response from the neighboring pair than simple songs.[563] The same sound has

Taxonomy Disputed. 3 subspecies,[962] including *H. a. albibarbis.*[315] Formerly a subspecies of *H. lar.*

Distinguishing Characteristics The coat color of dark-handed gibbons varies from buff to brown and black. The hands are dark. Males have white brows and cheeks; adult females have only white brows.[962]

Physical Characteristics Head and body length: 420–470mm *[16.5–18.5in].*[654] **Weight:** ♀ 5550–6400g *[12.2–14.1lb].*[654] **Intermembral index:** 121. **Adult brain weight:** 110g *[3.9oz].*[346]

Habitat Tropical lowland rain forest, swamp forest. Dark-handed gibbons prefer dry-ground forest but also use swamp forest.[258]

Diet Fruit, 58%; leaves, 39%; flowers, 3%; animal prey, 1%.[881]

Life History Weaning: NA. **Sexual maturity:** NA. **Estrus cycle:** NA. **Gestation:** NA. **Age 1st birth:** NA. **Birth interval:** 38mo.[564] **Life span:** 32y.[962]

Locomotion Brachiation for long-distance travel; bipedal walking on large branches and the ground.[289]

Social Structure 1 male–1 female monogamous pair with their offspring in a family group.[564] **Group size:** 4.4 (2–7).[564] **Home range:** 25–29ha.[881] **Day range:** 1335m (650–2200) *[4380ft (2133–7218)].*[881]

Behavior Diurnal and arboreal. Dark-handed gibbons

NATURAL HABITAT

R. WIRTH

Male.

Dark-handed or Agile Gibbon *continued*

In midswing from one branch to another.

Gibbons are bipedal when they are on the ground or on a large branch.

different meanings to the gibbons, depending on the context: *hoo* is friendly within the group but aggressive when 2 groups interact.[290] The great call does not vary from one geographic location to another; it is hypothe-

sized to be genetically determined.[290] **Sleeping site:** Large boughs of big trees.[289]

Malay Peninsula, Sumatra, Borneo, (Indonesia)

Kloss's Gibbon *Hylobates (Hylobates) klossii*

Kloss's gibbon is the smallest gibbon.

Taxonomy Monotypic.

Distinguishing Characteristics Male, female, and infant Kloss's gibbons are all black.[118] They have one-third less hair density than *H. lar*.[881]

Physical Characteristics Head and body length: 457mm *[18.0in]*.[986] **Weight:** 5.8kg *[12.8lb]*.[881] **Intermembral**

index: 126.[234] **Adult brain weight:** 91.1g *[3.2oz]*.[315]

Habitat Primary, evergreen lowland and hill forest.[898]

Diet Fruit, 70%; animal prey, 25%; leaves, 2%.[881]

Life History Infant: 24mo.[839] **Weaning:** 12mo.[315] **Juvenile:** 24–48mo.[839] **Subadult:** 48–>72mo.[839] **Sexual maturity:** NA. **Estrus cycle:** NA. **Gestation:** 210d.[315] **Age 1st birth:** NA. **Birth interval:** 40mo.[488] **Life span:** NA. **Births:** Year-round.[488]

Locomotion Brachiation.[881]

Social Structure Monogamous 1 male–1 female with offspring in a cohesive family group.[839]

Emigration: At about age 8, both male and female offspring are evicted from the family group.[839] **Group size:** 3.4 (2–6).[898] **Home range:** 7–11,[839] 34ha,[881] to 54.[488] **Day range:** 1514m (885–2150) *[4967ft (2904–7054)]*.[881]

Behavior Diurnal and arboreal. Kloss's gibbon parents may aid their subadult offspring by gaining

vacant territory or intimidating a neighboring pair. It is thought to be advantageous for subadults to remain near their natal range to learn the location of food trees and sleeping sites while still relying on the parental territory.[839] Relative to dark-handed and white-handed gibbons *(H. agilis, H. lar)*, Kloss's gibbons are active longer each day and have a greater day range, which they travel at the end of the day. All 3 species spend the same amount of time eating fruit, but Kloss's eats more insects and fewer leaves.[881] **Association:** Kloss's gibbons are dominant to and always displace Mentawai Island leaf monkeys *(Presbytis potenziani)* from food trees.[840] **Mating:** In 1 recorded incident, a son took over his family's territory and mated with his presumed mother after his father died.[839]

Vocalizations: The great call is different from that of other gibbons and is reported to be the most beautiful.[81] The pair does not duet. The male usually sings 1 hour before dawn. The female sings for 20 minutes about 1 hour after dawn and has a visual display with leaps, twists, turns, and

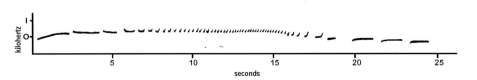

Kloss's Gibbon *continued*

leaf tearing. They do not sing every day and do not sing in the rain.[880] The male sings 7 times as much as the female.[882] The male's high-frequency trilling phrases may indicate his fitness and health.[118] **Sleeping site:** Tall emergent trees without lianas.[823] These gibbons do not sleep in the same tree 2 nights in a row, hypothetically to avoid predators and parasite infestations.[881] They leave the sleeping tree at 18 minutes after sunrise on average.[840]

Mentawai Islands
(Indonesia)

White-handed Gibbon *Hylobates (Hylobates) lar*

Taxonomy Disputed. 5 subspecies.[286] Formerly included *H. agilis*.[315]

Distinguishing Characteristics White-handed gibbons have a white face ring and white hands and feet. The coat color varies from cream to black and dark brown to red; the variation is not related to gender or age.[118]

Physical Characteristics Head and body length: ♀ 420–580mm *[16.5–22.8in]*, ♂ 435–585mm *[17.1–23.0in]*.[704] **Weight:** ♀ 4400–6800g *[9.7–15.0lb]*, ♂ 4970–7600g *[11.0–16.8lb]*.[704] **Intermembral index:** 129.7.[425] **Adult brain weight:** 107.7g *[3.8oz]*.[346] Males and females have equal-sized canines.

Habitat Primary and secondary, tropical dry deciduous and moist evergreen rain forest;[704] lowland to montane forest up to 2400m *[7874ft]* but usually at 250–500m *[820–1641ft]* elevation.[898]

Diet Fruit, 50%; leaves, 29%; insects, 13%; flowers, 7%;[682] new stems, shoots, buds.

Life History Infant: 0–24mo.[346] **Weaning:** 24mo.[346] **Juvenile:** 24–54mo.[708] **Subadult:** 54–72mo.[704] **Sexual maturity:** ♀ 108mo, ♂ 78mo.[346] **Estrus cycle:** 27d.[346] **Gestation:** 205d.[346] **Age 1st birth:** 112mo. **Birth interval:** 30mo.[346] **Life span:** 44y.[995]

Locomotion Brachiation. When on a large branch or the ground, these gibbons walk or hop bipedally with the arms raised over the head.[704]

Social Structure Monogamous 1 male–1 female with offspring in a cohesive family group. **Group size:** 5 (3–12).[898] **Home range:** 12–53.5ha.[854] **Day range:** 1490–1600m *[4889–5250ft]*.[681]

Behavior Diurnal and arboreal. The male white-handed gibbon defends his territory from other males; the female keeps away other females. The male and female groom each other 15 minutes per day. After a forest was selectively logged, the white-handed gibbons shifted and decreased their home range, used the lower forest levels, and ate more leaves.[417]

Association: Partially sympatric with siamangs *(H. syndactylus),* pileated gibbons *(H. pileatus),* and dark-handed gibbons *(H. agilis).* There is evidence of a hybrid zone between the last 2 species and *H. lar*.[83] White-handed gibbons compete with pig-tailed macaques *(Macaca nemestrina)* for food. The gibbons

Suspensory feeding.

ENDANGERED

USESA

211

White-handed Gibbon *continued*

have been observed to harass the macaques by loud hooting, chasing, and biting in order to protect a small tree with large fruit.
Vocalizations: The female starts the great call, and the male begins as she finishes. The great call is given at 8:16 a.m. on average. They may call more than once per day.[768]

China, Thailand, Laos, Burma, Malay Peninsula, Sumatra

Either gender of white-handed gibbon can be a light or dark form.

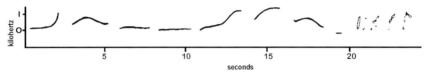

Silvery Javan Gibbon *Hylobates (Hylobates) moloch*

Female and infant.

Taxonomy Monotypic.[286]
Distinguishing Characteristics
Both male and female Javan gibbons are silvery gray with a dark cap. Some have a dark chest. Infants are a lighter color than adults.[996]
Physical Characteristics Head and body length: NA. **Weight:** 5.7kg *[12.6lb]*.[234] **Intermembral index:** 127.[234] **Adult brain weight:** 113.7g *[4.0oz]*.[346]

Habitat Primary and secondary forest and tropical[898] evergreen rain forest[429] from sea level to 1500m *[4922ft]*.[429]
Diet Fruit, 61%; leaves, 38%; flowers, 1%; buds;[430] animal prey, including caterpillars, termites, and other insects; honey.[430] These gibbons use 125 plant species.[430]
Life History Weaning: NA. **Sexual maturity:** NA. **Estrus cycle:** NA. **Gestation:** NA. **Age 1st birth:** NA. **Birth interval:** NA. **Life span:** 35y.[996]
Locomotion Brachiation.[429]
Social Structure Monogamous 1 male–1 female territorial family groups.[513] **Group size:** 3–4.[139] **Home range:** 17ha.[117] **Day range:** 1400m *[4593ft]*.[117]
Behavior Diurnal and arboreal. Silvery Javan gibbons are not found above 1500 meters *[4922ft]* because at that altitude there are fewer tree species, the trees are slender and unsuitable for brachiation,[429] and temperatures are lower. The density of these gibbons is greatest in lowland forest and decreases in montane forest.[429] They forage quietly in forest at 10 meters *[33ft]* or above and are detected only by branch movement and an occasional fruit drop.[430] This species has been protected by law since 1924 without success and is now critically endangered.
Vocalizations: Silvery gibbons do not duet.[431] They have 4 types of loud calls: the solo female song with which she defends the territory; the single male's song to attract a mate; the group call by all group members during border conflicts with another group; and the threat call by all group members to drive away predators such as leopards.[431]

E. THETFORD

Java (Indonesia)

Mueller's Bornean Gray Gibbon *Hylobates (Hylobates) muelleri*

size: 3.5 (2–5).[567] **Home range:** 38ha.[767] **Day range:** 890m (350–1520) *[2920ft (1148–4987)].*[767]

Behavior Diurnal and arboreal.[488] A female Mueller's Bornean gray gibbon reacts to playback of a taped female song by approaching and singing, and the male joins in the duet. When a tape of the male song is played, the male does not sing but leads the group to the song's source. In this experiment the female appears to exclude other females and enforce monogamy.[561] **Vocalizations:** A female dominates an interactive duet with a loud "bubbling" call.[488] The male produces lengthy solos.[561] A male's song comprises 6 notes that last 3 seconds each, and the song ends with a trill.[567] There are 4 songs per minute, and precise copies of the songs are rarely repeated. These gibbons start to sing at dawn, between 5 and 6:30 a.m.[567] A single unpaired male sings longer than mated pairs.[567] The calls are very different from those of the sympatric dark-handed gibbons *(H. agilis)*.[488] Captive hybrid offspring of *H. muelleri* and white-handed gibbons *(H. lar)* hint that gibbon songs may

Taxonomy Disputed. 3 subspecies.[286] Some taxonomists recognize this only as a subspecies of *H. agilis.* Others include the subspecies *H. m. albibarbis* in this species rather than in *H. agilis.*[315]

Distinguishing Characteristics Mueller's Bornean gray gibbons vary from brown to gray, with a dark cap and underparts. Males have a pale, incomplete face ring. Females can be darker than males.[118]

Physical Characteristics Head and body length: 420–470mm *[16.5–18.5in].*[654] **Weight:** 5.0–6.4kg *[11.0–14.1lb].*[654] **Intermembral index:** 129.[234] **Adult brain weight:** NA.

Habitat Primary and secondary, logged, dipterocarp forest up to 1500m *[4922ft].*[488]

Diet Fruit, 62%; leaves, 32%; flowers, 4%; animal prey, 2%.[488]

Life History Weaning: NA. **Sexual maturity:** NA. **Estrus cycle:** NA. **Gestation:** NA. **Age 1st birth:** NA. **Birth interval:** 36mo.[488] **Life span:** 47y.[998] Births: Year-round.[488]

Locomotion Brachiation.[488]

Social Structure Monogamous 1 male–1 female with offspring in a cohesive family group.[488] **Group**

Mueller's Bornean Gray Gibbon *continued*

be controlled genetically, as in birds and amphibians. The female hybrid gibbon's song was different from either parent's; the hybrid male's song had elements of both species' songs.[822] **Association:** Partially sympatric and occasionally hybridizing with dark-handed gibbons *(H. agilis)*.[488]

Borneo (Indonesia)

Gibbons walk bipedally on large tree branches.

Pileated or Capped Gibbon *Hylobates (Hylobates) pileatus*

Taxonomy Monotypic species.[286]

Distinguishing Characteristics
Male pileated gibbons are black, with white fingers, toes, and prepubital tuft. Males have thick white brows and white hair that encircles the face. Females are silver buff, with black cheeks and cap. Male and female infants have cream-colored fur. Subadult males are silver buff, like females, and change to black at maturity.[999]

Physical Characteristics Head and body length: NA. **Weight:** ♀ 6360–8640g *[14.0–19.0lb]*, ♂ 7860–10,450g *[17.3–23.0lb]*. **Intermembral index:** NA. **Adult brain weight:** 114.2g *[4.0oz]*.[346]

Male.

Habitat Primary, moist and dry, evergreen and deciduous, and montane forest.[898]

Diet Fruit, 71%; flowers, 15%; leaves, 11%; young shoots, 2%; insects (including termites and caterpillars) and leaf galls, 1%.[488]

Life History Weaning: NA. **Sexual maturity:** NA. **Estrus cycle:** NA. **Gestation:** NA. **Age 1st birth:** NA. **Birth interval:** NA. **Life span:** 39y.[999]

Locomotion Brachiation, 86%; leaping, 8.7%; climbing, 5.9%; bipedal, 1%.[767]

Social Structure Monogamous 1 male–1 female family groups.[767] **Group size:** 4 (2–6).[768] **Home range:** 15–50ha.[898] **Day range:** 833m (450–1350) *[2733ft (1476–4429)]*.[767]

Behavior Diurnal and arboreal. Pileated gibbons begin their activity at dawn (6 a.m.) and end in midafternoon (12:45–3:50 p.m.), for an average activity period of only 8.2 hours.[118] They spend 39% of feeding time in high canopy, 49% in midcanopy, and 2% in lower canopy.[118] Activity budget: Rest, 37%; feeding, 26%; travel, 25%; grooming, 5%; swinging, 4%; play, 3%.[118] **Association:** Sympatric with white-handed gibbons *(H. lar)* in Thailand.[767] **Vocalizations:** These gibbons give their loud call once a day at about

Female.

10 a.m. The female starts singing, and the male joins in halfway through the female's final trilling. Wind and rain may delay or inhibit the great call.[768] **Sleeping site:** Canopy or emergent trees (usually dipterocarps) over 25 meters *[82ft]* high. These gibbons sleep an average of 15.8 hours. The mother sleeps with her infant; the male and other offspring sleep in other trees.[767]

Thailand, Cambodia

kilohertz

5   10   15   20

seconds

Black Gibbon *Hylobates (Nomascus) concolor*

DIGITAL ILLUSTRATION

Taxonomy Disputed. 5 subspecies.[413] Formerly included *H. leucogenys* and *H. gabriellae,* which are now recognized as valid species. A new subspecies was reported from Viet Nam in 1989.[284]

Distinguishing Characteristics Male black gibbons are entirely black. Depending on the subspecies, the female changes at maturity from black to a brown buff, buff, or gray, with variable amounts of brown to black on the crown, chest, and belly. Infants are born yellow, with reddish facial skin. At about 6 months the fur of both genders changes to black.[987]

Physical Characteristics Head and body length: 457–635mm *[18.0–25.0in].*[118] **Weight:** 4.5–9.0kg *[9.9–19.8lb].*[497] **Intermembral index:** 140.[234] **Adult brain weight:** NA.

Habitat Mixed broadleaf, evergreen, deciduous, and semideciduous montane forest up to 2900m *[9515ft],* the highest recorded altitude for any gibbon.[60]

Diet Leaf buds and shoots, 61%; fruit, 21%; leaves, 11%; flowers, 7%.[147] Black gibbons use 53 plant species, including bamboo.[922]

Life History Weaning: 18mo.[497] **Juvenile:** 24–84mo.[497] **Subadult:** 84–108mo.[497] **Sexual maturity:** NA. **Estrus cycle:** NA. **Gestation:** NA. **Age 1st birth:** NA. **Birth interval:** NA. **Life span:** 36y.[987] By age 3, offspring are no longer carried.

Locomotion Brachiation.

Social Structure Monogamous family groups.[60] An early preliminary study reported this species to live in 1 male–multifemale groups with up to 4 females,[330] but later studies have reported monogamous pairs with larger group sizes during the dry season.[748] One hypothesis is that the scarcity of high-energy fruit in the dry season may relax territorial boundaries.[748] **Group size:** 2–5.[749] **Home range:** 300–500ha.[497] **Day range:** 90–750m *[295–2461ft].*[497]

Behavior Diurnal and arboreal. Black gibbons will come to the ground to eat bamboo shoots. Males lead the group and help with parental care. The subspecies *H. c. hainanus,* which lives on Hainan Island off China, is critically endangered and on the verge of extinction.[206]

Vocalizations: The great songs of the male and female are different and distinct. The male's call is a long, modulated whistle; the female's, a complex trill.[497] The morning song lasts 10 minutes.[147] Juveniles join in the parental duet, but their songs are often incomplete and may be the song of the opposite sex.[60] The songs and the male-female duet must be practiced.

China, Viet Nam, Laos

seconds

Golden-cheeked Gibbon *Hylobates (Nomascus) gabriellae*

Female.

Taxonomy Disputed. Monotypic. Elevated from a subspecies of *H. concolor* in 1989.[315]

Distinguishing Characteristics Golden-cheeked gibbon males are black, with a small amount of reddish golden hair on the cheeks and more black on the chin.[286] Females are buffy. Infants are whitish buff.[286]

Physical Characteristics Head and body length: NA. **Weight:** ♀ 5750g *[12.7lb].*[987] **Intermembral index:** NA. **Adult brain weight:** NA.

Habitat Tropical forest.
Diet NA.
Life History Weaning: NA. **Sexual maturity:** NA. **Estrus cycle:** NA. **Gestation:** NA. **Age 1st birth:** NA. **Birth interval:** NA. **Life span:** 46y.[987]
Locomotion Brachiation.
Social Structure NA. **Group size:** NA. **Home range:** NA. **Day range:** NA.
Behavior Diurnal and arboreal. As of 1995, golden-cheeked gibbons in their natural habitat had not been the subject of a long-term study. **Vocalizations:** In captivity the male begins the great call, and the female

Laos, Viet Nam, Cambodia

Juvenile male.

joins in. The male repeats his call after the female finishes.[164]

Chinese White-cheeked Gibbon *Hylobates (Nomascus) leucogenys*

Male.

Habitat Tropical broadleaf evergreen forest.[921]

Diet NA.

Life History Weaning: NA. **Sexual maturity:** ♀ 48mo.[285] **Estrus cycle:** NA. **Gestation:** 200–212d.[285] **Age 1st birth:** 54mo.[285] **Birth interval:** NA. **Life span:** 28y.[988]

Locomotion Brachiation.[305]

Social Structure Monogamous 1 male–1 female with offspring in a cohesive territorial family group.[305] **Group size:** 3.7 (2–6).[305] **Home range:** NA. **Day range:** NA.

Behavior Diurnal and arboreal. The behavior of Chinese white-cheeked gibbons, other than their vocalizations, has not been well studied in their natural habitat.

Vocalizations: The male and female duet is antiphonal (not in unison). The female begins a song, and the male answers with his. The male's song starts simply and gets more complex.[165] The male may display with branch shaking and brachiating while singing.[165] The male's song may indicate a pair's locale to neighboring gibbons. The female's song repulses neighbors.[165] The great call enhances intragroup cohesion.[165]

China, Viet Nam

Female.

Taxonomy Disputed. 2 subspecies.[286] Elevated from a subspecies of *H. concolor*.[315]

Distinguishing Characteristics Male Chinese white-cheeked gibbons are black, with white cheeks. Females are golden or buff. Neonates have a whitish buff coat.[988]

Physical Characteristics Head and body length: 457–635mm *[18.0–25.0in]*.[118] **Weight:** ♂ 5.6kg *[12.3lb]*, ♀ 5.8kg *[12.8lb]*.[490] **Intermembral index:** 140. **Adult brain weight:** NA.

Siamang *Hylobates (Symphalangus) syndactylus*

Taxonomy 2 subspecies. Classified in its own genus, *Symphalangus,* until 1972.[286]

Distinguishing Characteristics Both male and female siamangs are black and large, with a stocky build relative to other gibbons. Adults have a dark gray throat sac.[1000] The male has a long scrotal tuft that resembles a short tail.[118]

Physical Characteristics Head and body length: 737–889mm *[29.0–35.0in]*.[118] **Weight:** ♀ 10,000–11,140g *[22.0–24.6lb]*, ♂ 12,270–14,770g *[27.0–32.6lb]*.[490] **Intermembral index:** 147.[425] **Adult brain weight:** 121.7g *[4.3oz]*.[346]

Habitat Primary and secondary lowland and montane

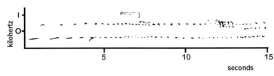

forest up to 3800 meters *[12,468ft]*. Siamangs are found at lower densities in secondary forest.[898]

Diet Leaves, 59%; fruit, 31%; flowers, 8%; animal prey, 3%.[145] Fig seeds are dispersed.[682]

Life History Weaning: NA. **Sexual maturity:** NA. **Estrus cycle:** NA. **Gestation:** 189–239d.[285] **Age 1st birth:** 108mo.[708] **Birth interval:** 36,[708] 59mo.[285] **Life span:** 35y.[708]

Locomotion Brachiation.[145]

Social Structure Monogamous 1 male–1 female with offspring in a family group.[145] **Group size:** 3.5[139] (2–10).[898] **Home range:** 47ha[681] (15–50).[898] **Day range:** 738–969m *[2421–3179ft]*.[488]

Siamang *continued*

Behavior Diurnal and arboreal. Siamangs feed mostly in the midcanopy (61%) but also in emergent trees (25%) and understory (14%).[145] The female leads the group and eats faster and longer than the male.[682] Because of their leafy diet and larger body size, siamangs are able to live at higher altitudes (which have fewer fruit trees) than other gibbons.[682] The lower amount of activity that characterizes siamangs has been hypothesized to reflect their less nutritionally rich diet of leaves or the higher cost of detoxifying secondary compounds.[145] The male siamang carries the infant after it is 8 months old and returns it to the female to nurse and to sleep. **Association:** Sympatric with the smaller-bodied white-handed and dark-handed gibbons *(H. lar, H. agilis)* but not in extensive competition for the same food. Siamangs eat more leaves from the more common trees; the gibbons seek out rare and high-quality fruits by traveling twice as far each day. Siamangs also associate with dusky leaf monkeys *(Trachypithecus obscurus),* but the dietary overlap of tree species used by both is only 16%.[145] **Vocalizations:** Siamangs inflate their throat sac to make their great call. The male screams his part of the call while the female produces a series of barks.[118] **Sleeping site:** Emergent trees.[682]

Siamangs have a throat sac that they inflate when they make their great call.

Female and infant.

Sumatra (Indonesia), Malay Peninsula

The scrotal tuft of the male.

Chimpanzee *Pan troglodytes*

Great Apes   *Families:* Pongidae and Hominidae

The great apes comprise the families Pongidae and Hominidae. They have larger bodies and bigger brains than other primates and are sexually dimorphic—the males are bigger than the females. Generally great apes are less arboreal and more terrestrial than lesser apes. Unlike lesser apes, great apes build nests and have no ischial callosities. The social structure of each genus of great ape is different. Orangutans are primarily solitary. Gorillas live in one male–multifemale troops. Chimpanzees live in fission-fusion communities. Humans have a variety of social structures, with monogamous pairs and one male–multifemale family groups occurring most often.

Great apes can recognize themselves in a mirror. In captive experiments, members of each great ape species have shown the ability to learn a sign language such as American Sign Language. An orangutan named Chantek, a lowland gorilla named Koko, and Washoe and many other chimpanzees use sign language to communicate with humans and other captive members of their species. Kanzi, a bonobo at the Language Research Center in Decatur, Georgia, can understand human speech and uses lexigrams to communicate with humans. He has even learned to make stone tools.[951]

Orangutans   *Family:* Pongidae

Orangutans are more arboreal and more solitary than the other apes. Though the males travel on the ground more than the females, they all live in the trees and use their grasping hands and feet to climb slowly and to suspend themselves from tree branches to feed. Adult females travel with their offspring, but adult males interact only with estrous females. Orangutans have one of the most prolonged developments of any mammal. Subadult male orangutans, though sexually mature, may not develop the fatty cheek pads that characterize a fully adult male until they are between 15 and 19 years old.[1016] The females have their first infant between the ages of 12 and 15 and give birth only every 7 to 8 years after that.[523] This low reproductive rate makes it difficult for orangutan populations to recover from the rapid habitat destruction that has caused them to decline precipitously in the last two decades.

Orangutans diverged from the other apes between 16 million and 10 million years ago. One of their recent ancestors was the largest known primate, called *Gigantopithecus blacki.* This species is estimated to have weighed up to 300 kilograms (about 660 pounds), and it lived in Asia during the Pleistocene epoch, 1 million to 2 million years ago. *Gigantopithecus* is believed to have been terrestrial and, like the giant panda, may have eaten bamboo. Its teeth were first discovered in a Chinese drugstore, where they were ground up to be used for folk medicine.[234]

Gorillas, Chimpanzees, and Humans   *Family:* Hominidae

Traditionally all great apes were placed in the family Pongidae, and humans were placed in a separate family—Hominidae. Recently, new fossils and biomolecular and genetic evidence have led most experts to conclude that the African apes diverged from each other about 5 million years ago and are more similar to each other than to orangutans. We *Homo sapiens* share 98.4% of our DNA with chimpanzees *(Pan troglodytes).* In contrast, the bird species red-eyed vireos and white-eyed vireos share only 97.1% of their genetic material.[1002]

Though we and African apes have many obvious morphological differences, gorillas *(Gorilla gorilla),* chimpanzees *(Pan troglodytes),* and bonobos, or pygmy chimpanzees *(Pan paniscus),* are our closest relatives. We are similar enough to belong to the same family. An extraterrestrial taxonomist using our current criteria would probably put us humans in the same genus as the chimpanzees.[1002]

Gorillas and chimpanzees walk quadrupedally and use their knuckles to carry the weight of their head and torso. The fossil record shows that our human ancestors diverged from the other apes over the last 5 million years in Africa by developing bipedal locomotion first and then an enlarged brain. They began to use flake tools about 2.5 million years ago. Human language and culture are believed to have developed relatively recently, with the most obvious advances being documented by archeologists as having occurred in the last 40,000 years.[1002]

In response to our greater understanding of our close similarity to great apes in terms of their capacity for self-recognition and their innate intelligence, a book entitled *The Great Ape Project,* edited by Peter Singer and Paola Cavalieri, calls for certain civil rights to be granted to all great apes, including the rights to life, liberty, and freedom from torture.[1015]

Sumatran Orangutan *Pongo abelii*

Female.

succulent fruits are preferred, and large fruits with a hard husk are eaten.[844]

Life History Weaning: 42mo.[346] **Sexual maturity:** ♀ 84mo, ♂ 114mo.[346] **Estrus cycle:** 30d.[346] **Gestation:** 260d.[346] **Age 1st birth:** 144–196mo.[1018] **Birth interval:** 96mo (72–144)[1018] **Life span:** >50y.[1018]

Locomotion Suspensory, quadrumanous climbing (using both hands and feet to grasp), and quadrupedal walking on the ground on clenched fists rather than on knuckles.

Social Structure Males are solitary, and females travel with their offspring.[898] Adolescent males and females

Taxonomy Disputed. Formerly a subspecies of *P. pygmaeus*. Recent genetic evidence suggests that this is a full species that diverged 1.5 million years ago.[112] Dermatoglyphic evidence (from fingerprints) also supports 2 *Pongo* species.[66]

Distinguishing Characteristics Sumatran orangutans are thinner than Borneo orangutans (*P. pygmaeus*) and have a paler red coat, longer hair, and a longer face. The Sumatran male's cheek

pads are covered with a fine white hair.[512]

Physical Characteristics Head and body length: NA. **Weight:** NA. **Intermembral index:** NA. **Adult brain weight:** 413.3g *[14.6oz]*.[346]

Habitat Primary lowland swamp to montane forest up to 4000 meters *[13,124ft]*, but altitudes below 500m *[1641ft]* are preferred.[898]

Diet Fruit, leaves, bark, flowers, and animal prey, including small mammals and termites. Over 200 plant species are eaten.[1007] Ripe

C. VAN SCHALK (NATURAL HABITAT)

Male.

This stick was modified and used by an orangutan to probe a tree hole to obtain social insects.

Sumatran Orangutan *continued*

Juvenile.

associate occasionally.[1007] **Group size:** 1–3. **Home range:** 200–1000ha. [898] **Day range:** 800–1200m *[2625–3937ft]*.[1007]

Behavior Diurnal and arboreal. One population of Sumatran orangutans found in a lowland swamp has been documented to modify and use sticks as tools to obtain honey and social insects from holes in trees.[852] They also use stick tools to remove seeds from fruits with stinging hairs.[1005] At the Ketambe study site, females have killed and eaten slow lorises *(Nycticebus coucang)*. It takes up to 2 hours for an orangutan to eat the whole animal.[845] **Mating:** The female initiates and is the more active partner in consorting with an adult male. The may consort up to 20 days, with more than 20 copulations in various positions, including face-to-face. Subadult males occasionally force adult females to copulate.[740]

Vocalizations: The male long call, a long series of reverberating grunts, carries over 1km *[3281ft]* through the forest to signal the resident male's location.

Sumatra
(Indonesia)

Borneo Orangutan *Pongo pygmaeus*

Fully adult males have large cheek pads.

Taxonomy Disputed. Formerly included 2 subspecies, but *P. abelii* has been elevated to a full species.

Distinguishing Characteristics Bornean orangutans have coarse, long hair that varies from orange to brown or maroon in adults. Males have a large throat pouch. The infant's pink face darkens with age.[112]

Physical Characteristics Head and body length: ♀ 780mm *[30.7in]*, ♂ 970mm *[38.2in]*.[112] **Weight:** ♀ 37.0kg (33.0–45.0) *[81.5lb (72.7–99.2)]*,[425] ♂ 77.5kg[425] (80.0–9.1)[869] *[170.8lb (176.3–200.6)]*.[869] **Intermembral index:** 139.[234] **Adult brain weight:** 413.3g *[14.6oz]*.[346] Cranial

capacity: ♂ 425 cu cm *[25.9 cu in]*, ♀ 370 cu cm *[22.6 cu in]*.[666] Arm span: 2250mm *[88.6in]*.[625] Penis length: 40mm *[1.6in]*.[300]

Habitat Primary rain forest from lowland swamp to upland forest up to 1500m *[4922ft]*.[315]

Diet Fruit, 60%; young leaves, shoots, bark, soil, and animal prey, including insects, eggs, baby birds, and squirrels.[512] Males eat more termites than females do. Over 400 plants and plant parts are eaten.[256]

Life History Infant: 48mo. **Weaning:** 42,[346] 60–84mo.[523] **Juvenile:** 48–108mo.[869] **Subadult:** 108–240mo.[869] **Sexual maturity:** ♀ 84mo, ♂ 114mo.[346] **Estrus cycle:** 30[346]–35d.[523] **Gestation:**

244d (223–267).[523] **Age 1st birth:** 144–180mo.[523] **Birth interval:** 32mo in captivity;[346] 84–96mo.[523] **Life span:** 59y.[625] Births: Year-round.[625] Development of the male's large cheek pads is influenced by social factors. In captivity, a subordinate male housed with a dominant adult male may not develop cheek pads until the 2 males are separated.[1016]

Locomotion Suspensory; quadrumanous climbing (using both hands and feet to grasp);[702] quadrupedal walking on the ground on clenched fists rather than on knuckles; occasional bipedal standing.[625]

Social Structure The male is solitary and his territory overlaps the territories of several separate females that travel with their offspring.[869] Two adult females occasionally travel together for up to 3 days. Several subadults may associate with each other.[625] Males emigrate; female offspring settle near their mother's home range.[869] **Group size:** 1–3. **Home range:** 42–777ha.[702] **Day range:** 305–800m *[1001–2625ft]*.[702]

Behavior Diurnal and arboreal. Bornean orangutans feed in the periphery of trees at 20–30m *[66–98ft]*, suspended by at least 2 limbs.[702] Males use the lower 14 meters *[46ft]* of the forest and the ground 4 times more than females.[702] The small food patches in Bornean forests, which cannot feed more than 1 orangutan, force the orangutans to remain solitary and limit their social interactions.[869] Orangutans in the care of humans have been taught American Sign Language;[935] 1 orangutan has learned 150 signs.[556] Activity budget: Feeding, 45.9%; rest, 39.2%; travel, 11.1%; nest building, 1%.[702] **Association:** In south Borneo these orangutans compete with and displace proboscis monkeys *(Nasalis larvatus)*.[924] **Mating:** Estrous females are attracted to the long call of the adult male and move toward him to initiate sex.[594] Matings are often prolonged and are sometimes face-to-face.[625] Subadult males may forcibly rape females. **Vocalizations:** 18.[966] The most notable is the long call of the adult male; it can be heard up to 1km *[3281ft]* away and tells other orangutans his location.[625] The

NATURAL HABITAT

Making a night nest.

Borneo Orangutan *continued*

Adult female and infant.

The grasping hands and feet of orangutans are used in quadrumanous climbing.
This juvenile is giving a "kiss-squeak" vocalization, a sign of annoyance.

"kiss squeak" is a sign of annoyance. **Sleeping site:**
Borneo orangutans construct a new leaf nest each
night by breaking branches of trees to form a plat-
form. They may also construct leaf shelters to protect
themselves from rain.[625]

Borneo
(Indonesia)

Orangutans occasionally stand bipedally.

The orangutan on the right, a former captive, learned sign language when she was a juvenile.
The male on the left is still a subadult.

Mountain Gorilla *Gorilla gorilla beringei*

NATURAL HABITAT

This silverback male named Rugabo, the leader of one of the groups habituated for tourism, was shot and killed in 1995 in the aftermath of the civil war in Rwanda.

Taxonomy 1 of 3 subspecies of *G. gorilla*.[333]

Distinguishing Characteristics
Mountain gorillas have longer hair than the other *G. gorilla* subspecies. They are black except for the adult male, which has a silvery white back and gray hips.[333]

Physical Characteristics Head and body length: NA. **Weight:** ♀ 97.7kg *[215.3lb]*, ♂ 159.2kg *[350.9lb]*.[425] **Intermembral index:** 116.[234] **Adult brain weight:** NA.
Habitat Montane and bamboo forest at 2800–3965m *[9187–13,009ft]* elevation.[413]
Diet Leaves, shoots, and stems, 85.8%; wood, 6.9%; roots, 3.3%; flowers, 2.3%; fruit, 1.7%; animal prey (including grubs, snails, and ants), 0.1%.[249] These

gorillas eat parts of 142 plant species, including wild celery, thistles, nettles, and bamboo,[869] but only 3 types of fruit.[687] All ages and both genders occasionally eat their own feces (coprophagy).[413]

Life History Infant: 36–48mo.[869] **Weaning:** NA. **Juvenile:** 48–84mo.[599] **Subadult:** 108–192mo.[869] **Sexual maturity:** NA. **Estrus cycle:** 28d (27–39).[413] **Gestation:** NA. **Age 1st birth:** 120–144mo.[869] **Birth interval:** 36–59mo.[413] **Life span:** 40–50y.[599] Births: Year-round. Females are receptive for 1–3 days per cycle.[413]

Locomotion Quadrupedal knuckle walking; climbing; limited bipedal standing.[980]

Social Structure 1 male–multifemale groups (60.7%). Less often (35.7% of sightings) there are 2 or more silverbacks (adult males) per group.[413] Troop territories overlap; the silverback defends the troop, not the territory.[979] **Emigration:** Most males and 60% of females emigrate.[869] Males leave at age 11 and may travel alone or with other males for 2–5 years before attracting a female to form a new group. Starting at age 8, females transfer to a new group.[413]

NATURAL HABITAT

Adult female.

Mountain Gorilla *continued*

Group size: 9. **Home range:** 400–800ha.[249] **Day range:** 400m (100–2500) *[1312ft (328–8203)]*.[249]

Behavior Diurnal and terrestrial. Adult mountain gorillas climb less than immature ones. The silverback is dominant to all other members and leads the direction of travel. There is no discernible female hierarchy.[979] Group cohesion is enhanced by the desire to remain close to the silverback, which protects juveniles and infants. Infants orphaned between ages 3 and 5 spend 2–3 times more time with the silverback.[869] The silverback's threat display is impressive, with hooting, chest beating and slapping, and throwing of vegetation.[413] Such a display is usually mild except when a female tries to transfer to a nearby group.[413] Infanticide and

NATURAL HABITAT

This silverback is taking the edible outer layer off a stem.

cannibalism of an infant have been reported.[367] These gorillas spend 25% of the day eating and rest during midday. In 1994 the population was estimated at about 600 individuals—before the civil unrest in Rwanda and Zaire, which has caused the death of an unknown number of gorillas.[24] **Mating:** Young (nulliparous) estrous females have a slight sexual swelling and initiate mating.[599] **Vocalizations:** 12.[347] The belch vocalization is a contact call and a sign of contentment made while foraging.[599] **Sleeping site:** Nests are built on the ground, using nonfood plants.[249]

Uganda, Rwanda,
Zaire

NATURAL HABITAT

This mountain gorilla is sitting in a tobacco field that was recently carved out of his forest habitat.

NATURAL HABITAT

A 4-year-old juvenile male.

Western Lowland Gorilla *Gorilla gorilla gorilla*

Habitat Dense primary and secondary forest, lowland swamp, and montane forest up to 3050m [10,007ft].[898]

Diet Fruit, 67%; seeds, leaves, stems, and pith, 17%; animal prey (including termites, caterpillars, and other insect larvae), 3%.[413] These gorillas are seasonal frugivores, eating fruits during the wet season and more herbs and bark in the dry season. They eat parts of at least 97 plant species.[687] They eat termites every day. The diet contains high levels of phenols and condensed tannins, but alkaloids and fatty fruits are avoided. Sugary fruits and pith are preferred, as well as protein-rich leaves and bark. The diet is similar to that of chimpanzees, but these gorillas eat larger fruits and mature leaves and stems.[701]

Life History Weaning: 52mo.[346] **Sexual maturity:** ♀ 78mo, ♂ 120mo.[346] **Estrus cycle:** 32d (25–42).[413] **Gestation:** 256d[346] (237–285).[413] **Age 1st birth:** 108–132mo.[666] **Birth interval:** 48mo.[346] **Life span:** 50y.[708]

Locomotion Quadrupedal knuckle walking; climbing; limited bipedal standing; suspensory (8%).[687] Males stay near the main trunk of a tree or on large branches, usually distributing their weight on several supports.[687]

Social Structure 1 male–multifemale groups.[413] Large groups may have 2 silverback males with several females and their offspring.[687] **Emigration:** Both males and females emigrate. Males may travel with other males or remain solitary; females transfer to another group.[413] **Group size:** 3–21.[413] **Home range:** 800–1800ha.[687] **Day range:** 2300m [7546ft].[687]

Behavior Diurnal and terrestrial; arboreal climbing. Western lowland gorillas climb more than mountain gorillas.[687] The chest beating display of these lowland gorillas is a threat display that may include hooting and throwing of vegetation.[413] Lone males rest more than gorillas in a troop and range farther in the wet season in search of fruit. Males spend more time on

Taxonomy This subspecies may be a separate species from the 2 eastern subspecies.[586]

Distinguishing Characteristics The top of the western lowland gorilla's head is brown. The rest of the body is brown gray. Adult males have an enlarged sagittal crest and a silvery color on the back that extends to the rump and thighs.[333]

Physical Characteristics Head and body length: ♀ 1500mm [59.1in], ♂ 1700mm [66.9in].[333] **Weight:** ♀ 71.5kg [157.6lb], ♂ 169.5kg [373.6lb].[425] **Intermembral index:** 115.6.[425] **Adult brain weight:** 505.9g [17.8oz].[346] Cranial capacity: ♂ 535 cu cm [32.6 cu in], ♀ 460 cu cm [28.1 cu cm].[666] Penis length: 30mm [1.2in].[300]

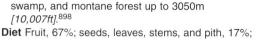

Nigeria to Zaire

Gorillas have grasping feet.

An infant rides on its mother's back.

Western Lowland Gorilla *continued*

Male lowland gorillas have a brown sagittal crest.

Gorillas walk on their knuckles, not their palms.

the ground eating herbs in the dry season.[687] Females feed higher in trees and eat more leaves than males.[687] **Mating:** Gorillas mate dorsoventrally and occasionally face-to-face. **Vocalizations:** Vocalizations and chest beating are heard more in the morning.[413] **Sleeping site:** These gorillas make a new nest each night.

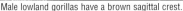

Eastern Lowland Gorilla *Gorilla gorilla graueri*

Taxonomy 1 of 3 subspecies of *G. gorilla*.[333]

Distinguishing Characteristics Eastern lowland gorillas have a longer face and a broader chest than western lowland gorillas. Easterns are black except for the adult male, which has a silvery, saddle-shaped back.[333]

Physical Characteristics Head and body length: NA. **Weight:** ♀ 80.0kg *[176.3lb]*, ♂ 175.2kg *[386.1lb]*.[234] **Intermembral index:** NA. **Adult brain weight:** NA.

Habitat Primary, secondary, and bamboo forest and marshes at 1000–2400m *[3281–7874ft]* elevation.[898]

Diet Fruit, leaves, bark, orchid bulbs, ants (6 species). These gorillas eat the most protein-rich parts of at least 104 plant species: 33 vines, 25 herbs, 20 trees, 13 shrubs, 6 grasses, 4 ferns, and 3 epiphytes.[298] They will raid banana, pea, maize, and taro crops.[898]

Life History NA.

Locomotion Quadrupedal knuckle walking; climbing; limited bipedal standing.[413]

Social Structure 1 male–multifemale groups. **Group size:** NA. **Home range:** 3200ha.[298] **Day range:** NA.

Behavior Diurnal, arboreal, and terrestrial. Eastern lowland gorillas have a larger home range than mountain gorillas and eat only 15 of the 56 species that are found in both ranges.[105] Eastern lowland gorillas reportedly have dietary traditions: groups do not eat the same foods as other groups if they have not been brought up on the foods from an early age.[105] During October and November these gorillas occupy bamboo forest, digging 200mm *[8in]* holes in the ground to reach young bamboo shoots.[105] At other times of the year they forage in secondary forest and in patches of elephant grass, whose pith is an important part of the diet.[613] **Sleeping site:** New nests are made on the ground each night.[730]

Zaire

H. & J. MORLAND (NATURAL HABITAT)

The eastern lowland gorilla eats more than 100 plant species.

Bonobo or Pygmy Chimpanzee *Pan paniscus*

Vernon is an adult male that lives in a zoo. Notice that he has groomed the hair off his arms, probably because of the boredom of captivity.

Taxonomy Monotypic. The genus name was changed from *Chimpansee* in 1969.[315] The common name "pygmy chimpanzee" is a misnomer because this species is the same size as some *P. troglodytes*.

Distinguishing Characteristics Bonobos have a more slender build than chimpanzees *(P. troglodytes),* with longer limbs and a narrower chest. The face is black from birth. The hair on top of the head appears to be parted, making the ears less visible. A white tail tuft is seen in adults and infants.[900]

Physical Characteristics Head and body length: ♀ 700–760mm *[27.6–29.9in]*, ♂ 730–830mm *[28.7–32.7in]*.[900] **Weight:** ♀ 31.0kg *[68.3lb]*, ♂ 39.0kg *[86.0lb]*.[900] **Intermembral index:** 102.7.[425] **Adult brain weight:** NA. Cranial capacity: ♂ 350 cu cm *[21.4 cu in]*, ♀ 345 cu cm *[21.0 cu in]*.[666] Canine length:

♀ 9mm *[0.4in]*, ♂ 11.1mm *[0.4in]*.[490]

Habitat Primary and secondary tropical lowland rain forest.[199]

Diet Fruit, leaves, terrestrial herbs, truffles, honey, stems, shoots, and animal prey, including earthworms, termites, ants, reptiles,[94] squirrels, duikers, and bats.[49] Bonobos have not been observed to hunt meat or eat monkeys.[724]

Life History Infant: 0–48mo.[982] **Weaning:** NA. **Juvenile:** 48–84mo.[982] **Subadult:** 84–132mo.[982] **Sexual maturity:** NA. **Estrus cycle:** 35–40d.[620] **Gestation:** 240d.[666] **Age 1st birth:** 168mo.[982] **Birth interval:** 54mo (48–72).[982] **Life span:** 40y. Adult females have swellings throughout their cycles.[620]

Locomotion Quadrupedal knuckle walking; climbing; suspensory; some bipedal.[981]

Social Structure Multimale-multifemale fission-fusion communities.[394] Bonobos travel in mixed-gender foraging groups.[900] Troops avoid each other and are not overtly hostile.[394] **Emigration:** Females emigrate;[394] males remain in their mother's group. **Group size:** Community, 50–200; foraging, 6–15.[900] **Home range:** 2200–5800ha.[982] **Day range:** 1200–2400m *[3937–7874ft]*.[982]

Behavior Diurnal, arboreal, and terrestrial. Bonobos come to the ground to feed on terrestrial herbaceous vegetation, which is

plentiful and may allow them to forage in larger and more stable groups than chimpanzees.[520] Bonobos are less aggressive and reconcile more often than chimpanzees.[157] They have fewer male–male grooming associations, more male–female associations, and strong female–female associations.[620] Male bonobos share food with females and share the least with other males.[620] Tool use is limited to leaf umbrellas and branch dragging during displays. Bonobos use their hands to dig for earthworms along stream banks and for truffles in secondary forest.[49] They wash roots and herbs in stream water.[49] In Zaire, humans hunt them for food and pets.

Mating: Bonobos commonly engage in sexual behavior that has no reproductive value for conception[157] but has social value, reducing tensions within the group caused by competition for food or by agonistic encounters within the group. Sexual behavior includes dorsoventral and face-to-face mating and female–female genital-to-genital rubbing and is exhibited by all combinations of ages and genders.[199] **Vocalizations:** 14,[347] including a high-pitched food peep, hooting, an alarm bark, and a copulatory scream.[644] The "high hoot" is a distance call often given by a whole group at any

Bonobos of all ages engage in sexual behavior, often face-to-face.

From puberty, female bonobos have a continuous estrous swelling.

Bonobo or Pygmy Chimpanzee *continued*

Bonobos are less aggressive than chimpanzees.

time of day or night in response to calls of other bonobos, to thunder, and to falling trees.[376] Humans can hear this call for half a kilometer *[1641ft]*.[841] The "high hoot" may help bring together dispersed foraging parties to a common sleeping site.[376] **Sleeping site:** A new nest is built in a tree every night.

Occasionally bonobos walk bipedally.

Zaire

Infant bonobos are born with a black face.

Bonobos often feed on the ground on herbaceous vegetation.

229

Chimpanzee *Pan troglodytes*

Taxonomy Disputed. 3 subspecies. There is genetic evidence that the western subspecies, *P. t. verus,* may be a separate species from the eastern subspecies, *P. t. troglodytes* and *P. t. schweinfurthii.*[23] The genus name was changed from *Chimpansee* in 1969.[315]

Distinguishing Characteristics Chimpanzees are black, but some old individuals have a gray back. Both genders often have short white beard. The ears are prominent. Infants have a white tail tuft and pink to brown facial skin, which darkens by adulthood.[900] Chimpanzee subspecies are not very distinctive in appearance;[900] few zoos are sure of which subspecies they exhibit.

Physical Characteristics Head and body length: 816mm (737–959) *[32.1in (29.0–37.8)].*[413] **Weight:** ♀ 32–47kg *[70.5–103.6lb],* ♂ 40–60kg *[88.2–132.2lb].*[869] **Intermembral index:** 103–106.[234] **Adult brain weight:** 410.3g *[14.5oz].*[346] Cranial capacity: ♂ 400 cu cm *[24.4 cu in],* ♀ 355 cu cm *[21.7 cu in].*[666] Canine length: ♀ 11.7mm *[0.5in],* ♂ 15mm *[0.6in].*[490] Penis length: 80mm *[3.1in].*[300] Weights and the intermembral index vary with the subspecies.[869]

Habitat Primary and secondary, dry woodland savanna, grassland, and tropical rain forest from lowland to montane up to >3000m *[>9843ft].*[413]

Diet Fruit, 45–76%; leaves, 12–45%; flowers, 1–18%; seeds, 1–11%; animal prey (including mammals, 10 bird species, ants, termites, wasps, and grubs) and galls, 0–5%.[300] The diet varies seasonally.[300] Over 250 food types are eaten.[899] Mammals eaten include other primates (12 species), bushbucks, bushpigs, duikers, rodents, and hyraxes.[300] Chimpanzees use some plants for what has been described as medicinal purposes.[904] Researchers have isolated an anti-tumor agent in 1 such plant.[470]

Chimpanzee knuckles are specialized for knuckle walking.

This infant is having a playful swing in the trees. Note the white tail tuft.

Chimpanzee *continued*

Life History Infant: 60mo.[869] **Weaning:** 48mo.[346] **Juvenile:** 60–96mo.[869] **Subadult:** 96–132mo.[869] **Sexual maturity:** ♀ 135mo, ♂ 156mo.[346] **Estrus cycle:** 36d.[346] **Gestation:** 240d.[666] **Age 1st birth:** 168–180mo.[869] **Birth interval:** 60mo.[346] **Life span:** 53y.[708] Mating and births: Year-round. Birth peak: Oct–Nov.[300]

Locomotion Quadrupedal knuckle walking and running; climbing; suspensory; leaping; limited brachiation; occasional bipedalism.[981]

Social Structure Multimale-multifemale, fission-fusion communities in which a core of related males patrol the community territory boundaries.[300] Females live a more solitary life, spending 65% of their time alone with their offspring within a core area[899] half the size of the males'. Males have a stable dominance hierarchy and are dominant over females.[300] **Emigration:** Females emigrate to other communities.[869] **Group size:** ♀ 7–25, ♂ 5–16,[582] up to 100;[869] foraging, 4.[903] **Home range:** 1250,[899] 3900–7800ha.[413] **Day range:** 3900m [12,796ft].[903]

Behavior Diurnal, arboreal, and terrestrial. Chimpanzees travel most on the ground but feed in trees during the day.[981] It has

Using a grass stem as a tool, this female *(P. t. schweinfurthii)* is extracting termites from a mound.

been hypothesized that chimpanzees in different areas have different cultural traditions (proto-culture) that are passed on from one generation to the next.[1011] Examples include cooperative hunting of adult western red colobuses *(Procolobus badius)* in Ivory Coast,[63] nut cracking in West Africa,[1004] termite fishing in Tanzania,[1001] and medicinal plant use in Uganda.[1010] Meat

sharing by males is a coalition strategy to reward friends and allies.[842] Chimpanzees modify and use objects in their environment as tools for termite fishing, ant dipping, and sponging water to drink and as hammers, anvils, and weapons.[300] In 1993, in the first documented case of a nonhuman primate using a tool to capture a mammal, a chimp was observed using a modified tool to capture a squirrel in a tree hole.[391] In West Africa, chimps carry stones to *Coula* and *Panda* nut trees in order to crack open the hard nuts; it takes a young chimp 4–5 years to learn this nut-cracking behavior.[1004] The males of one commu-

Chimpanzees have 34 vocalizations that humans can recognize.

A West African chimpanzee *(P. t. verus)* cracks palm nuts with a stone hammer on a rock anvil.

231

Chimpanzee *continued*

nity have stalked and killed members of another community in what was termed a war. Males form alliances and coalitions to further their position in the hierarchy. To intimidate rivals, they enact aggressive displays that include branch dragging, loud vocalizations, chasing, and attacking. A third male may mediate a dispute, bringing 2 rivals together to groom and thus reconcile.[159] Captive chimps have been observed to use deception by using the appearance of friendly reconciliation to draw in and then retaliate against an aggressor that escaped an earlier conflict.[159] Ten cases of cannibalistic infanticide have been reported from Mahale and Gombe.[367] Allomothering includes adoption of orphaned chimps.[300] Chimpanzees are hunted by humans for food and pets, and approximately 2000 are in U.S. medical research laboratories.[1014] **Mating:** A dominant male and an estrous female may consort away from other chimps during estrus, but promiscuous mating is common. One

female copulated 50 times in 1 day with 14 different males.[300] Meat is sometimes offered by a male to gain the sexual favors of an estrous female.[770] Both males and females have orgasms and specific mating vocalizations.[413] **Vocalizations:** 34 graded vocalizations, from soft grunts, whimpers, and lip smacks to barks, squeals, and screams.[300] The most notable vocalization is the pant-hoot, given in situations of social excitement. To announce a large supply of fruit to other community members, males drum on the buttresses of trees and give a long-distance food call.[899]

Guinea to Ghana, Nigeria, Cameroon, Gabon, Congo Republic, Central African Republic, Zaire, Uganda, Tanzania

Human *Homo sapiens*

Taxonomy Monotypic. This species shares 98.4% of its genetic materials with its sister species the chimpanzee *(Pan troglodytes)*.[1002]

Distinguishing Characteristics Humans are distinguished from other hominoids by having an upright bipedal posture, a lack of fur covering the body, a nonopposable hallux (big toe), and an extremely large bulbous brain.[234]

Physical Characteristics Head and body length: ♀ 1370–1774mm *[53.9–69.8in]*, ♂ 1450–1816mm *[57.1–71.5in]*.[425] **Weight:** ♀ 55,000g (37,000–76,600) *[121.2lb (81.5–168.8)]*,[425] ♂ 68,230g[425] (43,000–76,100)[209] *[150.4lb (94.8–167.7)]*. **Intermembral index:** 72.[234] **Adult brain weight:** 1250g *[44.1oz]*.[346] Cranial capacity: 1392 cu cm *[84.9 cu in]*.[780] Penis length: Mean 130mm *[5.1in]*.[300] Length and weight ranges are ranges of means.

Habitat All terrestrial habitats on earth except Antarctica,[234] up to 5250m *[17,225ft]* elevation.

Diet Opportunistic omnivores,[234] humans rely on agriculture to produce much of their food. They supplement that diet with hunting and fishing. The diet is partly determined by cultural tradition and local environment.

Life History Weaning: 720d.[346] **Sexual maturity:** ♀ 198mo.[346] **Estrus cycle:** 28d.[346] **Gestation:** 270d.[342] **Age 1st birth:** 192–240mo.[342] **Birth interval:** 10–48mo.[346] **Life span:** 80–90y.[342] Mating and births: Year-round. Infants are born at an immature stage of development relative to their size;[234] they cannot cling and must be carried.

Locomotion Habitually bipedal, which is unique among mammals.[234]

Social Structure More variable than in any other primate species.[234] 1 male–1 female monogamous pairs and 1 male–multifemale family groups are the most common. Humans are fiercely territorial and will go to war to defend or acquire resources. **Emigration:** Both males and females emigrate. **Group size:** Variable. **Home range:** Variable. **Day range:** Variable.

Behavior Diurnal and entirely terrestrial.[234] Humans are behaviorally flexible.[423] They can learn and teach complex ideas and abstract theories.[423] They communicate symbolically, using syntax and a system of grammar called language.[423] They devise and transmit a complex system of shared beliefs from one generation to the next that is referred to as culture.[423] Each culture has its own shared creation myth, language, values, customs, art, and rituals.[423] Humans make and habitually use tools, shelters, and fire.[423] The complexity of tools and technology has evolved to the point that human survival depends on them.[423] Humans have developed a system of food production known as agriculture, using domesticated plants and animals. This innovation has allowed cultures to increase their production, trading networks, population, social complexity, and social stratification and has given rise to the city-state and what is termed civilization.[423] The total weight (biomass) of all living humans is greater than for any other mammal.[423] Human population was 2 billion in 1930 and reached 5 billion in 1986. It will reach 6 billion before 1998.[1019] Humans are changing every aspect of the planet, including its climate and biodiversity. **Association:** Humans associate with several species of mammals

that they have domesticated, including dogs, cats, cows, horses, goats, sheep, pigs, and chickens. Many animals preferentially associate with humans; the most notable are rats, mice, and cockroaches. **Mating:** Human sexual behavior is varied and has other functions besides reproduction. Females are receptive throughout their cycle and show no external sign of ovulation. **Scent marking:** Humans have apocrine glands in the ears, armpits, navel, nipples, and anogenital region. Large numbers of eccrine glands secrete sweat for evaporative cooling.[780] Pheromones produced by humans may influence an individual's sexual attractiveness.[420] Menstrual synchrony among females is probably achieved by scent clues.[420] **Vocalizations:** The human vocal tract and muscles are specialized to increase the variety of sounds produced. Human vocalization, or speech, is organized by the brain into patterns of grammatical and symbolic communication. There are approximately 3000 human languages presently in use. **Sleeping site:** Humans make and habitually use shelters in which to sleep. The materials used vary from simple objects in the environment to complex structures that last hundreds of years.

Acknowledgments

This book would not have been possible without the long and dedicated efforts of all the primatologists, field assistants, trackers, and forest guards who in the last 35 years have studied, written about, and protected the subjects of their research. The primates subjects of this research must be acknowledged for enduring the human interference in their lives that research entails. In some cases they have lost their lives so that we humans can better understand our own biology and evolution.

I would like to thank Barbara Ruth, who planted the seed from which this book has grown. My deep appreciation goes to Patricia Wright, Russell A. Mittermeier, Stephen Nash, and Jane Goodall for their early encouragement and continual support of this book.

I am grateful to all the field researchers and field assistants who allowed me to visit and helped me photograph the primates they painstakingly habituated. I am in awe of their hard work and dedication. Thanks go to the following: Liz Balko, Louise Bartlett, G. Isaberia-Basuta, Bill Bleisch, Ramesh Boonratana, Le Xuan Canh, Colin Chapman, Deborah Forster, Liza Gadsby/Pandrillus, Birute Galdikas, Bettina and Andrew Grieser Johns, Lou Griffin, Clifford Jolly, Sara Kariko, Simon Macomber, Louise Martin, Shuichi Matsumura, Sebastiao Silva Ramos Neto, Ronald and Bette Noe and the staff of Institud d'Ecologic Tropical/Project Colobus (Florent, Yves, Ferdinand, Benjamin, Abdulaye, and Isaac), Claudio Pereira Nogueria, Lucio Peisoto de Oliveira, Vanessa Peerson, Jane Phillips-Conroy, Sheila and David Siddle, Kiki Strecker, Karen Strier, Shirley Strum, Chia Tan, Bernard Thierry, Dan Turk, Marc van Roosmalen, Edward Veado, Patricia C. Wright, and Jiahua Xie.

My sincere thanks go to the following people who helped with permissions and travel logistics: Ezeobi Amarka, Laurentis N. Ambu, Benjamin Andriamihaja, Nick Ashton-Jones, Nenny Babo/Sulawesi Natural Resources Conservation Information Center, Oronto Douglas, Earthwatch, Joe Erwin, Ken Glander, Philip and Francis Hall, Dang Huy Huynh/Institute for Ecology and Biological Resources, Dennis Ifeguere, Institute for the Conservation of Tropical Environments, Turner and Miriam Isoun/Niger Delta Wetlands Foundation, Peter Jenkins, Adam Koru/Gbanraun Wildlife Park Committee, Levantes Group, Clive Marsh, Shirley McGreal, Katharine Okoli, Lilan Ortega/Centre Suisse de Recherches Scientifique, Ramon Palete, Mary Pearl, Nguyen Nhu Phuong, Bruce Powell, Alan Rabinowitz, Emile Rajeriarson, Gervais Rakotoarivelo, Georges Rakotonirina, Loret Rasabo, Jamie Spencer, Cao Van Sung, Nguyen Mau Tai/Viet Nam Department of Forestry Protection, Pierre Telata, Vu Ngoc Thanh, Leonie Vejjajiva, and Vern Wietzel.

I wish to thank all the following people who kindly contributed their photographs to make this book as complete as possible: Jorg Adler, Stuart and Jeanne Altmann, Ron Austing/Cincinnati Zoo and Botanical Gardens, Suzanne Baker, R.A. Barnes, Simon Bearder, Quentin Bloxam, Sue Boinski, Ramesh Boonratana, CNRS–Ecotrop/Alain R. Devez, Carolyn Crockett, Gerald Curbit/Bruce Coleman Ltd., Anna Feistner, Stephen Ferrari, Jack Fooden, Augustine Fuentes, E. Gadsby/Pandrillus, Reg Gates, A. Gautier-Hion/Laboratory of Primatology—University of Rennes (France), Ken Glander, Colin Groves, John K. Hampton Jr. and Suzanne H. Hampton, Alison Hannah, David Haring, Charles H. Janson, Margaret F. Kinnaird, R.C. Kirkpatrick, Zig Koch, Charlie Kuntunidisz, Mark Leighton, Lysa Leland, Jean-Marc Lernould, L.C. Marigo, Clive Marsh, Judith Masters, Scott McGraw, William McGrew, Bernard Meier, Russell A. Mittermeier, John and Hilary Morland, Kristin Mosher/Jane Goodall Institute, Thomas Mutschler, Tilo Nadler, L.T. Nash, Stephen Nash, John Oates, Monica Olson, Peter Oxford/Natural Science Photos, Perth Zoo (Western Australia), L. Porter, H.L. Queiroz, Ed Regan, John Robinson, Peter Rodman, S.M. Rowe, George B. Schaller, Roland and Julia Seitre, Myron Shekelle, Craig B. Stanford, Tom Struhsaker, Zang Tai, Richard Tenaza, E. Thetford, Ronald Tilson, Ximena Valderama, C. van Schaik, Jan Vermer, John M. Walker, Bernard Walton/BBC Natural History Unit, Haven Wiley, R. Wirth, Andrew L. Young, and Qi-kun Zhao.

I would like to acknowledge E.H. Haimoff and Edinburgh University Press for the use of the sonograms from "Acoustic and Organizational Features of Gibbon Songs," which appeared in *The Lesser Apes,* edited by H. Preuschoft, D.J. Chivers, W.Y. Brockleman, and N. Creel (Edinburgh University Press, Edinburgh, 1984). Special thanks go to S.K. Bearder, P.E. Honess, L. Ambrose, and Plenum Press for the use of the sonograms from "Species Diversity among Galagos, with Special Reference to Mate Recognition," which appeared in *Creatures of the Dark: The Nocturnal Prosimians,* edited by L. Alterman, G.A. Doyle, and M.K. Izard (Plenum Press, New York, 1995).

My sincere thanks go to the following people who carefully reviewed the text of this book, often on short notice: Sylvia Atsalis *(Microcebus rufus);* Simon Bearder (Galagonidae); Bill Bleisch *(Trachypithecus* and *Pygathrix (Rhinopithecus));* Ramesh Boonratana *(Pygathrix avunculus);* Douglas Brandon-Jones (colobine taxonomy and characteristics); Diane Doran *(Gorilla* and *Pan);* Ardith Eudey *(Macaca);* John Fleagle *(Presbytis);* Augustine Fuentes *(Nasalis* and *Presbytis potenziani);* Paul Garber *(Saguinus* and *Leontopithecus);* Colin Groves (taxonomy in the introduction); Sharon Gursky (Tarsiidae); Paul Honess (Galagonidae); Charles H. Janson (Cebinae, Alouattinae, and Atelinae); Lois Lippold *(Pygathrix nemaeus* and *P. nigripes);* Scott McGraw (Cercopithecinae); Kelly McNeese *(Pan);* Alan Mootnick (Hylobatidae); L.T. Nash (Galagonidae); Marilyn Norconk *(Callicebus* and Pitheciinae); John Oates *(Colobus* and *Procolobus);* Leila Porter (Lepilemuridae); Anthony Rylands *(Callithrix);* Helga Schultz *(Loris tardigradus);* Gary Shapiro *(Pongo);* Carel van Schaik *(Pongo);* Karen Weisenseel (Loridae); and Patricia C. Wright *(Aotus, Callithrix,* and Lemuroidea). The author is solely responsible for any mistakes or misinterpretations found in this book.

I appreciate the efforts of Donna DiGiovanni, Ray Hamel, Jeffrey McMahon, Colin Mulcahy, Mike and Joann Palochak, Nancy Ruggeri, and Nicole Valenzuela, who helped during the preparation of the manuscript. I am especially grateful to Louis Budd Myers and Rosemary Sheffield for their invaluable editorial advice, and Lorraine Atherton, who read the proofs.

My sincerest thanks go to my parents, Stan and Louise Rowe, and my grandparents Dorothy and Stanley Rowe. They gave me the desire and the freedom to experience and appreciate many of the natural habitats of the world.

I am most indebted to Marc Myers for the unfailing friendship, dedication, and computer skills that he graciously provided throughout all stages of this book's evolution.

My deepest love and gratitude are for my wife, Abigail Barber, whose patience, hard work, and loving support have made it possible for me to complete this book.

About the Photographs

Primates in their natural habitat are difficult to photograph. The forests are dark, the monkeys are often high in the trees and backlit, and they are almost always moving, if not running away from us predatory humans. In this book the majority of the photos by the author were taken in zoos or under captive conditions in which the primate is still dependent on humans. The captive photographs were taken at the institutions listed below.

The photos of species in their natural habitat were usually taken in national parks and reserves or at research sites where the primates have been habituated to the presence of humans. Habituation is usually the long and difficult process of gaining the trust of a group of primates so that they don't run away from humans and their behavior can be studied. The photographs taken in the natural habitat are captioned as such on the side. A list of places visited in order to photograph primates is included below.

Several species described in this book have never been photographed, because they are so little known. Illustrations were produced from museum skins and taxonomic descriptions to show what these taxa look like. A handful of the photographs that are captioned on the side as "digitally composited" have enhanced backgrounds. For aesthetic reasons, the bars of cages were considered too intrusive to allow appreciation the species and were digitally removed from those photographs. Three species (Hylobates concolor, Presbytis comata, and Colobus satanas) have digital illustrations that clearly captioned. Rather than using conventional colored-pencil illustrations for each of these species, a computer program was used to modify a photograph of a closely related species: the coat color was altered according to the scientific description and after consultation with a taxonomist who has studied the species.

My basic camera for photographing primates under captive conditions is a Nikon F4 with an 80–200 f2.8 lens mounted on an Arca Swiss head on a Bogen 3232 tripod. I often use a Nikon SB-25 flash to light the face and to photograph nocturnal species. I use a 105mm macro lens for close-ups and a 500mm f4.0 telephoto lens for shooting through wire and for full-frame shots of primates in large cages. For photographing arboreal monkeys in their natural habitat, the telephoto lens is often the only possible choice, even though it is heavy and must be used with a tripod.

The photographs were taken with Fujichrome 100 ASA film, which is often push-processed to 200 ASA. In extremely low light conditions I use 400 ASA film, though it has more grain. Fujichrome Velvia film is rarely my choice for primates, because of its slow speed and high contrast.

Zoological institutions: Abidjan Zoo, Abidjan, Ivory Coast; Arashiyama West Institute, Texas Snow Monkey Sanctuary, Dilley, TX; Audubon Park and Zoological Garden, New Orleans; Banham Zoo and Monkey Sanctuary, Banham, Norwich, UK; Beardsley Zoological Gardens, Bridgeport, CT; Bristol, Clifton and West of England Zoological Society, Bristol, Clifton, UK; Central Park Conservation Park, New York; Centro de Primatología do Rio de Janeiro, Rio de Janeiro; Cercopan, Calabar, Nigeria; Chicago Zoological Park, Chicago; Chiengmai Zoo, Chiengmai, Thailand; Chimfunshi Wildlife Orphanage, Zambia; Cincinnati Zoo and Botanical Gardens, Cincinnati; Colchester Zoo, Colchester, Essex, UK; Cricket—the West Country Wildlife Park, Chard, Somerset, UK; Cuc Phong Endangered Primate Rescue Center, Viet Nam; Drill Ranch, Calabar, Nigeria; Duke University Primate Center, Durham, NC; Franklin Park Zoo, Boston; Greater Baton Rouge Zoo, Baton Rouge, LA; Hanoi Zoo, Hanoi, Viet Nam; Houston Zoological Gardens, Houston; Howletts Zoo, Bekesbourne, UK; Lincoln Park Zoological Gardens, Chicago; Los Angeles Zoo, Los Angeles; Louisiana Purchase Gardens and Zoo, Monroe, LA; Milwaukee County Zoological Park, Milwaukee; Monkey Hill, Hong Kong; National Zoological Park, Washington, DC; New York Wildlife Conservation Park/Bronx Zoo, Bronx, New York; North Carolina Zoological Park, Asheboro, NC; North of England Zoological Society, Chester, Cheshire, UK; Parc Zoologique de Paris, Paris; Parc Zoologique de Tsimbazaza, Antananarivo, Madagascar; Rio Primate Center, Rio de Janeiro; Royal Zoological Society of Antwerp, Antwerp, Belgium; San Diego Wild Animal Park, San Diego; San Diego Zoological Garden, San Diego; Santa Ana Zoo, Santa Ana, CA; Singapore Zoological Gardens, Singapore; Strasbourg Primate Research, Strasbourg, France; Strasbourg Zoo de l'Orangerie, Strasbourg, France; Taipei Zoo, Taipei, Taiwan; Trevor Teaching Zoo, Millbrook, NY; Twycross Zoo, East Midlands Zoological Society, Twycross, Warks, UK.

National parks and research sites: Analamazaotra Special Reserve (Perinet), Andasibe, Madagascar; Village of Apugeezi, Nigeria; Awash National Park, Ethiopia; Baboon Sanctuary, Belize; Berenty Private Reserve, Tolagara (Fort Dauphin), Madagascar; Caratinga Biological Research Station, Minas Gerais, Brazil; Cat Ba Island National Park, Cat Ba, Viet Nam; Cross River National Park, Nigeria; Danum Valley, Sabah, Malaysia; Debro Lebanus, Ethiopia; Village of Gbanraun, Nigeria; Gunung Leuser National Park, Sumatra, Indonesia; Karenta National Park, South Sulawesi, Indonesia; Kibale National Park, Fort Portal, Uganda; Na Hang Forest Reserve, Na Hang, Viet Nam; Okomo National Park, Nigeria; Parc National des Virunga, Zaire; Queen Elizabeth National Park, Uganda; Ranomafana National Park, Madagascar; Ranthambore National Park, India; Sariska National Park, India; Tabin Forest Reserve, Sabah, Malaysia; Taï National Park, Ivory Coast; Tangkoko Forest Reserve, Sulawesi, Indonesia; Tanjung Puting National Park, Kalimantan, Indonesia; Uaso Nyiro Baboon Project, Kenya; Wat Khao Tao Mo, Phetburi Province, Thailand; Wat Tham Pla, Chaing Rai Province, Thailand; Zoo Negara, Kuala Lumpur, Malaysia.

Biographical Information

About the Author
Noel Rowe is the director of Primate Conservation Inc., a nonprofit foundation that funds research and conservation projects on the least known and most endangered primates in their natural habitat. He has been a freelance photographer and a student of primatology for the last 10 years. He has traveled around the world to photograph primates in their natural habitats and in zoos.

About Jane Goodall, Ph.D.
Jane Goodall is world renowned for her more than 30 years of research on chimpanzees at Gombe Stream National Park in Tanzania. The first to document how similar chimpanzee behavior is to human behavior, she discovered that primates use tools, hunt and eat meat, and make war. She is the director of the Jane Goodall Institute for Wildlife Research and Conservation that was established to fund research and conservation on great apes. In addition, Dr. Goodall is working tirelessly to improve the living conditions and psychological well-being of chimpanzees in captivity.

About Russell A. Mittermeier, Ph.D.
Russell Mittermeier is currently the president of Conservation International and has been chairman of the Primate Specialist Group of the Species Survival Commission for the World Conservation Union (IUCN) for 19 years. He is the author of more than 200 articles and books, including *A Global Strategy for Primate Conservation* (1987), *Ecology and Behavior of Neotropical Primates* (vols. 1 and 2, 1981 and 1989), and *The Lemurs of Madagascar* (1994).

Glossary

Acacia **woodland** Scattered low, thorny trees of the *Acacia* genus found in Africa.

adaptation A genetic, morphological, or behavioral change that a species makes to help it survive in its environment.

adaptive radiation A closely related taxonomic group that has evolved different features that allow it to exploit different niches. The lemurs of Madagascar are a good example.

agouti A fur color in which each hair has alternating light and dark bands, giving the fur a streaked appearance.

alarm call A loud vocalization given by all ages and genders in response to a disturbance such as the appearance of a predator or another group.

allogrooming Reciprocal cleaning of the fur.

allomothering Care of an infant by a group member other than its mother; similar to baby-sitting.

allopatric Not sharing the same geographic range.

Anthropoidea In one classification system, the suborder that comprises monkeys and apes but not tarsiers; often called the higher primates or simians. *Compare* Haplorrhini.

antiphonal Alternating, as in a call and response.

apocrine glands Specialized sweat glands in the armpits and the perineum. The fluid secretion contains portions of the glandular cells and their contents. Bacteria act upon apocrine secretions, resulting in perspiration odor. *Compare* eccrine glands.

arboreal Adapted to life in trees.

association A grouping of two or more species that travel and/or feed together.

baculum A bone in the penis of some mammals.

bezoar stones Gallstones of primates that live in limestone-rich habitats; believed by some to possess magical powers.

bicornuate Having two crescent-shaped parts.

bilophodont Having two parallel ridges.

bipedal Using only two limbs for locomotion, as in humans and birds.

black-water river A river that appears black and clear because its water contains tannins and lacks silt.

brachial gland A gland on the arm or wrist.

brachiation Swinging with arms from one branch to another.

caatinga Brazilian rain forest characterized by low trees growing on sandy soils that lack nutrients.

canines Pointed teeth; also called eye-teeth. The two upper canines are usually the longest, most pointed teeth.

canopy The upper forest level, which receives the most sun. It may be continuous or have gaps or taller emergent trees.

canting or cantilevering Springing out from a branch to catch prey with the hands while the hind legs hold on to the branch.

carnivore An animal that eats the flesh of other animals as a major percentage of its diet. *Compare* faunivore, folivore, frugivore, graminivore, omnivore.

Catarrhini The infraorder that comprises Old World monkeys and apes.

cathemeral Active periodically throughout the day and night, as in lemurs.

cecum A saclike appendage between the small and large intestines.

cheek pouch A structure in the mouth used to store food temporarily; found in Old World monkeys but not colobines.

circumgenital Around the genitalia.

circumocular ring A ring around the eye.

clade All of the species that have evolved from one common ancestor.

coniferous forest Cone-bearing, usually evergreen trees, such as cedars, pines, and firs.

consort With regard to a male and an estrous female, to temporarily associate and travel together to enhance the male's chances of paternity. Consortion is found in baboons and chimpanzees.

conspecific Belonging to the same species.

coprophagy The ingestion of feces.

crepuscular Active in the twilight mostly at dawn and dusk.

cryptic Hard to detect. Cryptic primates often are quiet and prefer to be camouflaged by dense foliage.

day range The distance a group travels in a day, usually measured in meters.

deciduous forest Forest with trees that lose their leaves during one season of the year.

dental formula A formula expressing the number and kinds of teeth in half the mouth (incisors first, then canines, premolars, and molars), multiplied by 2 to give the total. The formula for great apes is $\frac{2.1.2.3}{2.1.2.3} \times 2 = 32$.

derived feature A specialized adaptation (behavioral or morphological characteristic) that differs from that of the taxon's ancestors.

dermatoglyphic Having skin patterns on the fingers, palms, or feet.

dipterocarp forest Southeast Asian forest with a large portion of tall trees of the family Dipterocarpaceae.

display A conspicuous behavior involving posture, action, and/or vocalization that usually conveys a message to members of the same species.

diurnal Active during the day.

dominant Higher ranking in a hierarchy, as determined by the displacement of other troop members for priority of access to food, mates, or resting places.

eccrine glands Sweat glands whose

secretions are by ducts and do not contain portions of the secretory cells themselves. Eccrine secretions are important to body heat regulation. *Compare* apocrine glands.

ecological niche The specific role of an organism in its environment and community that allows it to survive and reproduce.

ecology The study of how organisms are affected by and interact with other organisms in a particular environment.

ecosystem The interrelationships among the living organisms and the chemical and physical components of a given environment.

emergent trees Trees that grow above the top of the canopy.

emigration The departure of an individual from a group to travel alone or join another group, usually at adulthood.

endemic Found only in one region or country.

epigastric gland A gland in the stomach region.

epitheliochorial placenta A more ancestral placenta in which the epithelium of the fetus is in contact with the entire uterus, and the fetal circulation is separated from the mother's by a few layers of tissue.

estrus The period around ovulation when females are willing to mate, as indicated by perineal swellings or pheromonal or behavioral signals.

exudate Ooze from a tree's vascular system in the form of sap, gum, or resin.

eye ring A ring of colored fur around the eye.

faunivore An animal that eats other animals as a major percentage of its diet. *Compare* carnivore, folivore, frugivore, graminivore, omnivore.

fermentation The process by which microorganisms break down complex molecules. Colobines have bacteria in their cecum that enable them to digest their high-cellulose leafy diet.

fission-fusion A social structure in which individuals of a troop split into small subgroups for foraging and re-form as a large group when food resources permit.

folivore An animal that eats leaves as a major percentage of its diet. *Compare* carnivore, faunivore, frugivore, graminivore, omnivore.

frugivore An animal that eats fruit as a major percentage of its diet. *Compare* carnivore, faunivore, folivore, graminivore, omnivore.

gall, leaf gall A sphere-shaped swelling of plant tissue caused by insect or bacterial damage.

gallery Forest found along streams and rivers.

genus The taxonomic classification below family and above species; indicated by italics and capitalization (e.g., *Cebus*).

gestation The period between conception and birth.

graminivore An animal that eats grass

seeds and grains as a major percentage of its diet. *Compare* carnivore, faunivore, folivore, frugivore, omnivore.

great call The territorial song produced by gibbons.

grooming An interaction between two primates, with one cleaning the other's fur.

habituated troop Animals that have gradually learned to trust researchers and can be followed and studied.

hallux The first toe of the foot.

Haplorrhini In one classification system, the suborder that includes the tarsiers with the higher primates. *Compare* Anthropoidea.

hierarchy A rank-ordered system defined by interactions among individuals, with dominants and subordinates.

higher primates *See* Anthropoidea.

home range The area used by a troop over an annual cycle; usually expressed in hectares (ha). 1 ha = about 2.5 acres.

hyoid bone A bone or bone complex in the throat that supports the larynx and trachea.

igapo Forest in a black-water river basin that floods seasonally.

inclusive fitness A measure of an individual's total genetic representation in the next generation. In this system, two of your sister's offspring are equivalent to one of your own.

infanticide The deliberate killing of infants by members of their own species.

insular Isolated from others of one's type, as on an island.

intergroup Between or among different groups.

intermembral index A number expressing the relative lengths of an animal's forelimbs and hind limbs; calculated by multiplying the sum of the radius and humerus lengths by 100, then dividing by the sum of the femur and tibia lengths.

intragroup Within a group.

IPS Congress International Primatological Society Congress, a conference held in even-numbered years.

ischial callosities The two pads on the rump, upon which Old World monkeys and gibbons sit.

juvenile An individual between infancy and sexual maturity.

karst A region of eroded limestone formations with caves and cracks.

karyotype The shape, size, and number of somatic chromosomes.

keystone foods A small number of foods upon which a species depends to survive the harshest conditions of the year.

knuckle walking A type of four-limbed locomotion used by gorillas and chimpanzees in which the weight of the head and torso is borne on the knuckles.

lexigram A picture that represents a word or a concept.

lip smacking A gesture that many primate species use as a friendly signal when approaching another individual.

long call A call made by male orangutans,

consisting of a series of loud grunts that carry for more than 1 kilometer (3281 feet).

loud call A long-distance vocalization usually given by a male in a one-male social system to indicate his presence and/or the group's location.

lowland forest Forest below 1 kilometer (3281 feet) in altitude. Lowland forest often has the most biodiversity.

malar Of the cheek or the cheekbone.

matrifocal Pertaining to a social structure in which females are the center of the group and males are peripheral.

matrilineal Pertaining to kinship or descent through female lines.

monogamy A type of social structure centered on one male and one female that mate only with each other.

monotypic Pertaining to a taxon with only one representative, such as a genus with only one species or a species with no subspecies.

montane forest Forest found at higher altitudes and characterized by shorter trees covered with mosses and lichens.

morphology The structure and form of an organism.

natal troop The group into which an individual is born.

natural selection The evolutionary process by which better-adapted individuals have more offspring that survive and reproduce than less well adapted individuals do. In the long term this process leads to genetic changes in a population.

Neotropics The tropical regions of Central and South America.

nocturnal Active at night.

noyau A social structure in which a male's territory overlaps the smaller territories of several females; found in several nocturnal primates.

nulliparous Pertaining to a female that has not had offspring. A nulliparous female has small nipples and can usually be distinguished from a parous female, which has had offspring and consequently has long, distended nipples.

omnivore An animal that eats both animal and vegetable material as major percentages of its diet. *Compare* carnivore, faunivore, folivore, frugivore, graminivore.

order A taxonomic classification level below class and above family. The order Primates is in the class Mammalia.

os clitoris A bone found in the clitoris.

paleontologist One who studies fossil plants and animals.

pelage The hair or fur of a mammal.

perineal swelling Inflation of the tissue between the anus and the vulva with fluid, signaling the onset of estrus and, at its fullest, ovulation.

pheromones Chemicals secreted by an animal that communicate information to other individuals of the same species.

photoperiod The relative lengths of daylight and darkness, whose seasonal changes affect certain organisms in particular ways.

phylogeny The evolutionary history of an organism or a taxon.

Platyrrhini The infraorder that comprises New World monkeys.

polyandry A social structure in which the one breeding female of the group mates with more than one male during a given mating season.

polygyny A social structure in which one breeding male mates with more than one female in the group during a given mating season.

postorbital closure The full enclosure of the eye socket by bone; found in anthropoid primates, including humans. Prosimians have open eye sockets.

prehensile Able to grasp by wrapping around; often refers to the tail of some larger Neotropical monkeys.

primary forest Forest that has matured and remained undisturbed for several centuries.

primitive feature A behavioral or morphological feature shared by a species and its ancestors.

proceptive Pertaining to behavior that indicates an individual's interest in sex.

procumbent Lying flat and forward, as the prosimian incisors that form the tooth comb.

promiscuous Mating with more than one partner per breeding season.

Prosimii In one classification system, the suborder that comprises lorises, bush babies, lemurs, and tarsiers. *Compare* Strepsirrhini.

protoculture The first signs of a cultural tradition.

quadrumanous Using both the hands and the feet for grasping and climbing.

quadrupedal Using four limbs for locomotion.

rain forest A botanically diverse forest in the tropics or subtropics with plentiful year-round rainfall.

receptive Ready to mate; usually corresponds to "estrous."

relict An isolated population that has survived only in one small area.

reproductive strategy The physiological and behavioral limitations and choices available to an individual to reproduce its genes. Rodents, for example, have many offspring that receive a minimum of parental care, whereas most primates have only one offspring and carry it and care for it.

reproductive success The amount of genes passed on to the next generation.

restinga A moist habitat in Brazil, consisting of trees with many bromeliads.

rhinarium The moist part around the nose that enhances smell in some animals, including dogs and prosimians.

sacculated Divided into small chambers.

sagittal crest A bone ridge on top of the cranium that is most prominent in male gorillas.

salado A mineral deposit in South American forests similar to a salt lick.

savanna Open, flat tropical grassland with varying amounts of trees and seasonal rainfall.

scent gland An odor-producing organ.

scent mark An olfactory message communicated by rubbing glandular secretions or leaving urine or feces at a specific site.

sclerocarpic frugivore An animal that eats hard-shelled fruits and seeds as a major percentage of its diet.

scrub Woody, low-growing shrubs that are often found in arid regions or in areas of continual human deforestation.

secondary compounds Often-toxic chemicals produced by plants to discourage animals from eating the plant.

secondary forest Forest that is regenerating from a natural or human disturbance; usually characterized by numerous small trees and often thick viny growth.

seed disperser An animal that passes seeds through its digestive tract and deposits them unharmed. Some seeds germinate faster after having passed through an animal's digestive tract.

seed predator An animal that eats and destroys the seeds of plants.

semideciduous forest Forest in which some but not all of the tree species lose their leaves during one season of the year.

sexual dichromatism Color pattern variation between genders of the same species.

sexual dimorphism Size, color, or weight variation between genders of the same species.

simian Apelike. *See* Anthropoidea.

solitary Spending a majority of time alone, not in a group or a pair.

species The basic biological unit of classification in which like members are similar in appearance, can interbreed, and are different from other species, with which they do not normally interbreed. The species epithet is the uncapitalized word that follows the genus name, as in *Cebus apella*.

spiny desert An arid forest type found in southern Madagascar made up of many endemic thorny plant species.

sternal Pertaining to the chest region near the sternum, or breastbone.

Strepsirrhini In one classification system, the suborder that includes lorises, bush babies, and lemurs but not tarsiers. *Compare* Prosimii.

subadult An individual between sexual maturity and full adulthood; similar to a human adolescent.

subfossil A bone that has not been buried long enough to fossilize and is usually

much younger than a fossil.

subordinate Lower ranking in a hierarchy and therefore giving way to dominant individuals.

subspecies A classification level for a species subgroup that lives in a different geographical region and is morphologically different. The second uncapitalized word after the genus name designates the subspecies, as in *Cebus apella nigritus*.

subtropical In latitudes bordering the tropical and temperate zones.

suprapubic Above the pubic region.

suspensory behavior Hanging, climbing, or moving below tree branches.

sympatric Sharing the same geographic area.

systematics The scientific study of how organisms are related and classified.

tapetum A layer of the eye that enhances night vision; found in some mammals, including cats and prosimians.

taxon A named group of related organisms in a classification system, such a subspecies or species.

taxonomy The descriptive science of classification and naming of organisms.

terra firma forest Lowland forest that is not flooded seasonally.

terrestrial Adapted to life on the ground.

territory The area an animal or group actively defends from other members of its species.

thorn scrub A low-growing community of thorny trees and shrubs found in an arid environment.

tooth comb A prosimian's lower incisors, which are inclined forward like a comb and are used for grooming.

transverse In a crosswise direction.

tropical The geographical region between the tropic of Cancer (23.5° north latitude) and the tropic of Capricorn (23.5° south latitude).

understory The lowest forest level, between the ground and 10 meters (33 feet).

varzea A forest on a white-water river that floods every year.

ventroventral Belly-to-belly.

ventrum The belly region.

vertical clinging and leaping A type of locomotion in which the primate posture is parallel to the trunk of a tree when clinging and leaping from one trunk to another. Primates using this locomotion usually have longer hind limbs than forelimbs.

vestigial Rudimentary, as in a structure that is reduced or not visible.

white-water river A river that contains much eroded silt, so that it is very cloudy and light brown.

woodland Scattered short trees interspersed with grassland.

Resources and Bibliography

Popular Books

General

African Silences, by P. Matthiessen. Random House, New York, 1991.

The Aye-aye and I: A Rescue Mission in Madagascar, by G. Durrell. Arcade, New York, 1992.

Chimpanzee Politics: Power and Sex among Apes, by F. de Waal. Johns Hopkins Press, Baltimore, 1990.

The Deluge and the Ark: A Journey into Primate Worlds, by D. Peterson. Avon, New York, 1989.

Female Choices: Sexual Behavior of Female Primates, by M. F. Small. Cornell University Press, Ithaca, NY, 1993.

Good Natured: The Origins of Right and Wrong in Humans and Other Animals, by F. de Waal. Harvard University Press, Cambridge, 1996.

The Great Ape Project, by P. Singer and P. Cavalieri. St. Martin's Press, New York, 1993.

The Multimedia Guide to the Non-human Primates, by F. D. Burton and M. Eaton. Prentice-Hall Canada, Scarborough, Ont., 1995.

Peacemaking among Primates, by F. de Waal. Harvard University Press, Cambridge, 1989.

The Third Chimpanzee: The Evolution and Future of the Human Animal, by J. Diamond. Harper Perennial, New York, 1992.

Visions of Caliban: On Chimpanzees and People, by D. Peterson and J. Goodall. Houghton Mifflin, Boston, 1993.

Walking with the Great Apes: Jane Goodall, Dian Fossey, Birute Galdikas, by S. Montgomery. Houghton Mifflin, Boston, 1991.

The Woman That Never Evolved, by S. B. Hrdy. Harvard University Press, Cambridge, 1983.

Rain Forests

Diversity and the Tropical Rain Forest, by J. Terborgh. Scientific American Library, New York, 1992.

The Last Rain Forests: A World Conservation Atlas, by M. Collins, ed. Oxford University Press, New York, 1990.

The Primary Source: Tropical Forests and Our Future, by N. W. W. Myers. Norton and Co., New York, 1992.

Field Studies

Almost Human: A Journey into the World of Baboons, by S. C. Strum. Random House, New York, 1987.

Faces in the Forest: The Endangered Muriqui Monkeys of Brazil, by K. Strier. Oxford University Press, New York, 1990.

Gorillas in the Mist, by D. Fossey. Houghton Mifflin, Boston, 1983.

In Quest of the Sacred Baboon: A Scientist's Journey, by H. Kummer. Princeton University Press, Princeton, NJ, 1995.

In the Shadow of Man, by J. Goodall. Houghton Mifflin, Boston, 1988.

The Langurs of Abu: Female and Male Strategies of Reproduction, by S. B. Hrdy. Harvard University Press, Cambridge, 1980.

Reflections of Eden: My Years with the Orangutans of Borneo, by B. M. F. Galdikas. Little, Brown, Boston, 1995.

Through a Window: My Thirty Years with the Chimpanzees of Gombe, by J. Goodall. Houghton Mifflin, Boston, 1990.

Ape Language Studies

The Education of Koko, by F. Patterson and E. Linden. Holt, Rinehart and Winston, New York, 1981.

Kanzi: The Ape at the Brink of the Human Mind, by S. Savage-Rumbaugh and R. Lewin. John Wiley and Sons, New York, 1994.

Silent Partners: The Legacy of the Ape Language Experiments, by E. Linden. Times Books, New York, 1986.

Field Guides

The Behavior Guide to African Mammals, by D. E. Estes. University of California Press, Berkeley and Los Angeles, 1992.

The Lemurs of Madagascar, by R. A. Mittermeier. Conservation International, Washington, DC, 1994.

Neotropical Rainforest Mammals: A Field Guide, by L. H. Emmons. University of Chicago Press, Chicago, 1990.

For Children

Among the Orangutans: The Birute Galdikas Story, by E. Gallardo. Chronicle Books, San Francisco, 1993.

Monkeys and Apes Coloring Book, by J. Green. Dover, New York, 1988.

World Wide Web Sites

African Primates at Home
http://www.indiana.edu/~primate/primates.html
Information about an African field study.

American Society of Primatologists
http://www.asp.org/asp/index.htm
Introduces the society and its conservation awards.

The Dian Fossey Gorilla Fund
http://deathstar.rutgers.edu/projects/gorilla/gorilla.html
Mountain gorilla conservation.

Global SchoolNet Supports the Jane Goodall Institute
http://www.gsn.org/gsn/proj/jgi/index.html
Chimpanzee conservation.

The Great Ape Project
http://envirolink.org/arrs/gap/gaphome.html
Promoting greater legal consideration for great apes.

International Primate Protection League
http://www.sims.net/organizations/ippl/ippl.html
Stopping the illegal trade of endangered primate species.

Net Vet Primates
http://netvet.wusttl.edu/primates.html
Many links to various primate related resources.

Orangutan Foundation International
http://www.ns.net/orangutan/
Orangutan conservation.

Pan Africa News
http://inrui.zool.kyoto-u.ac.jp/PAN/home.html
A newsletter about chimpanzees in Africa.

Primate Conservation Inc.
http://www.primate.org
Matching grants for research and conservation projects.

Primate Info Net
http://www.primate.wisc.edu/pin
An excellent resource from the University of Wisconsin.

Primate Society of Great Britain
http://www.ana.ed.ac.uk/PSGB/
Introduces the society.

South Texas Primate Observatory
http://www.wscape.com/stpo
Information about the Japanese macaques in Texas.

Bibliography

AJP = Am J Primatol
AJPA = Am J Phys Anthropol
IJP = Int J Primatol
IPS = International Primatological Society
IUCN = The World Conservation Union (formerly the International Union for Conservation of Nature)

1. Abbott DH. 1993. Social conflict and reproductive suppression in marmoset and tamarin monkeys. In Mason WH, Mendoza SP (eds), *Primate Social Conflict,* 331–372. State Univ of New York Press, Albany.

2. Abbott DH, Barrett J, George LM. 1993. Comparative aspects of the social suppression of reproduction in marmosets and tamarins. In Rylands AB (ed), *Marmosets and Tamarins: Systematics, Behaviour, Ecology,* 152–162. Oxford Univ Press, Oxford.

3. Aggimarangsee N. 1992. Survey for semi-tame colonies of macaques in Thailand. *Nat Hist Bull Siam Soc* 40:103–166.

4. Agoramoorthy G. 1994. An update on the long term field research on red howler monkeys *Alouatta seniculus* at Hato Masaguaral, Venezuela. *Neotrop Primates* 2(3):7–8.

5. Aich H, Moos-Heilen R, Zimmermann E. 1990. Vocalizations of adult gelada baboons *(Theropithecus gelada):* acoustic structure and behavioural context. *Folia Primatol* 55:109–132.

6. Aimi M, Bakar A. 1992. Taxonomy and distribution of *Presbytis melalophus* group in Sumatra, Indonesia. *Primates* 33(2):191–206.

7. Aimi M, Hardhasasmita HS, Sjarmidi A, Yuri D. 1986. Geographical distribution of Aygula-group of the genus *Presbytis* in Sumatra. *Kyoto Univ Overs Res Rep Stud Asian Non-human Primates* 5:45–58.

8. Alberts S. 1987. Parental care in captive siamangs *(Hylobates syndactylus).* *Zoo Biol* 6:401–406.

9. Alberts SC, Sapolsky RM, Altmann S. 1992. Behavioral endocrine and immunological correlates of immigration by an aggressive male into a natural primate group. *Horm Behav* 26(2):167–178.

10. Albignac R, Justin, Meier B. 1991. Study of the first behavior of *Allocebus trichotis* Gunther 1875 (hairy-eared dwarf lemurs): prosimian lemur rediscovered in the northeast of Madagascar (Biosphere Reserve of Mananara-Nord). In Ehara A, Kimura T, Takenaka O, Iwamonto M (eds), *Primatology Today*. Elsevier Science, Amsterdam.

11. Alfred JRE, Sati JP. 1987. Sexual behavior in the *Hylobates hoolock* [abstract]. *IJP* 8(5):530.

12. Alfred JRE. 1992. The hoolock gibbon—*Hylobates hoolock. Primate Rep* 32:65–69.

13. Allman J, McLaughlin T, Hakeem A. 1993. Brain weight and life-span in primate species. *Proc Natl Acad Sci USA* 90:118–122.

14. Alperin R. 1993. *Callithrix argentata* (Linnaeus, 1771): considerações taxonômicas e descrição de subespécie nova. *Bol Mus Para Emilio Goeldi Zool* 9(2):317–328.

15. Altmann J, Alberts SC, Dubach J, et al. 1994. What-you-see-is-what-you-get: social structure and observed mating behavior predict paternity in a savannah baboon group. Paper presented at IPS Congress 15, Kuta, Bali, Indonesia. [Abstract available: 1994. *15th Congress IPS Handbook and Abstracts*: 98.]

16. Altmann J. 1980. *Baboon Mothers and Infants*. Harvard Univ Press, Cambridge.

17. Altmann SA, Altmann J. 1970. *Baboon Ecology*. Univ of Chicago Press, Chicago.

18. Anderson C. 1993. Talk presented at New York Consortium in Evolutionary Primatology/New York Regional Primatology Colloquium, City Univ of New York, New York.

19. Anderson JR, Roeder J. 1989. Responses of capuchin monkeys *(Cebus apella)* to different conditions of mirror-image stimulation. *Primates* 30(4):581–587.

20. Anderson MJ. 1995. Comparative morphology and speciation in galagos. Poster presented at the International Conference on the Biology and Conservation of Prosimians, The North of England Zoological Society, Chester, UK. [Abstract available: 1996. *Folia Primatol*. In press.].

21. New species of squirrel monkey from Brazilian Amazonia. 1988. *Primate Conserv* 9:10.

22. *Sustainable Agriculture and the Environment in the Humid Tropics*. 1993. National Research Council, National Academy Press, Washington, DC.

23. Morin PA, Moore JJ, Chakraborty R, Li J. 1994. Kin selection, social structure, gene flow, and the evolution of chimpanzees. *Science* 265(5176):1193–1201.

24. Earth almanac: epitaph for two gorillas. 1995. *Natl Geogr* 188(5).

25. Ansorge V, Hammerschmidt K, Todt D. 1992. Communal roosting and formation of sleeping clusters in Barbary macaques *(Macaca sylvanus)*. *AJP* 28:271–280.

26. Anzenberger G. 1992. Monogamous social systems and paternity in primates. In Martin RD, Dixson AF, Wickings EJ (eds), *Paternity in Primates: Genetic Tests and Theories*, 203–224. Karger, Basel, Switzerland.

27. Anzenberger G. 1993. Social conflict in two monogamous New World primates: pairs and rivals. In Mason WH, Mendoza SP (eds), *Primate Social Conflict*, 291–330. State Univ of New York Press, Albany.

28. Aquino R, Encarnación F. 1994. Owl monkey populations in Latin America: field work and conservation of owl monkeys. In Baer J, Weller RE,

Kakoma I (eds), Aotus: *The Owl Monkey*, 60–96. Academic Press, New York.

29. Aquino R, Encarnación F. 1986. Population structure of *Aotus nancymai* (Cebidae: Primates) in Peruvian Amazon lowland forest. *AJP* 11:1–7.

30. Armada JLA, Barreso CML, Lima M, Muniz JAPC. 1987. Chromosome studies in *Alouatta belzebul. AJP* 13(3):283–296.

31. Atsalis S. In prep. Seasonal patterns in diet, distribution, and reproduction of *Microcebus rufus* (family Cheirogaleidae) in Ranomafana National Park. PhD diss, City Univ of New York, New York.

32. Atsalis S. 1996. Pers com. City College of New York, New York, Jan.

33. Ayres JM. 1986. The conservation status of the white uakari. *Primate Conserv* 7:22–25.

34. Balakrishnan M, Easa PS. 1986. Habitat preferences of the larger mammals in Parambikulam Wildlife Sanctuary, Kerala, India. *Biol Conserv* 37:191–200.

35. Baldwin JD, Baldwin JI. 1981. The squirrel monkeys, genus *Saimiri*. In Coimbra-Filho AF, Mittermeier RA (eds), *Ecology and Behavior of Neotropical Primates*, vol 1, 277–330. Academia Brasileira de Ciências, Rio de Janeiro.

36. Bartechi U, Heyman EW. 1987. Sightings of red uakaris, *Cacajao calvus rubicundus*, at the Rio Blanco, Peruvian Amazonia. *Primate Conserv* 8:34–35.

37. Barton R. 1985. Grooming site preferences in primates and their functional implications. *IJP* 6(5):519–531.

38. Bauers KA, de Waal FBM. 1991. "Coo" vocalizations in stumptailed macaques: a controlled functional analysis. *Behaviour* 119(1–2):151–158.

39. Bearder S, Honess P, Ambrose L. 1994. Speciation among nocturnal primates [abstract]. *Folia Primatol* 62:222.

40. Bearder SK, Martin RD. 1979. The social organization of a nocturnal primate revealed by radio tracking. In Amlaner CJ Jr, Macdonald DW (eds), *A Handbook on Biotelemetry and Radio Tracking*. Pergamon Press, Oxford.

41. Bearder SK. 1987. Lorises, bushbabies, and tarsiers: diverse societies in solitary foragers. In Smuts BB, Cheney DL, Seyfarth RM, Wrangham RW, Struhsaker TT (eds), *Primate Societies*, 11–24. Univ of Chicago Press, Chicago.

42. Bearder SK, Doyle GA. 1979. Ecology of *Galago senegalensis* and *Galago crassicaudatus* with some notes on their behaviour in the field. In Martin RD, Doyle GA, Walker AC (eds), *Prosimian Biology*, 109–130. Univ of Pittsburgh Press, Pittsburgh.

43. Bearder SK, Honess PE, Ambrose L. 1995. Species diversity among galagos with special reference to mate recognition. In Alterman L, Doyle GA, Izard MK (eds) *Creatures of the Dark: The Nocturnal Prosimians*. Plenum Press, New York.

44. Beauchamp G, Canaba G. 1990. Group size variability in primates. *Primates* 31(2):171–182.

45. Bennett EL. 1986. Environmental correlates of ranging behaviour in the banded langur, *Presbytis melalophos. Folia Primatol* 47:26–38.

46. Bennett EL, Davies AG. 1994. The ecology of Asian colobines. In Davies AG, Oates JF (eds), *Colobine Monkeys: Their Ecology, Behaviour, and Evolution*, 129–172. Cambridge Univ Press, Cambridge.

47. Bennett EL, Sebastian AC. 1988. Social organization and ecology of proboscis monkeys *(Nasalis larvatus)* in mixed coastal forest of Sarawak. *IJP* 9(3):233–255.

48. Berger G, Tylinek E. 1985. *Monkeys and Apes*. Arco, New York.

49. Bermejo M, Illera G, Sabater Pi J. 1994. Animals and mushrooms consumed by bonobos *(Pan*

paniscus): new records from Lilungu (Ikela) Zaire. *IJP* 15(6):879–898.

50. Bernstein I, Williams L, Ramsay M. 1983. The expression of aggression in Old World monkeys. *IJP* 4(2):113–125.

51. Bernstein IS. 1968. The latong of Kuala Selangor. *Behaviour* 32:1–16.

52. Bernstein IS, Baker S. 1988. Activity pattern in a captive group of Celebes black apes *(Macaca nigra)*. *Folia Primatol* 51:61–75.

53. Bernstein IS, Balcaen P, Dresdale L, et al. 1976. Differential effects of forest degradation on primate populations. *Primates* 17(3):401–411.

54. Bertrand M. 1969. *The Behavioral Repertoire of the Stumptailed Macaque: A Descriptive and Comparative Study*. Karger, Basel, Switzerland.

55. Bicca-Marques JC. 1992. Drinking behaviour in the black howler monkey *(Alouatta caraya)*. *Folia Primatol* 58:107–111.

56. Bicca-Marques JC, Calegaro-Marques C. 1994. Twins or adoption? *Neotrop Primates* 2(3):6–7.

57. Bleisch W, Cheng A, Ren X, Xie J. 1993. Preliminary results from a field study of wild Guizhou snub-nosed monkeys *(Rhinopithecus brelichi)*. *Folia Primatol* 60:72–82.

58. Bleisch W, Xie J. 1994. Ecology and behavior of Guizhou golden monkeys *(Rhinopithecus brelichi)*. Paper presented at IPS Congress 15, Kuta, Bali, Indonesia. [Abstract available: 1994. *15th Congress IPS Handbook and Abstracts:* 279.]

59. Bleisch WV. 1994. Management recommendations for conservation of the Guizhou golden monkey and the Fanjing Mountain Reserve. Unpub report to Chinese Ministry of Forestry, Wildlife Conservation Society, The Bronx, New York.

60. Bleisch WV, Chen N. 1991. Ecology and behavior of wild black-crested gibbons *(Hylobates concolor)* in China with a reconsideration of evidence of polygyny. *Primates* 32(2):539–548.

61. Boccia ML, Laudenslager M, Reite M. 1988. Food distribution, dominance, and aggressive behaviors in bonnet macaques. *AJP* 16:123–130.

62. Bocian C. In prep. Niche separation in sympatric population of black and white colobus. PhD diss, City Univ of New York, New York.

63. Boesch C. 1994. Hunting strategies of Gombe and Tai chimpanzees. In Wrangham RW, McGrew WC, de Waal FBM, Heltne PG (eds), *Chimpanzee Culture*, 77–92. Harvard Univ Press, Cambridge.

64. Boesch C. 1994. Chimpanzees–red colobus monkeys: a predator–prey system. *Anim Behav* 47:1135–1148.

65. Boese GLS. 1975. Social behavior and ecological considerations of west African baboons *(Papio papio)*. In Tuttle RH (ed), *Socioecology and Psychology of Primates*, 205–230. Mouton, The Hague.

66. Boestani AN, Smits W. 1994. Dermatoglyphics of the Bornean and Sumatran orangutan. Paper presented at IPS Congress 15, Kuta, Bali, Indonesia. [Abstract available: 1994. *15th Congress IPS Handbook and Abstracts:* 335.]

67. Boinski S. 1989. Why don't *Saimiri oerstedii* and *Cebus capucinus* form mixed-species groups? *IJP* 10(2):103–114.

68. Boinski S. 1991. The coordination of spatial position: a field study of the vocal behavior of adult squirrel monkeys. *Anim Behav* 41:89–102.

69. Boinski S. 1994. Affiliation patterns among male Costa Rican squirrel monkeys. *Behaviour* 130(3–4):191–211.

70. Boinski S, Chapman CA. 1995. Predation on primates: where are we and what's next. *Evol Anthropol* 4(1):1–3.

71. Boinski S, Mitchell CL. 1994. Male residence and association patterns in Costa Rican squirrel monkeys *(Saimiri oerstedii)*. AJP 34:157–169.

72. Boinski S, Mitchell CL. 1995. Wild squirrel monkey *(Saimiri sciureus)* "caregiver" calls: context and acoustic structure. AJP 35:129–137.

73. Boinski S, Newman JD. 1988. Preliminary observations on squirrel monkey *(Saimiri oerstedii)* vocalizations in Costa Rica. AJP 14:329–343.

74. Boonratana R, Canh LX. 1994. Conservation of the Tonkin snub-nosed monkey in Vietnam. Paper presented at IPS Congress 15, Kuta, Bali, Indonesia. [Abstract available: 1994. *15th Congress IPS Handbook and Abstracts:* 278.]

75. Boonratana R, Canh LX. 1994. *A Report on the Ecology, Status, and Conservation of the Tonkin Snub-nosed Monkey* (Rhinopithecus avunculus) *in Northern Vietnam.* Wildlife Conservation Society, Bronx, New York.

76. Boubli JP. 1994. The black uakari monkey in the Pico Da Neblina National Park. *Neotrop Primates* 2(3):11–12.

77. Bourlière F, Hunkeler C, Bertrand M. 1970. Ecology and behavior of Lowe's guenon *(Cercopithecus campbelli lowei)* in the Ivory Coast. In Napier JR, Napier PH (eds), *Old World Monkeys: Evolution, Systematics, and Behavior,* 297–350. Academic Press, New York.

78. Brain C. 1992. Deaths in a desert baboon troop. *IJP* 13(6):593–599.

79. Branch LC. 1983. Seasonal and habitat differences in the abundance of primates in the Amazon (Tapajós) National Park, Brazil. *Primates* 24(3):424–431.

80. Brandon-Jones D. 1985. Colobus and leaf monkeys. In Macdonald D (ed), *Primates,* 102–112. Torstar Books, New York.

81. Brandon-Jones D. 1994. The primates of the Mentawai Islands: a conservation imperative. Paper presented at IPS Congress 15, Kuta, Bali, Indonesia. [Abstract available: 1994. *15th Congress IPS Handbook and Abstracts:* 129.]

82. Brandon-Jones D. 1995. Pers com. Natural History Museum, London, Dec.

83. Brockelman WY, Gittins SP. 1984. Natural hybridization in the *Hylobates lar* species group: implications of speciation in gibbons. In Preuschoft H, Chivers DJ, Brockelman WY, Creel N (eds), *The Lesser Apes: Evolutionary and Behavioural Biology,* 498–532. Edinburgh Univ Press, Edinburgh.

84. Brotoisworo E, Dirgayusa IWA. 1991. Ranging and feeding behavior of *Presbytis cristata* in the Pangandaran Nature Reserve, West Java, Indonesia. In Ehara A, Kimura T, Takenaka O, Iwamoto M (eds), *Primatology Today.* Elsevier Science, Amsterdam.

85. Brown CH. 1989. The active space of blue monkey and grey cheeked mangabey vocalizations. *Anim Behav* 37(6):1023–1034.

86. Bruce KE, Estep DQ. 1992. Interruption of and harassment during copulation by stumptailed macaques, *Macaca arctoides. Anim Behav* 44:1029–1044.

87. Buchanan DB, Mittermeier RA, van Roosmalen MGM. 1981. The saki monkey, genus *Pithecia.* In Coimbra-Filho AF, Mittermeier RA (eds), *Ecology and Behavior of Neotropical Primates,* vol 1, 391–418. Academia Brasileira de Ciências, Rio de Janeiro.

88. Buchanan-Smith HM. 1991. A field study on the red-bellied tamarin, *Saguinus l. labiatus,* in Bolivia. *IJP* 12(3):259–277.

89. Burton FD, Eaton M. 1995. *The Multimedia Guide to the Non-human Primates.* Prentice-Hall Canada, Scarborough, Ontario, Canada.

90. Burton FD, Snarr KA, Harrison SE. 1995. Preliminary report on *Presbytis francoisi leucocephalus. IJP* 16(2):311–329.

91. Busse CD, Gordon TP. 1983. Attacks on neonates by a male mangabey *(Cercocebus atys). AJP* 5:345–356.

92. Busse CD, Gordon TP. 1984. Infant carrying by adult male mangabeys *(Cercocebus atys). AJP* 6:133–141.

93. Butynski T. 1982. Harem-male replacement and infanticide in the blue monkey *(Cercopithecus mitis stuhlmanni)* in Kibale Forest, Uganda. *AJP* 3:1–22.

94. Butynski TM. 1982. Vertebrate predation by primates: a review of hunting patterns and prey. *J Hum Evol* 11:421–430.

95. Butynski TM. 1982. Blue monkey *(Cercopithecus mitis stuhlmanni)* predation on galago. *Primates* 23(4):563–566.

96. Butynski TM. 1988. Guenon birth seasons and correlates with rainfall and food. In Gautier-Hion A, Bourlière F, Gautier JP, Kingdon J (eds), *A Primate Radiation: Evolutionary Biology of the African Guenons,* 284–322. Cambridge Univ Press, Cambridge.

97. Butynski TM. 1990. Comparative ecology of blue monkeys *(Cercopithecus mitis)* in high- and low-density populations. *Ecol Monogr* 60(1):1–26.

98. Bynum EL. 1994. Population characteristics and habitat use in hybrid *Macaca tonkeana* and *Macaca hecki* in central Sulawesi, Indonesia. Paper presented at IPS Congress 15, Kuta, Bali, Indonesia. [Abstract available: 1994. *15th Congress IPS Handbook and Abstracts:* 302.]

99. Byrne G, Suomi SJ. 1995. Development of activity patterns, social interactions, and exploratory behavior in infant tufted capuchins *(Cebus apella). AJP* 35:255–270.

100. Byrne RW, Whiten A, eds. 1988. *Machiavellian Intelligence: Social Expertise and the Evolution of Intellect in Monkeys, Apes, and Humans.* Oxford Univ Press, Oxford.

101. Caine NG. 1993. Flexibility and cooperation as unifying themes in *Saguinus* social organization and behaviour: the role of predation pressure. In Rylands AB (ed), *Marmosets and Tamarins: Systematics, Behaviour, Ecology,* 200–219. Oxford Univ Press, Oxford.

102. Caldecott JO. 1986. *An Ecological and Behavioral Study of the Pig-tailed Macaque.* Karger, Basel, Switzerland.

103. Caro TM, Sellen DW, Parish A, et al. 1995. Termination of reproduction in nonhuman and human female primates. *IJP* 16(2):205–221.

104. Cartmill M. 1979. *Daubentonia, Dactylopsila,* woodpeckers, and klinorhynchy. In Martin RD, Doyle GA, Walker AC (eds), *Prosimian Biology,* 655–672. Univ of Pittsburgh Press, Pittsburgh.

105. Casimir M. 1975. Feeding ecology and nutrition of an eastern gorilla group in the Mt. Kahuzi region (République du Zaïre). *Folia Primatol* 24:81–136.

106. Chapman C. 1987. Flexibility in diets of three species of Costa Rican primates. *Folia Primatol* 49:90–105.

107. Chapman C. 1990. Ecological constraints on group size in three species of Neotropical primates. *Folia Primatol* 55:1–9.

108. Chapman C, Chapman LJ. 1990. Dietary variability in primate populations. *Primates* 31(1):121–128.

109. Charles-Dominique P; Martin RD, trans. 1977. *Ecology and Behavior of Nocturnal Primates: Prosimians of Equatorial West Africa.* Columbia Univ Press, New York.

110. Charles-Dominique P. 1985. Bush babies, lorises, and pottos. In Macdonald D (ed), *Primates,* 36–41. Torstar Books, New York.

111. Charles-Dominique P, Petter JJ. 1980. Ecology and social life of *Phaner furcifer.* In Charles-Dominique P, Cooper HM, Hladik A, et al (eds), *Nocturnal Malagasy Primates: Ecology, Physiology, and Behavior,* 75–96. Academic Press, New York.

112. Chemnick L, Ryder O. 1994. Cytological and molecular divergence of orangutan subspecies. In Ogden JJ, Perkins LA, Sheeran L (eds), *Proceedings of the International Conference on Orangutans: The Neglected Ape,* 74–78. Zoological Society of San Diego, San Diego.

113. Cheney RM, Seyfarth RM. 1990. *How Monkeys See the World.* Univ of Chicago Press, Chicago.

114. Chism J, Olson DK, Rowell TE. 1983. Diurnal births and perinatal behavior among wild patas monkeys: evidence of an adaptive pattern. *IJP* 4(2):167–184.

115. Chism J, Rowell T, Olson D. 1984. Life history patterns of female patas monkeys. In Small MF (ed), *Female Primates,* 175–192. AR Liss, New York.

116. Chivers DJ. 1977. The feeding behaviour of siamangs *(Symphalangus syndactylus).* In Clutton-Brock TH (ed), *Primate Ecology: Studies of Feeding and Ranging Behaviour in Lemurs, Monkeys, and Apes,* 335–383. Academic Press, London.

117. Chivers DJ. 1984. Feeding and ranging in gibbons: a summary. In Preuschoft H, Chivers DJ, Brockelman WY, Creel N (eds), *The Lesser Apes: Evolutionary and Behavioural Biology,* 267–284. Edinburgh Univ Press, Edinburgh.

118. Chivers DJ. 1985. Gibbons. In Macdonald D (ed), *Primates,* 118–123. Torstar Books, New York.

119. Choudhury A. 1989. Ecology of the capped langur *(Presbytis pileatus)* in Assam, India. *Folia Primatol* 52:88–92.

120. Choudhury A. 1990. Population dynamics of hoolock gibbons *(Hylobates hoolock)* in Assam, India. *AJP* 20:37–41.

121. Abegg C, Thierry B, Kaumanns W. 1994. Reconciliation in two groups of lion-tailed macaques *(Macaca silenus).* Paper presented at IPS Congress 15, Kuta, Bali, Indonesia. [Abstract available: 1994. *15th Congress IPS Handbook and Abstracts:* 241.]

122. Clark AB. 1988. Interspecific differences and discrimination of auditory and olfactory signals of *Galago crassicaudatus* and *Galago garnettii. IJP* 9(6):557–571.

123. Clarke MR, Glander KE. 1984. Female reproductive success in a group of free ranging howling monkeys *(Alouatta palliata)* in Costa Rica. In Small MF (ed), *Female Primates,* 111–126. AR Liss, New York.

124. Clutton-Brock TH. 1975. Feeding behaviour of red colobus and black and white colobus in East Africa. *Folia Primatol* 23:165–207.

125. Coates A, Poole TB. 1983. The behavior of the callitrichid monkey, *Saguinus labiatus labiatus,* in the laboratory. *IJP* 4(4):339–371.

126. Colquhoum IC. 1993. The socioecology of *Eulemur macaco:* a preliminary report. In Kappeler PM, Ganzhorn JU (eds), *Lemur Social Systems and Their Ecological Basis,* 11–24. Plenum, New York.

127. Colyn M. 1994. Donnees ponderales sur les primates Cercopithecidae d'Afrique centrale (Bassin du Zaire/Congo). *Mammalia* 58(3):483–487.

128. Colyn M, Gautier-Hion A, Thys van den Audenaerde D. 1991. *Cercopithecus dryas* Schwarz 1932 and *C. salongo* Thys van den Audenaerde 1977 are the same species with an

age-related coat pattern. *Folia Primatol* 56:167–170.

129. Cooper HM. 1980. Ecological correlates of visual learning in nocturnal prosimians. In Charles-Dominique P, Cooper HM, Hladik A, et al (eds), *Nocturnal Malagasy Primates: Ecology, Physiology, and Behavior*, 191–204. Academic Press, New York.

130. Corbet GB, Hill JE. 1992. Primate. In *The Mammals of the Indomalayan Region: A Systematic Revision*, 42–52, 161–185, 419–421. Oxford Univ Press, Oxford.

131. Cordeiro NJ. 1994. Opportunist killers: blue monkeys feed on forest birds. *Folia Primatol* 63:84–87.

132. Cords M. 1984. Mating patterns and social structure in redtail monkeys *(Cercopithecus ascanius)*. *7 Tierpsychol* 64:313–329.

133. Cords M. 1987. Forest guenons and patas monkeys: male-male competition in one-male groups. In Smuts BB, Cheney DL, Seyfarth RM, Wrangham RW, Struhsaker TT (eds), *Primate Societies*, 98–111. Univ of Chicago Press, Chicago.

134. Cords M. 1988. Mating systems of forest guenons: a preliminary review. In Gautier-Hion A, Bourlière F, Gautier JP, Kingdon J (eds), *A Primate Radiation: Evolutionary Biology of the African Guenons*, 323–339. Cambridge Univ Press, Cambridge.

135. Cords M. 1990. Mixed-species association of east African guenons: general pattern or specific examples? *AJP* 21:101–114.

136. Cords MJ, Aureli F. 1993. Patterns of reconciliation among juvenile long-tailed macaques. In Pereira ME, Fairbanks LA (eds), *Juvenile Primates: Life History, Development, and Behavior*, 271–284. Oxford Univ Press, New York.

137. Costello RK, Dickinson C, Rosenberger AL, Boinski S, Szalay FS. 1993. Squirrel monkey (genus *Saimiri*) taxonomy: a multidisciplinary study of the biology of species. In Kimbel WH, Martin LB (eds), *Species, Species Concepts, and Primate Evolution*, 177–210. Plenum, New York.

138. Cowgill UM, States SJ, States KJ. 1989. A twenty-five-year chronicle of a group of captive nocturnal prosimians *(Perodicticus potto)*. *Mammal Rev* 9(2):83–89.

139. Cowlishaw G. 1992. Song function in gibbons. *Behaviour* 121(1–2):131–153.

140. Crandlemire-Sacco J. 1988. An ecological comparison of two sympatric primates: *Saguinus fuscicollis* and *Callicebus moloch* of Amazonia Peru. *Primates* 29(4):465–475.

141. Crockett CM, Eisenberg JF. 1987. Howlers: variation in group size and demography in primate societies. In Smuts BB, Cheney DL, Seyfarth RM, Wrangham RW, Struhsaker TT (eds), *Primate Societies*, 54–68. Univ of Chicago Press, Chicago.

142. Crockett CM, Pope TR. 1993. Consequences of sex differences in dispersal for juvenile red howler monkeys. In Pereira ME, Fairbanks LA (eds), *Juvenile Primates: Life History, Development, and Behavior*, 104–118. Oxford Univ Press, New York.

143. Crockett CM, Sekulic R. 1982. Gestation length in red howler monkeys. *AJP* 3:291–294.

144. Crompton RH, Andah PM. 1986. Location and habitat utilization in free-ranging *Tarsius bancanus*: a preliminary report. *Primates* 27(3):337–355.

145. Curtin SH, Chivers DJ. 1978. Leaf-eating primates of peninsular Malaysia: the siamang and the dusky leaf-monkey. In Montgomery GG (ed), *The Ecology of Arboreal Folivores*, 441–464. Smithsonian Institution Press, Washington, DC.

146. da Cunha C, Barnett AA. 1990. Sightings of the golden-backed uacari *(Cacajao melanocephalus*

ouakary) on the upper Rio Negro, Amazonas, Brazil. *Primate Conserv* 11:8–11.

147. Lan DY. 1993. Feeding and vocal behaviours of black gibbons *(Hylobates concolor)* in Yunnan: a preliminary study. *Folia Primatol* 60:94–105.

148. Dasilva GA. 1989. The ecology of western black and white colobus *(Colobus polykomos polykomos* Zimmerman 1780) on a riverine island in southeastern Sierra Leone. PhD thesis, Univ of Oxford, Oxford, UK.

149. Dasilva GA. 1992. The western black-and-white colobus as a low-energy strategist: activity budgets, energy expenditure, and energy intake. *J Anim Ecol* 61:79–91.

150. Dasilva GA. 1994. Diet of *Colobus polykomos* on Tiwai Island: selection of food in relation to its seasonal abundance and nutritional quality. *IJP* 15(5).655–680.

151. Davies AG. 1991. Seed-eating by red leaf monkeys *(Presbytis rubicunda)* in dipterocarp forest of northern Borneo. *IJP* 12(2):119–140.

152. Davies AG, Bennett EL, Waterman PG. 1988. Food selection by two south-east Asian colobine monkeys *(Presbytis rubicunda* and *Presbytis melalophos)* in relation to plant chemistry. *Biol J Linn Soc* 34:33–56.

153. Davies CR, Ayres JM, Dye C, Deane LM. 1991. Malaria infection rate of Amazonian primates increases with body weight and group size. *Funct Ecol* 5:655–662.

154. Davies AG. 1995. A new baby is born. *Radio Times* 13.

155. de la Torre S. 1994. Feeding habits of *Saguinus nigricollis graellsi* in northeastern Ecuador. Paper presented at IPS Congress 15, Kuta, Bali, Indonesia. [Abstract available: 1994. *15th Congress IPS Handbook and Abstracts:* 82.]

156. de Waal F. 1989. *Peacemaking among Primates.* Harvard Univ Press, Cambridge.

157. de Waal FBM. 1988. The communicative repertoire of captive bonobos *(Pan paniscus)* compared to that of chimpanzees. *Behaviour* 106:183–251.

158. de Waal FBM. 1992. Appeasement, celebration, and food sharing in the two *Pan* species. In Nishida T, McGrew WC, Marler P, Pickford M, de Waal FBM (eds), *Topics in Primatology: Human Origins*, vol 1, 37–50. Univ of Tokyo Press, Tokyo.

159. de Waal FBM. 1993. Reconciliation among primates: a review of empirical evidence and unresolved issues. In Mason WH, Mendoza SP (eds), *Primate Social Conflict*, 111–144. State Univ of New York Press, Albany.

160. de Waal FBM. 1993. Codevelopment of dominance relations and affiliative bonds in rhesus monkeys. In Pereira ME, Fairbanks LA (eds), *Juvenile Primates: Life History, Development, and Behavior*, 259–270. Oxford Univ Press, New York.

161. Decker BS. 1994. Effects of habitat disturbance on the behavioral ecology and demographics of the Tana River red colobus *(Colobus badius rufomitratus)*. *IJP* 15(5):703–737.

162. Defler TR. 1989. The status and some ecology of primates in the Colombian amazon. *Primate Conserv* 10:51–55.

163. Defler TR. 1995. The time budget of a group of wild woolly monkeys *(Lagothrix lagotricha)*. *IJP* 16(1):107–120.

164. Demars C, Berthomier, Goustard M. 1983. The great call of *Hylobates concolor hainanus*: comparison with the homologous emission of *H. concolor gabriellae* and *H. concolor leucogenys*. In Seth PK (ed), *Perspectives in Primate Biology*, 99–102. Today and Tomorrow's Printers and Publishing, New Delhi.

165. Deputte BL. 1982. Dueting in male and female songs of the white-cheeked gibbon *(Hylobates concolor leucogenys)*. In Snowdon CT, Brown CH,

Petersen MR (eds), *Primate Communication*, 67–93. Cambridge Univ Press, Cambridge.

166. Deputte BL. 1991. Reproduction parameters of captive grey-cheeked mangabeys. *Folia Primatol* 57:57–69.

167. Dew L, Wright PC. 1994. Conservation implications of seed dispersal by primates in a Malagasy rainforest (Ranomafana National Park). Paper presented at IPS Congress 15, Kuta, Bali, Indonesia. [Abstract available: 1994. *15th Congress IPS Handbook and Abstracts:* 124.]

168. Dewar RE. 1984. Extinction in Madagascar: the loss of the subfossil fauna. In Martin PS, Klein RG (eds), *Quaternary Extinction: A Prehistoric Revolution*, 574–593. Univ of Arizona Press, Tucson.

169. Dewsbury DA, Pierce JD Jr. 1989. Copulatory patterns of primates as viewed in broad mammalian perspective. *AJP* 17:51–72.

170. Permadi D, Tumbelaka LI, Yusuf TL. 1994. Reproductive patterns of *Tarsius* spp. in captive breeding. Poster presented at IPS Congress 15, Kuta, Bali, Indonesia. [Abstract available: 1994. *15th Congress IPS Handbook and Abstracts:* 421.]

171. Dietz JM, Baker AJ, Miglioretti D. 1994. Seasonal variation in reproduction, juvenile growth, and adult body mass in golden lion tamarins *(Leontopithecus rosalia)*. *AJP* 34:115–132.

172. Dittus WPJ. 1975. Population dynamics of the toque monkey, *Macaca sinica*. In Tuttle RH (ed), *Socioecology and Psychology of Primates*, 125–152. Mouton, The Hague.

173. Dittus WPJ. 1977. The social regulation of population density and age-sex distribution in the toque monkey. *Behaviour* 63:281–322.

174. Dixson AF. 1987. Baculum length and copulatory behavior in primates. *AJP* 13:51–60.

175. Dixson AF. 1989. Sexual selection, genital morphology, and copulatory behavior in male galagos. *IJP* 10(1):47–55.

176. Dixson AF. 1994. Reproductive biology of the owl monkey. In Baer J, Weller RE, Kakoma I (eds), *Aotus: The Owl Monkey*, 113–133. Academic Press, New York.

177. Dixson AF, Bossi T, Wickings EJ. 1993. Male dominance and genetically determined reproductive success in the mandrill *(Mandrillus sphinx)*. *Primates* 34(4):525–532.

178. Djuwantoko. 1994. Habitat and conservation of ebony leaf monkey in deciduous forests (teak), in central Java. Paper presented at IPS Congress 15, Kuta, Bali, Indonesia. [Abstract available: 1994. *15th Congress IPS Handbook and Abstracts:* 83.]

179. Doran D. 1995. Pers com. State Univ of New York Stony Brook, Dec.

180. Doyle GA. 1979. Development of behavior in prosimians with special reference to the lesser bushbaby, *Galago senegalensis moholi*. In Doyle GA, Martin RD (eds), *The Study of Prosimian Behavior*, 158–206. Academic Press, New York.

181. Duckworth JW. 1994. Field sightings of the pygmy loris, *Nycticebus pygmaeus*, in Laos. *Folia Primatol* 63:99–101.

182. Dunbar RIM, Nathan MF. 1972. Social organization of the Guinea baboon *(Papio papio)*. *Folia Primatol* 17:321–334.

183. Dunbar RIM. 1977. Feeding ecology of gelada baboons: a preliminary report. In Clutton-Brock TH (ed), *Primate Ecology: Studies of Feeding and Ranging Behaviour in Lemurs, Monkeys, and Apes*, 251–275. Academic Press, London.

184. Dunbar RIM. 1991. Functional significance of social grooming in primates. *Folia Primatol* 57:121–131.

185. Dunbar RIM, Dunbar EP. 1974. Ecology and population dynamics of *Colobus guereza* in Ethiopia. *Folia Primatol* 21:188–208.

186. Dutrillaux B, Dutrillaux AM, Lombard M, et al. 1988. The karyotype of *Cercopithecus solatus* Harrison 1988, a new species belonging to *C. lhoesti*, and its phylogenetic relationship with other guenons. *J Zool Lond* 215(4):611–617.

187. Eames JC, Robson CR. 1993. Threatened primates in southern Vietnam. *Oryx* 27(3):146–154.

188. Easley SP. 1982. Ecology and behavior of *Callicebus torquatus*, Cebidae, Primates. PhD diss, Washington Univ, St Louis.

189. Easley SP, Coelho AM Jr. 1991. Lipsmacking: an indication of social status in baboons. *Folia Primatol* 56:190–201.

190. Egler SG. 1992. Feeding ecology of *Saguinus bicolor bicolor* (Callitrichidae: Primates) in a relict forest in Manaus, Brazilian Amazonia. *Folia Primatol* 59:61–76.

191. Ehardt CL. 1988. Affiliative behavior of adult female sooty mangabeys *(Cercocebus atys). AJP* 15:115–127.

192. Ehardt CL. 1988. Absence of strongly kin-preferential behavior by adult sooty mangabeys *(Cercocebus atys). AJPA* 76:233–243.

193. Ehrlich A, Macbride L. 1989. Mother-infant interactions in captive slow lorises *(Nycticebus coucang). AJP* 19:217–228.

194. Eisenberg JF. 1973. General notes. *J Mammal* 54(4):955–957.

195. Eisenberg JF. 1973. Reproduction in two species of spider monkeys, *Ateles fusciceps* and *Ateles geoffroyi. J Mammal* 54(21):955–957.

196. Eisenberg JF. 1989. Order Primates. In *The Northern Neotropics: Panama, Colombia, Venezuela, Guyana, Suriname, French Guiana*, vol. 1 of *Mammals of the Neotropics*, 233–261. Univ of Chicago Press, Chicago.

197. Elliot DG. 1910. Dr. D. G. Elliot on new species of monkeys. *Ann Mag Nat Hist* 5:81.

198. Emmons LH, Feer F. 1990. Monkeys (Primates). *Neotropical Rainforest Mammals: A Field Guide*, 134–153. Univ of Chicago Press, Chicago.

199. Enomoto T. 1991. Flexibility in activities of solicitations for heterosexual intercourse in the bonobo *(Pan paniscus)*. In Ehara A, Kimura T, Takenaka O, Iwamonto M (eds), *Primatology Today*. Elsevier Science, Amsterdam.

200. Epple C, Belcher AM, Küderling I, et al. 1993. Making sense out of scents: species differences in scent glands, scent-marking behaviour, and scent-mark composition in the Callitrichidae. In Rylands AB (ed), *Marmosets and Tamarins: Systematics, Behaviour, Ecology*, 123–151. Oxford Univ Press, Oxford.

201. Epple G, Alveario MC, Katz Y. 1982. The role of chemical communication in aggressive behavior and its gonadal control in the tamarin *(Saguinus fuscicollis)*. In Snowdon CT, Brown CH, Petersen MR (eds), *Primate Communication*, 279–302. Cambridge Univ Press, Cambridge.

202. Erickson CJ. 1995. Feeding sites for extractive foraging by the aye-aye, *Daubentonia madagascariensis. AJP* 35:235–240.

203. Erwin J. 1994. Travels and travails in south Sulawesi. *Sulawesi Primate Newsl* 2(2).

204. Estes RD. 1991. Part 4, Primates. In *The Behavior Guide to African Mammals: Including Hoofed Mammals, Carnivores, Primates*, 447–558. Univ of California Press, Berkeley and Los Angeles.

205. Estrada A. 1984. Resource use by howler monkeys *(Alouatta palliata). IJP* 5(2):105–131.

206. Eudey AA, comp. 1987. *Action Plan for Asian Primate Conservation: 1987–1991*. IUCN, Gland, Switzerland.

207. Eudey AA. 1991. Macaque habitat preference in west-central Thailand and quaternary glacial events. In Ehara A, Kimura T, Takenaka O, Iwamonto M (eds), *Primatology Today*. Elsevier Science, Amsterdam.

208. Eudey AA. 1996. Pers com. Upland, CA, Jan.

209. Eveleth PB, Tanner JM, eds. 1990. *Worldwide Variation in Human Growth*. 2d ed. Cambridge Univ Press, Cambridge.

210. Fa JF. 1985. Baby care in Barbary macaque. In Macdonald D (ed), *Primates*, 92–93. Torstar Books, New York.

211. Fa M. 1989. The genus *Macaca:* a review of taxonomy and evolution. *Mammal Rev* 19(2):45–81.

212. Fairbanks LA. 1993. Juvenile vervet monkeys: establishing relationships and practicing skills for the future. In Pereira ME, Fairbanks LA (eds), *Juvenile Primates: Life History, Development, and Behavior*, 211–227. Oxford Univ Press, New York.

213. Fedigan L, Fedigan LM. 1988. *Cercopithecus aethiops:* a review of field studies. In Gautier-Hion A, Bourlière F, Gautier JP, Kingdon J (eds), *A Primate Radiation: Evolutionary Biology of the African Guenons*, 389–411. Cambridge Univ Press, Cambridge.

214. Fedigan LM, Fedigan L, Chapman C, Glander K. 1988. Spider monkey home ranges: a comparison of radio telemetry and direct observation. *AJP* 16:19–29.

215. Feistner ATC. 1991. Scent marking in mandrills, *Mandrillus sphinx. Folia Primatol* 57:42–47.

216. Feistner ATC. 1994. Biology and conservation of *Hapalemur griseus alaotrensis* at Lac Alaotra. Paper presented at IPS Congress 15, Kuta, Bali, Indonesia. [Abstract available: 1994. *15th Congress IPS Handbook and Abstracts:* 122.]

217. Fernandez MEB. 1994. Notes on the geographic distribution of howling monkeys in the Marajó Archipelago, Pará, Brazil. *IJP* 15(6):919–927.

218. See 217.

219. Ferrari S. 1988. The behaviour and ecology of the buffy-headed marmoset. PhD thesis, Univ College of London, London.

220. Ferrari S. 1991. An observation of western black spider monkeys, *Ateles paniscus chamek*, utilizing an arboreal water source. *Biotropica* 23(3):307–308.

221. Ferrari SF. 1987. Food transfer in a wild marmoset group. *Folia Primatol* 48:203–206.

222. Ferrari SF. 1991. Preliminary report on a field study of *Callithrix flaviceps*. In Rylands AB, Bernardes AT (eds), *A Primatologia no Brasil-3*, 159–171. Fundação Biodiversitas, Belo Horizonte, Brazil.

223. Ferrari SF. 1992. The care of infants in a wild marmoset *(Callithrix flaviceps)* group. *AJP* 26:109–118.

224. Ferrari SF. 1993. Ecological differentiation in the Callitrichidae. In Rylands AB (ed), *Marmosets and Tamarins: Systematics, Behaviour, Ecology*, 314–328. Oxford Univ Press, Oxford.

225. Ferrari SF. 1993. An update on the black-headed marmoset, *Callithrix nigriceps* Ferrari and Lopes 1992. *Neotrop Primates* 1(4):11–13.

226. Ferrari SF, Ferrari MAL. 1989. A re-evaluation of the social organization of the Callitrichidae, with reference to the ecological differences between genera. *Folia Primatol* 52:132–147.

227. Ferrari SF, Ferrari MAL. 1990. Predator avoidance in the buffy-headed marmoset, *Callithrix flaviceps. Primates* 33(3):323–338.

228. Ferrari SF, Lopes MA. 1992. A new species of marmoset genus *Callithrix* Erxleben 1777 Callitrichidae Primates, from western Brazilian Amazon. *Goeldiana Zool* 12:1–13.

229. Ferrari SF, Lopes MA, Krause EAK. 1993. Brief communication: gut morphology of *Callithrix nigriceps* and *Saguinus labiatus* from western Amazonia. *AJPA* 90(4):487–493.

230. Ferrari SF, Queiroz HL. 1994. Two new Brazilian primates discovered, endangered. *Oryx* 28(1):31–36.

231. Fimbel C. 1994. Ecological correlates of species success in modified habitats may be disturbance- and site-specific: the primates of Tiwai Island. *Conserv Biol* 8(1):106–113.

232. Fischer JO, Geissmann T. 1990. Group harmony in gibbons: comparison between white-handed gibbon *(Hylobates lar)* and siamang *(H. syndactylus). Primates* 31(4):481–494.

233. Fleagle JG. 1978. Locomotion, posture, and habitat utilization in two sympatric Malaysian leaf-monkeys *(Presbytis obscura* and *Presbytis melalophos)*. In Montgomery GG (ed), *The Ecology of Arboreal Folivores*, 243–253. Smithsonian Institution Press, Washington, DC.

234. Fleagle JG. 1988. *Primate Adaptation and Evolution*. Academic Press, New York.

235. Foerg R. 1982. Reproduction in *Cheirogaleus medius. Folia Primatol* 39:49–62.

236. Fogden MPI. 1979. A preliminary field study of the western tarsier, *Tarsius bancanus* Horsefield. In Martin RD, Doyle GA, Walker AC (eds), *Prosimian Biology*, 151–166. Univ of Pittsburgh Press, Pittsburgh.

237. de Fonseca GAB, Herrman G, Leite YLR, Mittermeier RA, Rylands AB. 1995. Lista anotada dos mamiferos do Brasil. Occasional Paper in Conservation Biology 3. Conservation International, Washington, DC.

238. Fontaine R. 1981. The uakaris, genus *Cacajao*. In Coimbra-Filho AF, Mittermeier RA (eds), *Ecology and Behavior of Neotropical Primates*, vol 1, 443–494. Academia Brasileira de Ciências, Rio de Janeiro.

239. Fooden J. 1969. *Taxonomy and Evolution of the Monkeys of Celebes (Primates: Cercopithecidae)*. Karger, Basel, Switzerland.

240. Fooden J. 1976. Provisional classification and key to living species of macaques (Primates: *Macaca). Folia Primatol* 25:225–236.

241. Fooden J. 1976. Primates obtained in peninsular Thailand June–July 1973 with notes on the distribution of continental Southeast Asian leaf monkeys *(Presbytis). Primates* 17(1):95–118.

242. Fooden J. 1981. Taxonomy and evolution of the *sinica* group of macaques: part 2, species account of the Indian bonnet macaque, *Macaca radiata. Fieldiana Zool*, ns, 9:1–52.

243. Fooden J. 1982. Taxonomy and evolution of the *sinica* group of macaques: part 3, species account of *Macaca assamensis. Fieldiana Zool*, ns, 10:1–51.

244. Fooden J. 1983. Taxonomy and evolution of the *sinica* group of macaques: part 4, species account of *Macaca thibetana. Fieldiana Zool*, ns, 17:1–20.

245. Ford SM. 1994. Taxonomy and distribution of the owl monkey. In Baer J, Weller RE, Kakoma I (eds), *Aotus: The Owl Monkey*, 1–59. Academic Press, New York.

246. Ford SM. 1994. Evolution of sexual dimorphism in body weight in platyrrhines. *AJP* 34:221–244.

247. Ford SM, Davis LC. 1992. Systematics and body size: implications for feeding adaptation in New World monkeys. *AJPA* 88:415–468.

248. Forman MF, French JA. 1992. Scent marking and food availability in golden-lion tamarins *(Leontopithecus rosalia)* and Wied's black tufted-ear marmoset *(Callithrix kuhli)* [abstract]. *AJP* 27(1):28.

249. Fossey D, Harcourt AH. 1977. Feeding ecology of free ranging mountain gorillas *(Gorilla gorilla beringei)*. In Clutton-Brock TH (ed), *Primate*

Ecology: Studies of Feeding and Ranging Behaviour in Lemurs, Monkeys, and Apes, 415–449. Academic Press, London.

250. Fox R. 1995. *The Challenge of Anthropology: Old Encounters and New Excursions*. Transaction Publishers, New Brunswick, NJ.

251. Freese CH, Oppenheimer JR. 1981. The capuchin monkey, genus *Cebus*. In Coimbra-Filho AF, Mittermeier RA (eds), *Ecology and Behavior of Neotropical Primates*, vol 1, 331–390. Academia Brasileira de Ciências, Rio de Janeiro.

252. Froehlich JW. 1994. Pers com. Univ of New Mexico, Albuquerque, NM, Aug.

253. Froehlich JW, Froelich PH. 1987. The status of Panama's endemic howling monkey. *Primate Conserv* 8:58–62.

254. Fuentes A. 1994. Social organization of the Mentawai langur *(Presbytis potenziani)*. AJPA 18(suppl):90.

255. Fuentes A. 1994. Social organization of the Mentawai langur *(Presbytis potenziani)*. Paper presented at IPS Congress 15, Kuta, Bali, Indonesia. [Abstract available: 1994. *15th Congress IPS Handbook and Abstracts: 261*.]

256. Galdikas BMF. 1994. Pers com. Simon Fraser Univ, Burnaby, BC, Canada, Aug.

257. Galdikas BMF. 1995. *Reflections of Eden: My Years with the Orangutans of Borneo*. Little, Brown, Boston.

258. Galdikas BMF, Shapiro GL. 1994. *A Guidebook to Tanjung Puting National Park*. PT Gramedia Pustaka Utama, Jakarta, Indonesia.

259. Galdikas BMF, Wood JW. 1990. Birth spacing patterns in humans and apes. *AJPA* 83:185–191.

260. Ganzhorn JU. 1985. Habitat separation of semi-free-ranging *Lemur catta* and *Lemur fulvus*. *Folia Primatol* 45:76–88.

261. Ganzhorn JU. 1993. Flexibility and constraints of *Lepilemur* ecology. In Kappeler PM, Ganzhorn JU (eds), *Lemur Social Systems and Their Ecological Basis*, 153–166. Plenum, New York.

262. Ganzhorn JU, Wright PC. 1994. Temporal patterns in primate leaf eating: the possible role of leaf chemistry. *Folia Primatol* 63:203–208.

263. Garber PA. 1991. A comparative study of positional behavior in three species of tamarin monkeys. *Primates* 32(2):219–230.

264. Garber PA. 1992. Vertical clinging, small body size, and evolution of feeding adaptations in the Callitrichinae. *AJPA* 88:469–482.

265. Garber PA. 1993. Feeding ecology and behaviour of the genus *Saguinus*. In Rylands AB (ed), *Marmosets and Tamarins: Systematics, Behaviour, Ecology*, 273–295. Oxford Univ Press, Oxford.

266. Garber PA. 1994. Phylogenetic approach to the study of tamarin and marmoset social systems. *AJP* 34:199–219.

267. Garcia JE, Braza F. 1987. Activity rhythms and use of space of a group of *Aotus azarae* in Bolivia during the rainy season. *Primates* 28(3):337–342.

268. Garcia JE, Braza F. 1993. Sleeping sites and lodge trees of the night monkey *(Aotus azarae)* in Bolivia. *IJP* 14(3):467–477.

269. Gartlan JS. 1970. Preliminary notes on the ecology and behavior of the drill *Mandrillus leucophaeus* Ritgen, 1824. In Napier JR, Napier PH (eds), *Old World Monkeys: Evolution, Systematics, and Behavior*, 445–480. Academic Press, New York.

270. Gaulin SJP, Gaulin CK. 1982. Behavioral ecology of *Alouatta seniculus* in Andean cloud forest. *IJP* 3(1):1–33.

271. Gaulin SJP, Sailer LD. 1984. Sexual dimorphism in weight among primates: the relative impact of allometry and sexual selection. *IJP* 5(6):515–536.

272. Gautier JP. 1988. Interspecific affinities among guenons as deduced from vocalizations. In Gautier-Hion A, Bourlière F, Gautier JP, Kingdon J (eds), *A Primate Radiation: Evolutionary Biology of the African Guenons*, 194–226. Cambridge Univ Press, Cambridge.

273. Gautier JP. 1988. New species of *Cercopithecus* in Gabon needs protection. *Primate Conserv* 9:27–28.

274. Gautier JP, Gautier-Hion A. 1982. Vocal communication within a group of monkeys: an analysis by biotelemetry. In Snowdon CT, Brown CH, Petersen MR (eds), *Primate Communication*, 5–39. Cambridge Univ Press, Cambridge.

275. Gautier JP, Loireau JN, Moysan F. 1986. Distribution, ecology, behaviour, and phylogenetic affinities of a new species of guenon, *Cercopithecus (l'hoesti) solatus* (Harrison, 1986) [abstract]. *Primate Rep* 14:106.

276. Gautier-Hion A. 1980. Seasonal variations of diet related to species and sex in a community of *Cercopithecus* monkeys. *J Anim Ecol* 49:237–269.

277. Gautier-Hion A. 1986. Paternal behaviour in arboreal guenons [abstract]. *Primate Rep* 14:31.

278. See 277.

279. Gautier-Hion A. 1988. The diet and dietary habits of forest guenons. In Gautier-Hion A, Bourlière F, Gautier JP, Kingdon J (eds), *A Primate Radiation: Evolutionary Biology of the African Guenons*, 257–283. Cambridge Univ Press, Cambridge.

280. Gautier-Hion A. 1988. Polyspecific associations among forest guenons: ecological, behavioural, and evolutionary aspects. In Gautier-Hion A, Bourlière F, Gautier JP, Kingdon J (eds), *A Primate Radiation: Evolutionary Biology of the African Guenons*, 452–476. Cambridge Univ Press, Cambridge.

281. Gautier-Hion A, Maisels F. 1994. Mutualism between a leguminous tree and large African monkeys as pollinators. *Behav Ecol Sociobiol* 34:203–210.

282. Gautier-Hion A, Quris R, Gautier JP. 1983. Monospecific vs. polyspecific life: a comparative study of foraging and antipredatory tactics in a community of *Cercopithecus* monkeys. *Behav Ecol Sociobiol* 12:325–335.

283. Gautier-Hion A, Tutin CEG. 1988. Simultaneous attack by adult males of a polyspecific troop of monkeys against a crowned hawk eagle. *Folia Primatol* 51:149–151.

284. Geissmann T. 1989. A female black gibbon, *Hylobates concolor* subspecies, from northwestern Vietnam. *IJP* 10(5):455–476.

285. Geissmann T. 1991. Reassessment of age of sexual maturity in gibbons (*Hylobates* spp.). *AJP* 23:11–22.

286. Geissmann T. 1994. Systematik der gibbons. *Z Koelner Zoo* 37(2):65–77.

287. Geissmann T. 1995. Captive management and conservation of gibbons in China and Vietnam with special reference to crested gibbons (*Hylobates concolor* group). *Primate Rep* 42:29–41.

288. Gevaerts H. 1992. Birth seasons of *Cercopithecus, Cercocebus*, and *Colobus* in Zaire. *Folia Primatol* 59:105–113.

289. Gittins SP. 1983. Use of the forest canopy by the agile gibbon. *Folia Primatol* 40:134–144.

290. Gittins SP. 1984. The vocal repertoire and song of the agile gibbon. In Preuschoft H, Chivers DJ, Brockelman WY, Creel N (eds), *The Lesser Apes: Evolutionary and Behavioural Biology*, 354–375. Edinburgh Univ Press, Edinburgh.

291. Gittins SP. 1984. The distribution and status of the hoolock gibbon in Bangladesh. In Preuschoft H, Chivers DJ, Brockelman WY, Creel N (eds), *The Lesser Apes: Evolutionary and Behavioural Biology*, 13–15. Edinburgh Univ Press, Edinburgh.

292. Gittins SP, Tilson RL. 1984. Notes on the ecology of the hoolock gibbon. In Preuschoft H, Chivers DJ, Brockelman WY, Creel N (eds), *The Lesser Apes: Evolutionary and Behavioural Biology*, 258–266. Edinburgh Univ Press, Edinburgh.

293. Glander KE, Wright PC, Seigler DS, Randrianasolo V, Randrianasolo B. 1989. Consumption of cyanogenic bamboo by a newly discovered species of bamboo lemur. *AJP* 19:119–124.

294. Glaston AR. 1983. Olfactory communication in the lesser mouse lemur *(Microcebus murinus)*. In Seth PK (ed), *Perspectives in Primate Biology*. Today and Tomorrow's Printers and Publishing, New Delhi.

295. Goldfoot DA. 1982. Multiple channels of sexual communication in rhesus monkeys: role of olfactory cues. In Snowdon CT, Brown CH, Petersen MR (eds), *Primate Communication*, 413–428. Cambridge Univ Press, Cambridge.

296. Goldizen AW. 1987. Tamarins and marmosets: communal care of offspring. In Smuts BB, Cheney DL, Seyfarth RM, Wrangham RW, Struhsaker TT (eds), *Primate Societies*, 34–43. Univ of Chicago Press, Chicago.

297. Goldstein SJ, Richard AF. 1989. Ecology of rhesus macaques *(Macaca mulatta)* in northwest Pakistan. *IJP* 10(6):531–567.

298. Goodall AG. 1977. Feeding and ranging behaviour of a mountain gorilla group *(Gorilla gorilla. beringei)* in the Tshibinda-Kahuzi region (Zaire). In Clutton-Brock TH (ed), *Primate Ecology: Studies of Feeding and Ranging Behaviour in Lemurs, Monkeys, and Apes*, 450–480. Academic Press, London.

299. Goodall J. 1992. Unusual violence in the overthrow of an alpha male chimpanzee at Gombe. In Nishida T, McGrew WC, Marler P, Pickford M, de Waal FBM (eds), *Topics in Primatology: Human Origins*, vol 1, 131–142. Univ of Tokyo Press, Tokyo.

300. Goodall J. 1996. *The Chimpanzees of Gombe: Patterns of Behavior*. Belknap Press of Harvard Univ Press, Cambridge.

301. Goodman SM. 1989. Predation by the grey leaf monkey *Presbytis hosei* of the contents of a bird's nest at Mt. Kinabalu Park, Sabah. *Primates* 30(1):127–128.

302. Goodman SM, O'Connor S, Langrand O. 1993. A review of predation on lemurs: implications for the evolution of social behavior in small, nocturnal primates. In Kappeler PM, Ganzhorn JU (eds), *Lemur Social Systems and Their Ecological Basis*, 51–66. Plenum, New York.

303. Gordon TP, Gust DA, Busse CD, Wilson ME. 1991. Behavior associated with postconception perineal swelling in sooty mangabeys, *Cercocebus torquatus atys*. *IJP* 12(6):585–597.

304. Gormus BJ, Wolf RH, Baskin GB, et al. 1988. A second sooty mangabey monkey with naturally acquired leprosy: first reported possible monkey to monkey transmission. *Int J Lepr Other Mycobact Dis* 56(1):61–65.

305. Goustard M. 1984. Patterns of loud calls in the concolor gibbon. In Preuschoft H, Chivers DJ, Brockelman WY, Creel N (eds), *The Lesser Apes: Evolutionary and Behavioural Biology*, 404–415. Edinburgh Univ Press, Edinburgh.

306. Grant JWA, Chapman CA, Richardson KS. 1992. Defended vs. undefended home range size of carnivores, ungulates, and primates. *Behav Ecol Sociobiol* 31:149–161.

307. Graves GR, O'Neil JP. 1980. Notes on the yellow-tailed woolly monkey *(Lagothrix flavicauda)* of Peru. *J Mammal* 61(2):345–347.

Bibliography *continued*

308. Griesser B. 1992. Infant development and paternal care in two species of sifakas. *Primates* 33(3):305–314.

309. Groves C. 1970. The forgotten leaf-eaters and the phylogeny of the Colobinae. In Napier JR, Napier PH (eds), *Old World Monkeys: Evolution, Systematics, and Behavior*, 555–588. Academic Press, New York.

310. Groves C, Angst R, Westwood C. 1993. The status of *Colobus polykomos dollmani* Schwarz. *IJP* 14(4):573–588.

311. Groves CP. 1973. Notes on the ecology and behaviour of the Angola colobus (*Colobus angolensis*, P. L. Sclater 1860) in N. E. Tanzania. *Folia Primatol* 20:12–26.

312. Groves CP. 1980. Speciation in *Macaca:* the view from Sulawesi. In Lindburg DG (ed), *The Macaques: Studies in Ecology, Behavior, and Evolution*, 84–124. Van Nostrand Reinhold, New York.

313. Groves CP. 1989. *A Theory of Human and Primate Evolution*. Oxford Univ Press, New York.

314. Groves CP. 1992. Book review: Titis, New World Monkeys of the Genus *Callicebus*. *IJP* 13(1):111–112.

315. Groves CP. 1993. Order Primates. In Wilson DE, Reader DM (eds), *Mammalian Species of the World: A Taxonomic and Geographic Reference*, 2d ed, 243–277. Smithsonian Institution Press, Washington, DC.

316. Groves CP. 1994. Book review: Primates of the World. *IJP* 15(5):789–790.

317. Groves CP. 1995. Pers com. Australian National Univ, Canberra, Dec.

318. Gurmaya KJ. 1986. Ecology and behavior of *Presbytis thomasi* in northern Sumatra. *Primates* 27(2):151–172.

319. Gurmaya KJ. 1994. Ecology and sociology of the all-male group of Thomas' leaf monkey *P. thomasi*. Paper presented at IPS Congress 15, Kuta, Bali, Indonesia. [Abstract available: 1994. *15th Congress IPS Handbook and Abstracts:* 80.]

320. Gursky SL. 1994. Infant care in spectral tarsier *(Tarsius spectrum)*, Sulawesi, Indonesia. *IJP* 15(6):843–853.

321. Gursky SL. 1995. Spectral tarsier population density in Tangkoko Nature Reserve, Sulawesi. Paper presented at the International Conference on the Biology and Conservation of Prosimians, The North of England Zoological Society, Chester, UK. [Abstract available: 1996. *Folia Primatol.* In press.]

322. Gursky SL. 1995. Pers com. State Univ of New York at Stony Brook, Nov.

323. Gursky SL. 1996. Group size and composition in the spectral tarsier, *Tarsius spectrum:* implications for social organization. *Trop Biodiversity* 3(1):57–62.

324. Gust DA. 1991. Male age and reproductive behavior in sooty mangabeys, *Cercocebus torquatus atys. Anim Behav* 46(2):277–283.

325. Gust DA. 1994. Alpha-male sooty mangabeys differentiate between females' fertile and their postconception maximal swellings. *IJP* 15(2):289–301.

326. Gust DA, Gordon TP. 1993. Conflict resolution in sooty mangabeys. *Anim Behav* 46:685–694.

327. Gust DA, Gordon TP. 1994. The absence of matrilineally based dominance system in sooty mangabeys, *Cercocebus torquatus atys. Anim Behav* 47:589–594.

328. Hager M. 1993. Deep in the flooded forest. *Wildl Conserv* 96(1):53–63.

329. Haimoff EH. 1985. The organization of song in Mueller's gibbon *(Hylobates muelleri). IJP* 6(2):173–192.

330. Haimoff EH, Yang XJ, He SJ, Chen N. 1987. Preliminary observation of wild black crested gibbons *(Hylobates concolor concolor)* in Yunnan Province, PRC. *Primates* 28(3):319–335.

331. Hamilton WJ III. 1982. Baboon sleeping site preferences and relationships to primate grouping patterns. *AJP* 3:41–53.

332. Hammerschmidt K, Ansorge V. 1989. Birth of a Barbary macaque *(Macaca sylvanus):* acoustic and behavioural features. *Folia Primatol* 52:78–87.

333. Harcourt AH. 1985. Gorilla. In Macdonald D (ed), *Primates*, 136–143. Torstar Books, New York.

334. Harcourt C. 1991. Diet and behaviour of a nocturnal lemur, *Avahi laniger*, in the wild. *J Zool Lond* 223:667–674.

335. Harcourt C, Thornbeck J. 1990. *Lemurs of Madagascar and the Comoros: The IUCN Red Data Book*. IUCN, Gland, Switzerland.

336. Harcourt CS. 1987. Brief trap/retrap of the brown mouse lemur *(Microcebus rufus). Folia Primatol* 49:209–211.

337. Harcourt CS, Bearder SK. 1989. Comparison of *Galago moholi* in southern Africa with *Galago zanzibaricus* in Kenya. *IJP* 10(1):35–45.

338. Harcourt CS, Nash LT. 1986. Species differences in substrate use and diet between sympatric galagos in two Kenyan coastal forests. *Primates* 27(1):41–52.

339. Harding RSO. 1984. Primates of the Killini area, northwest Sierra Leone. *Folia Primatol* 42:96–114.

340. Harding RSO. 1984. Primates of the Killini area, northwest Sierra Leone. *Primate Conserv* 4:32–34.

341. Harrington J. 1979. Olfactory communication in *Lemur fulvus*. In Martin RD, Doyle GA, Walker AC (eds), *Prosimian Biology*, 331–346. Univ of Pittsburgh Press, Pittsburgh.

342. Harrison GA, Tanner JM, Pilbeam DR, Baker PT, eds. 1988. *Human Biology: An Introduction to Human Evolution, Variation, Growth, and Adaptability*, 3d ed. Oxford Univ Press, Oxford.

343. Harrison M. 1988. New guenon from Gabon. *Oryx* 22(4):190–191.

344. Harrison MJS. 1984. Optimal foraging strategies in the diet of the green monkey, *Cercopithecus sabaeus*, at Mt. Assirik, Senegal. *IJP* 5(5):435–471.

345. Harrison MJS, Hladik CM. 1986. Un primate granivore: le colobe noir dans la foret du Gabon; potentialite d'evolution du comportement alimentaire. *Rev Ecol Terre Vie* 41(4):281–298.

346. Harvey PH, Martin RD, Clutton-Brock TH. 1987. Life histories in comparative perspective. In Smuts BB, Cheney DL, Seyfarth RM, Wrangham RW, Struhsaker TT (eds), *Primate Societies*, 181–196. Univ of Chicago Press, Chicago.

347. Hauser M. 1993. The evolution of the non-human primate vocalizations: effects of phylogeny body weight and social context. *Am Nat* 142(3):528–543.

348. Held JR, Wolfe TL. 1994. Imports: current trends and usage. *AJP* 34:85–94.

349. Heltne PJ, Wojcik JF, Pook AG. 1981. Goeldi's monkey, genus *Callimico*. In Coimbra-Filho AF, Mittermeier RA (eds), *Ecology and Behavior of Neotropical Primates*, vol 1, 169–210. Academia Brasileira de Ciências, Rio de Janeiro.

350. Henzi SP. 1985. Genital signaling and the coexistence of male vervet monkeys *(Cercopithecus aethiops pygerythrus). Folia Primatol* 45:129–147.

351. Hershkovitz P. 1977. *Living New World Monkeys (Platyrrhini): With an Introduction to Primates.* Vol 1. Univ of Chicago Press, Chicago.

352. Hershkovitz P. 1979. The species of saki genus *Pithecia* (Cebidae, Primates) with notes on sexual dimorphism. *Folia Primatol* 31:1–22.

353. Hershkovitz P. 1983. Two new species of night monkey, genus *Aotus* (Cebidae, Platyrrhini): a preliminary report on *Aotus* taxonomy. *AJP* 4:209–243.

354. Hershkovitz P. 1984. Taxonomy of squirrel monkeys genus *Saimiri* (Cebidae, Platyrrhini): a preliminary report with description of a hitherto unnamed form. *AJP* 6:257–312.

355. Hershkovitz P. 1985. A preliminary taxonomic review of the South American bearded saki monkeys of the genus *Chiropotes* (Cebidae, Platyrrhini) with a description of a new saki species. *Fieldiana Zool* 27:1–46.

356. Hershkovitz P. 1987. The taxonomy of South American sakis, genus *Pithecia* (Cebidae, Platyrrhini): a preliminary report and critical review with a description of a new species and a new subspecies. *AJP* 12:387–468.

357. Hershkovitz P. 1990. Titis New World monkeys of the genus *Callicebus* (Cebidae, Platyrrhini): a preliminary taxonomic review. *Fieldiana Zool* 55:1–109.

358. Heymann EW. 1990. Further field notes on red uacaris, *Cacajao calvus ucayalii*, from the Quebrada Blanco, Amazonian Peru. *Primate Conserv* 11:7–8.

359. Heymann EW, Hartmann G. 1991. Geophagy in moustached tamarins, *Saguinus mystax* (Platyrrhini: Callitrichidae), at the Rio Blanco, Peruvian Amazonia. *Primates* 32(4):533–537.

360. Hill CM. 1994. The role of female diana monkeys *(Cercopithecus diana). Anim Behav* 47:425–431.

361. Hill D. 1994. Affiliative behaviour between adult males of the genus *Macaca. Behaviour* 130(3–4):293–308.

362. Hill D, van Hooff JARAM. 1994. Affiliative relationships between males in groups of non-human primates: a summary. *Behaviour* 130(3–4):143–149.

363. Hill WCO. 1953. *Primates: Comparative Anatomy and Taxonomy.* Vol 1, *Strepsirrhini.* Edinburgh Univ Press, Edinburgh.

364. Hill WCO. 1960. *Primates: Comparative Anatomy and Taxonomy.* Vol 4, *Cebidae, Part A.* Interscience, New York.

365. Hill WCO. 1962. *Primates: Comparative Anatomy and Taxonomy.* Vol 5, *Cebidae, Part B.* Interscience, New York.

366. Hill WCO. 1966. *Primates: Comparative Anatomy and Taxonomy.* Vol 6, *Catarrhini Cercopithecoidea Cercopithecinae.* Interscience, New York.

367. Hiraiwa-Hasegawa M. 1992. Cannibalism among non-human primates. In Elgar MA, Crespi BJ (eds), *Cannibalism: Ecology and Evolution among Diverse Taxa*, 323–338. Oxford Univ Press, Oxford.

368. Hladik C, Charles-Dominique P, Petter JJ. 1980. Feeding strategies of five nocturnal prosimians in the dry forest of the west coast of Madagascar. In Charles-Dominique P, Cooper HM, Hladik A, et al (eds), *Nocturnal Malagasy Primates: Ecology, Physiology, and Behavior*, 41–74. Academic Press, New York.

369. Hladik CM. 1977. A comparative study of the feeding strategies of two sympatric species of leaf monkey: *Presbytis senex* and *Presbytis entellus*. In Clutton-Brock TH (ed), *Primate Ecology: Studies of Feeding and Ranging Behaviour in Lemurs, Monkeys, and Apes*, 324–354. Academic Press, London.

370. Hladik CM. 1979. Diet and ecology of prosimians. In Doyle GA, Martin RD (eds), *The Study of Prosimian Behavior*, 307–358. Academic Press, New York.

371. Hladik CM, Charles-Dominique P. 1979. The behavior and ecology of the sportive lemur *Lepilemur mustelinus* in relation to its dietary peculiarities. In Martin RD, Doyle GA, Walker AC (eds), *Prosimian Biology*, 23–38. Univ of Pittsburgh Press, Pittsburgh.

372. Hohmann G. 1988. Analysis of loud calls provides new evidence for hybridization between two Asian leaf monkeys *(Presbytis johnii, Presbytis entellus)*. Folia Primatol 51:209–213.

373. Hohmann G. 1989. Vocal communication of wild bonnet macaques *(Macaca radiata)*. Primates 30(3):325–345.

374. Hohmann G. 1989. Group fission in Nilgiri langurs *(Presbytis johnii)*. IJP 10(5):441–445.

375. Hohmann G. 1991. Comparative analyses of age and sex specific patterns of vocal behaviour in four species of Old World monkeys. *Folia Primatol* 56:133–136.

376. Hohmann G, Fruth B. 1994. Structure and use of distance calls in wild bonobos *(Pan paniscus)*. IJP 15(5):767–783.

377. Hohmann G, Sunderraj FSW. 1990. Survey of Nilgiri langurs and lion-tail macaques in Tamil Nadu, south India. *Primate Conserv* 11:49–53.

378. Hohmann G, Vogl L. 1991. Loud calls of male Nilgiri langur *(Presbytis johnii)*: age, individual, and population specific differences. *IJP* 12(5):503–524.

379. Hohmann GM, Herzog MO. 1985. Vocal communication in lion-tailed macaques *(Macaca silenus)*. Folia Primatol 45:148–178.

380. Homewood AM, Rodgers WA. 1981. A previously undescribed mangabey from southern Tanzania. *IJP* 2(1):47–57.

381. Honess PE. 1993. Phylogenetic and population genetic studies of nocturnal primates in African forest communities. Progress Report 1. Oxford Brookes Univ, Oxford, UK. Unpub.

382. Horn AD. 1987. The socioecology of the black mangabey *(Cercocebus aterrimus)* near Lake Tumba, Zaire. *AJP* 12:165–180.

383. Horn AD. 1987. Taxonomic assessment of the allopatric gray-cheeked mangabey *(Cercocebus albigena)* and black mangabey *(C. aterrimus)*: comparative socioecological data and the species concept. *AJP* 12:181–187.

384. Horrocks JA, Hunte W. 1993. Interactions between juveniles and adult males in vervets: implications for adult male turnover. In Pereira ME, Fairbanks LA (eds), *Juvenile Primates: Life History, Development, and Behavior*, 228–244. Oxford Univ Press, New York.

385. Horwich RA. 1974. Development of behaviors in a male spectacled langur. *Primates* 15(2–3):151–178.

386. Horwich RA. 1983. Breeding behaviors in the black howler monkey *(Alouatta pigra)* of Belize. *Primates* 24:222–230.

387. Horwich RA, Gebhard K. 1983. Roaring rhythms in black howler monkeys *(Alouatta pigra)* of Belize. *Primates* 24(2):290–296.

388. Hoshino J. 1985. Feeding ecology of mandrills *(Mandrillus sphinx)* in Campo Animal Reserve, Cameroon. *Primates* 26(3):248–271.

389. Hrdy SB. 1980. *The Langurs of Abu: Female and Male Strategies of Reproduction.* Harvard Univ Press, Cambridge.

390. Hrdy SB, Whitten PL. 1987. Patterning of sexual activity. In Smuts BB, Cheney DL, Seyfarth RM, Wrangham RW, Struhsaker TT (eds), *Primate Societies*, 370–384. Univ of Chicago Press, Chicago.

391. Huffman MA, Kalunde MS. 1993. Tool-assisted predation on a squirrel by a female chimpanzee in the Mahale Mountains, Tanzania. *Primates* 34(1):93–98.

392. Hunkeler C, Bourlière F, Bertrand M. 1972. Le comportement social de la mone de Lowe *(Cercopithecus campbelli lowei)*. Folia Primatol 17:218–236.

393. Indani G. 1986. Seed dispersal by pygmy chimpanzees *(Pan paniscus):* a preliminary report. *Primates* 27(4):441–447.

394. Indani G. 1991. Cases of interunit group encounters in pygmy chimpanzees at Wamba, Zaire. In Ehara A, Kimura T, Takenaka O, Iwamonto M (eds), *Primatology Today*. Elsevier Science, Amsterdam.

395. Isbell LA. 1994. The vervets' year of doom. *Nat Hist* 103(8):49–55.

396. Islam MA, Husain KZ. 1982. A preliminary study on the ecology of the capped langur. *Folia Primatol* 39:145–159.

397. Iwano T, Iwakawa C. 1988. Feeding behaviour of the aye-aye *(Daubentonia madagascariensis)* on nuts of ramy *(Canarium madagascariensis)*. Folia Primatol 50:136–142.

398. Izard MK, Rassmussen DT. 1985. Reproduction in the slender loris *(Loris tardigradus malabaricus)*. AJP 8:153–165.

399. Izard MK, Weisenseel KA, Ange RL. 1988. Reproduction in the slow loris *(Nycticebus coucang)*. AJP 16:331–339.

400. Izawa K. 1975. Foods and feeding behavior of monkeys in the upper Amazon Basin. *Primates* 16(3):295–316.

401. Izawa K. 1976. Group size and composition of monkeys in the upper Amazonian Basin. *Primates* 17(3):367–399.

402. Izawa K. 1978. Frog-eating behavior of wild black capped *(Cebus apella)*. Primates 19(4):633–642.

403. Izawa K. 1978. A field study of the ecology and behavior of the black mantled tamarin *(Saguinus nigricollis)*. Primates 19(2):241–274.

404. Izawa K. 1993. Soil eating by *Alouatta* and *Ateles*. IJP 14(2):229–241.

405. Jablonski NG. 1995. The phyletic position and systematics of the douc langurs of Southeast Asia. *AJP* 35:185–205.

406. Jablonski NG, Yan-Zhang P. 1993. The phylogenetic relationship and classification of the douc and snub-nosed langurs of China and Vietnam. *Folia Primatol* 60:36–55.

407. Janson CH. 1984. Female choice and mating system of the brown capuchin monkey, *Cebus apella* (Primates: Cebidae). *Z Tierpsychol* 65:177–200.

408. Janson CH. 1986. The mating system as a determinant of social evolution. In Else JG, Lee PC (eds), *Primate Ecology and Conservation*, vol 2, 169–182. Cambridge Univ Press, Cambridge.

409. Janson CH, Boinski S. 1992. Morphological and behavioral adaptations for foraging in generalist primates: the case of the cebines. *AJPA* 88:483–498.

410. Janson CH, van Schaik CH. 1993. Ecological risk aversion in juvenile primates: slow and steady wins the race. In Pereira ME, Fairbanks LA (eds), *Juvenile Primates: Life History, Development, and Behavior*, 57–76. Oxford Univ Press, New York.

411. Jenamejeyan A, Ram Manchar B, Thiyagesan K. 1991. Food preference and diversity of bonnet macaques at Point Calimere, Tamil Nadu. *J Anim Morphol Physiol* 38(1–2):1–8.

412. Jenkins PA. 1987. *Catalogue of Primates in the British Museum (Natural History) and Elsewhere in the British Isles, Part 4: Suborder Strepsirrhini, Including the Subfossil Madagascaran Lemurs and Family Tarsiidae.* British Museum, London.

413. Jenkins PA. 1990. *Catalogue of Primates in the British Museum (Natural History) and Elsewhere in the British Isles, Part 5.* Natural History Museum Publications, London.

414. Ji W, Bleisch W. 1994. Conservation of snub-nosed monkeys *(Rhinopithecus* spp.) in China. Paper presented at IPS Congress 15, Kuta, Bali, Indonesia. [Abstract available: 1994. *15th Congress IPS Handbook and Abstracts:* 277.]

415. Johns AD. 1985. Current status of the southern bearded saki *Chiropotes satanas satanas.* Primate Conserv 5:28.

416. Johns AD. 1986. Notes on the ecology and current status of the buffy saki *Pithecia albicans.* Primate Conserv 7:26–28.

417. Johns AD. 1986. Effects of selective logging on the behavioral ecology of west Malaysian primates. *Ecology* 67(3):684–694.

418. Jolly A. 1994. Pers com. Princeton Univ, Princeton, NJ, Aug.

419. Jolly A. 1994. Female dominance and social structure in lemurs. Paper presented at IPS Congress 15, Kuta, Bali, Indonesia. [Abstract available: 1994. *15th Congress IPS Handbook and Abstracts:* 285.]

420. Jolly A. 1985. *The Evolution of Primate Behavior.* Macmillan, New York.

421. Jolly A, Rasamimanana HR, Kinnaird MF, et al. 1993. Territoriality in *Lemur catta* groups during the birth season at Berenty, Madagascar. In Kappeler PM, Ganzhorn JU (eds), *Lemur Social Systems and Their Ecological Basis*, 85–110. Plenum, New York.

422. Jolly CJ. 1993. Species, subspecies, and baboon systematics. In Kimbel WH, Martin LB (eds), *Species, Species Concepts, and Primate Evolution*, 67–109. Plenum, New York.

423. Jolly CJ, Plog F, eds. 1987. *Physical Anthropology and Archeology.* 4th ed. Knopf, New York.

424. Julliot C, Sabatier D. 1993. Diet of the red howler monkey *(Alouatta seniculus)* in French Guiana. IJP 14(4):527–550.

425. Jungers WL. 1985. Allometry of primate limb proportions size and scaling. In Jungers WL (ed), *Size and Scaling in Primate Biology*, 345–382. Plenum, New York.

426. Jürgens U. 1982. A neuroethological approach to the classification of vocalization in the squirrel monkey. In Snowdon CT, Brown CH, Petersen MR (eds), *Primate Communication*, 50–62. Cambridge Univ Press, Cambridge.

427. Kadam KM, Swayamprabha MS. 1980. Parturition in the slender loris *(Loris tardigradus lydekkerianus)*. Primates 21(4):567–571.

428. Kano T. 1990. The bonobos' peaceable kingdom. *Nat Hist* 99(11):62–71.

429. Kappeler M. 1984. The gibbon in Java. In Preuschoft H, Chivers DJ, Brockelman WY, Creel N (eds), *The Lesser Apes: Evolutionary and Behavioural Biology*, 19–31. Edinburgh Univ Press, Edinburgh.

430. Kappeler M. 1984. Diet and feeding behaviour of the moloch gibbon. In Preuschoft H, Chivers DJ, Brockelman WY, Creel N (eds), *The Lesser Apes: Evolutionary and Behavioural Biology*, 228–241. Edinburgh Univ Press, Edinburgh.

431. Kappeler M. 1984. Vocal bouts and territorial maintenance in moloch gibbon. In Preuschoft H, Chivers DJ, Brockelman WY, Creel N (eds), *The Lesser Apes: Evolutionary and Behavioural Biology*, 379–389. Edinburgh Univ Press, Edinburgh.

432. Kappeler PM. 1995. Scramble competition polygyny and the social organization of *Mirza coquereli.* Paper presented at the International Conference on the Biology and Conservation of Prosimians, The North of England Zoological Society, Chester, UK. [Abstract available: 1996. *Folia Primatol.* In press.]

433. Kappeler PM. 1987. Reproduction in the crowned lemur *(Lemur coronatus)* in captivity. *AJP* 12:497–503.

434. Kappeler PM. 1988. A preliminary study of olfactory behavior of captive *Lemur coronatus* during the breeding season. *IJP* 9(2):135–146.

435. Kappeler PM. 1990. The evolution of sexual size dimorphism in prosimian primates. *AJP* 21:201–214.

436. Kappeler PM. 1991. Patterns of sexual dimorphism in body weight among prosimian primates. *Folia Primatol* 57:132–146.

437. Kappeler PM. 1993. Sexual selection and lemur social systems. In Kappeler PM, Ganzhorn JU (eds), *Lemur Social Systems and Their Ecological Basis*, 223–240. Plenum, New York.

438. Kappeler PM, Ganzhorn JU. 1993. The evolution of primate communities and societies in Madagascar. *Evol Anthropol* 2(5):159–171.

439. Kavanagh M. 1978. The social behavior of doucs *(Pygathrix nemaeus nemaeus)* at San Diego Zoo. *Primates* 19(1):101–114.

440. Kavanagh M. 1983. *A Complete Guide to Monkeys, Apes, and Other Primates*. Oregon Press, London.

441. Kavanagh M, Dresdale L. 1975. Observations on the woolly monkey *(Lagothrix lagotricha)* in northern Colombia. *Primates* 16(3):285.

442. Kay RF, Plavcan JM, Glander KE, Wright PC. 1988. Sexual selection and canine dimorphism in New World monkeys. *AJPA* 77:385–397.

443. Kendrick KM, Dixson AF. 1984. A qualitative description of copulatory and associated behaviors of captive marmosets *(Callithrix jacchus)*. *IJP* 5(3):199–213.

444. Kingdon J. 1974. *Primates*. In *East African Mammals: An Atlas of Evolution in Africa*, vol 1, 99–328. Univ of Chicago Press, Chicago.

445. Kingdon J. 1988. What are face patterns and do they contribute to reproductive isolation in guenons? In Gautier-Hion A, Bourlière F, Gautier JP, Kingdon J (eds), *A Primate Radiation: Evolutionary Biology of the African Guenons*, 227–245. Cambridge Univ Press, Cambridge.

446. Kinnaird MF. 1990. Pregnancy, gestation, and parturition in free-ranging Tana River crested mangabeys *(Cercocebus galeritus galeritus)*. *AJP* 22:285–289.

447. Kinnaird MF. 1992. Variable resource defense by the Tana River crested mangabey. *Behav Ecol Sociobiol* 31:115–122.

448. Kinnaird MF. 1992. Competition for a forest palm: use of *Phoenix reclinata* by human and non-human primates. *Conserv Biol* 6(1):101–107.

449. Kinnaird MF. 1994. Intergroup interactions in *Macaca nigra*: a random walk or resource defense? Paper presented at IPS Congress 15, Kuta, Bali, Indonesia. [Abstract available: 1994. *15th Congress IPS Handbook and Abstracts*: 301.]

450. Kinzey WG. 1981. The titi monkeys, genus *Callicebus*. In Coimbra-Filho AF, Mittermeier RA (eds), *Ecology and Behavior of Neotropical Primates*, vol 1, 240–276. Academia Brasileira de Ciências, Rio de Janeiro.

451. Kinzey WG. 1977. Diet and feeding behaviour of *Callicebus torquatus*. In Clutton-Brock TH (ed), *Primate Ecology: Studies of Feeding and Ranging Behaviour in Lemurs, Monkeys, and Apes*, 127–152. Academic Press, London.

452. Kinzey WG. 1992. Dietary and dental adaptations in the Pitheciinae. *AJPA* 88:499–514.

453. Kinzey WG, Becker M. 1983. Activity patterns of the masked titi monkey, *Callicebus personatus*. *Primates* 24(3):337–343.

454. Kinzey WG, Cunningham EP. 1994. Variability in platyrrhine social organization. *AJP* 34:185–198.

455. Kinzey WG, Norconk MA. 1993. Physical and chemical properties of fruit and seeds eaten by *Pithecia* and *Chiropotes* in Surinam and Venezuela. *IJP* 14(2):207–227.

456. Kinzey WG, Rosenberger AL, Heisler PS, Prowse DL, Trilling JS. 1977. A preliminary field investigation of the yellow handed titi monkey, *Callicebus torquatus torquatus*, in northern Peru. *Primates* 18(1):159–181.

457. Kinzey WG, Wright PC. 1982. Grooming behavior in the titi monkey *(Callicebus torquatus)*. *AJP* 3:267–275.

458. Kirkevold BC, Crockett CM. 1987. Behavioral development and proximity patterns in captive DeBrazza's monkeys. In Zucker EL (ed), *Comparative Behavior of African Monkeys*, 39–66. AR Liss, New York.

459. Kirkpatrick C. 1994. The natural history of the Yunnan snub-nosed monkey (Colobinae: *Rhinopithecus bieti*). Paper presented at IPS Congress 15, Kuta, Bali, Indonesia. [Abstract available: 1994. *15th Congress IPS Handbook and Abstracts*: 281.]

460. Kirkpatrick RC, Long YC. 1994. Altitudinal ranging and terrestriality in the Yunnan snub-nosed monkey *(Rhinopithecus bieti)*. *Folia Primatol* 63:102–106.

461. Klein LL. 1972. The ecology and social organization of the spider monkey, *Ateles belzebuth*. PhD diss, Univ of California, Berkeley.

462. Klein LL, Klein DB. 1977. Feeding behaviour of the Colombian spider monkey. In Clutton-Brock TH (ed), *Primate Ecology: Studies of Feeding and Ranging Behaviour in Lemurs, Monkeys, and Apes*, 153–182. Academic Press, London.

463. Kleinman DG. 1981. *Leontopithecus rosalia*. *Mammal Species* 148:1–7.

464. Klopfer PH, Boskoff KJ. 1979. Maternal behavior in prosimians. In Doyle GA, Martin RD (eds), *The Study of Prosimian Behavior*, 123–157. Academic Press, New York.

465. Kobayashi S, Langguth AL. 1994. New titi monkey from Brazil. Paper presented at IPS Congress 15, Kuta, Bali, Indonesia. [Abstract available: 1994. *15th Congress IPS Handbook and Abstracts*: 399.]

466. Koenig A. 1995. Group size, composition, and reproductive success in wild common marmosets. *AJP* 35:311–317.

467. Kool KM. 1992. Food selection by the silver leaf monkey, *Trachypithecus auratus sondaicus*, in relation to plant chemistry. *Oecologia* 90:527–533.

468. Kool KM. 1992. The status of endangered primates in Gunung Halimun Reserve, Indonesia. *Oryx* 26(1):29–33.

469. Kool KM. 1993. The diet and feeding behavior of the silver leaf monkey *(Trachypithecus auratus sondaicus)* in Indonesia. *IJP* 14(5):667–700.

470. Koshimizu K, Ohigashi H, Huffman MA, Nishida T, Takasaki H. 1993. Physiological activities and the active constituents of potentially medicinal plants used by wild chimpanzees of the Mahale Mountains, Tanzania. *IJP* 14(2):345–352.

471. Krishnamani R. 1994. Diet composition of the bonnet macaque *(Macaca radiata)* in a tropical dry evergreen forest in southern India. *Trop Biodiversity* 2(2):285–301.

472. Kubzdela KS, Richard AF, Pereira ME. 1992. Social relations in semi free-ranging sifakas *(P. v. coquereli)* and the question of female dominance. *AJP* 28:139–145.

473. Kummer H. 1968. *Social Organization of Hamadryal Baboons: A Field Study*. Univ of Chicago Press, Chicago.

474. Kummer H. 1990. The social system of hamadryas baboons and its presumable evolution. In de Mello MT, Whiten A, Byrne RW (eds), *Baboons: Behaviour and Ecology, Use and Care*, 43–60. Brasília, Brazil.

475. Kummer H. 1995. *In Quest of the Sacred Baboon: A Scientist's Journey*. Princeton Univ Press, Princeton, NJ.

476. Kurup GU, Kumar A. 1993. Time budget and activity patterns of the lion-tailed macaque *(Macaca silenus)*. *IJP* 14(1):27–39.

477. Kuruvilla GP. 1980. Ecology of the bonnet macaque *(Macaca radiata* Geoffroy) with special reference to feeding habits. *J Bombay Nat Hist Soc* 75:976–988.

478. Kyes RC. 1988. Grooming with a stone in sooty mangabey *Cercocebus atys*. *AJP* 16(2):171–175.

479. Lahm S. 1986. Diet and habitat preference of *Mandrillus sphinx* in Gabon: implication of foraging strategy. *AJP* 11:9–26.

480. Lahm SA. 1985. Mandrill ecology and status of Gabon's rainforests. *Primate Conserv* 6:32–33.

481. Lambert F. 1990. Some notes on fig-eating by arboreal mammals in Malaysia. *Primates* 31(3):453–458.

482. Lee LL, Lin YS. 1990. Status of Formosan macaques in Taiwan. *Primate Conserv* 11:18–20.

483. Lee PC, Brennan EJ, Else JG, Altmann J. 1986. Ecology and behaviour of vervet monkeys in a tourist lodge habitat. In Else JG, Lee PC (eds), *Primate Ecology and Conservation*, vol 2, 229–236. Cambridge Univ Press, Cambridge.

484. Lee PC, Thornback J, Bennett EL. 1988. *Threatened Primates of Africa: The IUCN Red Data Book*. IUCN, Gland, Switzerland.

485. Lee RJ. 1994. Effects of hunting and live capture of wild populations of the crested black macaque *(Macaca nigra)* in northern Sulawesi, Indonesia. Paper presented at IPS Congress 15, Kuta, Bali, Indonesia. [Abstract available: 1994. *15th Congress IPS Handbook and Abstracts*: 125.]

486. Lehman SM, Robertson KL. 1994. Preliminary survey of *Cacajao melanocephalus melanocephalus* in southern Venezuela. *IJP* 15(6):927–934.

487. Lehman SM, Robertson KL. 1994. Survey of Humboldt's black head uakari *(Cacajao melanocephalus melanocephalus)* in southern Amazonas, Venezuela [abstract]. *AJP* 33(3):223.

488. Leighton DR. 1987. Gibbons: territoriality and monogamy. In Smuts BB, Cheney DL, Seyfarth RM, Wrangham RW, Struhsaker TT (eds), *Primate Societies*, 135–145. Univ of Chicago Press, Chicago.

489. Lemos de Sa RM, Pope TR, Struhsaker TT, Glander KE. 1993. Sexual dimorphism in canine length of woolly spider monkeys *(Brachyteles arachnoides*, E. Geoffroy 1806). *IJP* 14(5):755–763.

490. Leutenegger W, Cheverud J. 1982. Correlates of sexual dimorphism in primates: ecological and size variables. *IJP* 3(4):387–403.

491. Lindburg DG. 1977. Feeding behaviour and diet of rhesus monkey *(Macaca mulatta)* in Siwalik forest of north India. In Clutton-Brock TH (ed), *Primate Ecology: Studies of Feeding and Ranging Behaviour in Lemurs, Monkeys, and Apes*, 223–250. Academic Press, London.

492. Lippold L. 1977. The douc langur: a time for conservation. In Rainier Prince, Bourne GH (eds), *Primate Conservation*, 513–538. Academic Press, New York.

493. Lippold L. 1994. Status, distribution, and conservation of the douc langur *(Pygathrix nemaeus)* in Vietnam. Paper presented at IPS Congress 15, Kuta, Bali, Indonesia. [Abstract available: 1994. *15th Congress IPS Handbook and Abstracts*: 136.]

494. Lippold L. 1995. Pers com. San Diego State Univ, San Diego, Oct.

Bibliography *continued*

495. Lippold L. 1995. Distribution and conservation status of douc langurs in Vietnam. *Asian Primates* 4(4):4–6.

496. Lippold L. 1996. A new variety of douc langur. *Asian Primates*. In press.

497. Liu Z, Zhang Y, Jiang H, Southwick C. 1989. Population structure of *Hylobates concolor* in Bawanglin Nature Reserve, Hainan, China. *AJP* 19:247–254.

498. Loireau JN, Gautier-Hion A. 1988. Olfactory marking behaviour in guenons and its implications. In Gautier-Hion A, Bourlière F, Gautier JP, Kingdon J (eds), *A Primate Radiation: Evolutionary Biology of the African Guenons*, 246–254. Cambridge Univ Press, Cambridge.

499. Long Y. 1994. Time budget of Yunnan snub-nosed monkey *(Rhinopithecus bieti)*. Paper presented at IPS Congress 15, Kuta, Bali, Indonesia. [Abstract available: 1994. *15th Congress IPS Handbook and Abstracts: 282.*]

500. Lopes MA, Ferrari SF. 1994. Foraging behavior of a tamarin group *(Saguinus fuscicollis weddelli)* and interactions with marmosets *(Callithrix emiliae)*. *IJP* 15(3):373–388.

501. Lowen C, Dunbar RIM. 1994. Territory size and defendability in primates. *Behav Ecol Sociobiol* 35:347–354.

502. Loy J. 1987. The sexual behavior of African monkeys and the question of estrus. In Zucker EL (ed), *Comparative Behavior of African Monkeys*, 197–234. AR Liss, New York.

503. Loy J. 1989. Patas monkey copulations one mount, repeat if necessary. *AJP* 18:57–62.

504. Loy J, Argo B, Nestell GL, Vallett S, Wanamaker G. 1993. A reanalysis of patas monkeys' "grimace and gecker" display and a discussion of their lack of formal dominance. *IJP* 14(6):879–894.

505. Loy KM, Loy JD. 1987. Sexual differences in early social development among captive patas monkeys. In Zucker EL (ed), *Comparative Behavior of African Monkeys*, 23–38. AR Liss, New York.

506. Lucas PW, Hails CJ, Corlett RT. 1988. Status of banded langur in Singapore. *Primate Conserv* 9:136–137.

507. Luna ML. 1980. First field study of the yellow-tailed woolly monkey. *Oryx* 15(3):386–387.

508. Luna ML. 1987. Primate conservation in Peru: a case study of the yellow-tailed woolly monkey. *Primate Conserv* 8:122–124.

509. Macdonald D, ed. 1985. *Primates*. Torstar Books, New York.

510. Macedonia JM. 1993. Adaptation and phylogenetic constraints in the antipredation behavior of ring-tailed and ruffed lemurs. In Kappeler PM, Ganzhorn JU (eds), *Lemur Social Systems and Their Ecological Basis*, 67–84. Plenum, New York.

511. Mack DS, Kleiman DG. 1978. Distribution of scent marks in different contexts in captive lion tamarins *Leontopithecus rosalia* (primates). In Rothe H, Wolters HJ, Hearn JP (eds), *Biology and Behaviour of Marmosets*, 181–190. Eigenverlag Rothe, Göttingen, Germany.

512. MacKinnon J. 1985. Orang-utan. In Macdonald D (ed), *Primates*, 132–135. Torstar Books, New York.

513. MacKinnon JR, MacKinnon KS. 1984. Territoriality, monogamy, and song in gibbons and tarsiers. In Preuschoft H, Chivers DJ, Brockelman WY, Creel N (eds), *The Lesser Apes: Evolutionary and Behavioural Biology*, 291–297. Edinburgh Univ Press, Edinburgh.

514. MacLarnon AM, Chivers DJ, Martin RD. 1986. Gastro-intestinal allometry in primates and other mammals, including new species. In Else JG, Lee PC (eds), *Primate Ecology and Conservation*, vol 2, 73–86. Cambridge Univ Press, Cambridge.

515. Macleod MC. 1994. The woolly monkey—another patrilineal primate? Paper presented at IPS Congress 15, Kuta, Bali, Indonesia. [Abstract available: 1994. *15th Congress IPS Handbook and Abstracts: 262.*]

516. Maestripieri D. 1994. Social structure, infant handling, and mothering styles in group-living Old World monkeys. *IJP* 15(4):531–551.

517. Maisels F. 1993. Gut passage rates in guenons and mangabeys: another indicator of a flexible feeding niche. *Folia Primatol* 61:35–37.

518. Maisels F, Gautier-Hion A, Gautier JP. 1994. Diets of two sympatric colobines in Zaire: more evidence on seed-eating in forests on poor soils. *IJP* 15(5):681–703.

519. Maisels FG, Gautier JP, Cruickshank A, Bosefe JP. 1993. Attacks by crowned hawk eagles *(Stephanoaetus coronatus)* on monkeys in Zaire. *Folia Primatol* 61:157–159.

520. Malenky RK, Wrangham RW. 1994. A quantitative comparison of terrestrial herbaceous food consumption by *Pan paniscus* in the Lomako Forest, Zaire, and *Pan troglodytes* in the Kibale Forest, Uganda. *AJP* 32:1–12.

521. Malik I. 1986. Time budgets and activity patterns in free ranging rhesus monkeys. In Else JG, Lee PC (eds), *Primate Ecology and Conservation*, vol 2, 105–114. Cambridge Univ Press, Cambridge.

522. Manley GH. 1979. Function of the external genital glands of *Perodicticus* and *Arctocebus*. In Martin RD, Doyle GA, Walker AC (eds), *Prosimian Biology*, 313–330. Univ of Pittsburgh Press, Pittsburgh.

523. Markham R. 1994. Doing it naturally: reproduction in captive orangutans *(Pongo pygmaeus)*. In Ogden JJ, Perkins LA, Sheeran L (eds), *Proceedings of the International Conference on Orangutans: The Neglected Ape*, 166–170. Zoological Society of San Diego, San Diego.

524. Marsh C. 1983. Food availability diet and ranging behavior in Tana River red colobus, *Colobus badius rufomitratus*. In Seth PK (ed), *Perspectives in Primate Biology*. Today and Tomorrow's Printers and Publishing, New Delhi.

525. Marsh CW. 1981. Diet choice among red colobus *(Procolobus badius rufomitratus)* on the Tana River, Kenya. *Folia Primatol* 35:147–178.

526. Marsh CW. 1981. Ranging behaviour and its relation to diet selection in Tana River red colobus. *J Zool Lond* 195(4):473–492.

527. Martin RD. 1979. The study of prosimian behaviour. In Martin RD, Doyle GA, Walker AC (eds), *Prosimian Biology*, 4–16. Univ of Pittsburgh Press, Pittsburgh.

528. Martin RD. 1992. Female cycles in relation to paternity in primate societies. In Martin RD, Dixson AF, Wickings EJ (eds), *Paternity in Primates: Genetic Tests and Theories*, 238–274. Karger, Basel, Switzerland.

529. Wright PC, Martin LB. 1995. Predation, pollination, and torpor in two nocturnal prosimians: *Cheirogaleus major* and *Microcebus rufus* in the rain forest of Madagascar. In Alterman L, Doyle DA, Izard K (eds), *Creatures of the Dark: The Nocturnal Prosimians*, 45–60. Plenum, New York.

530. Martins ES, Ayres JM, Ribeiro do Valle MB. 1988. On the status of *Ateles belzebuth marginatus* with notes on other primates of the Iriri River basin. *Primate Conserv* 9:87–90.

531. Masataka N, Kohda M. 1988. Primate play vocalizations and their functional significance. *Folia Primatol* 50:152–156.

532. Masataka N, Thierry B. 1993. Vocal communication of Tonkean macaques in confined environments. *Primates* 34(2):169–180.

533. Masters JC. 1991. Loud calls of *Galago crassicaudatus* and *G. garnettii*, and their relation to habitat structure. *Primates* 32(2):153–167.

534. Masters JC, Lumsden WHR, Young DA. 1988. Reproduction and dietary parameters in wild greater galago populations. *IJP* 9(6):573–592.

535. Masters JC, Rayner RJ, Ludewick H, et al. 1994. Phylogenetic relationships among the Galaginae. *Primates* 35(2):177–190.

536. Masui K, Narita Y, Tanaka S. 1986. Information on the distribution of Formosan monkeys *Macaca cyclopis*. *Primates* 27(3):383–392.

537. Mathur R, Manohar BR. 1991. Departure of juvenile male *Presbytis entellus* from the natal group. *IJP* 12(1):39–44.

538. Matsumura S. 1993. Female reproductive cycles and sexual behavior of moor macaques *(Macaca maura)* in their natural habitat, south Sulawesi, Indonesia. *Primates* 34(1):99–103.

539. McGraw S. 1994. Census, habitat preference, and polyspecific association of six monkeys in the Lomako Forest, Zaire. *AJP* 34:295–307.

540. McGraw WS. In prep. The positional behavior and habitat use of six monkeys in the Tai Forest, Ivory Coast. PhD diss, State Univ of New York at Stony Brook.

541. McKey D, Waterman PG. 1982. Ranging behaviour of a group of black colobus *(Colobus satanas)* in the Douala-Edea Reserve, Cameroon. *Folia Primatol* 39:264–304.

542. McKey DB, Gartlan JS, Waterman PG, Choo GM. 1981. Food selection by black colobus monkeys *(Colobus satanas)* in relation to plant chemistry. *Biol J Linn Soc* 16:115–116.

543. Medway L. 1970. The monkeys of Sundaland: ecology and systematics of the cercopithecids of a humid equatorial environment. In Napier JR, Napier PH (eds), *Old World Monkeys: Evolution, Systematics, and Behavior*, 513–554. Academic Press, New York.

544. Mehlman PT. 1988. Food resources of the wild Barbary macaque *(Macaca sylvanus)* in high altitude fir forest, Ghomaran Rif, Morocco. *J Zool Lond* 214:469–490.

545. Mehlman PT. 1989. Comparative density, demography, and ranging behavior of Barbary macaques *(Macaca sylvanus)* in marginal and prime conifer habitats. *IJP* 10(4):269–293.

546. Meier B, Albignac R. 1989. Hairy-eared dwarf lemur rediscovered *(Allocebus trichotis)*. *Primate Conserv* 10:27.

547. Meier B, Albignac R. 1991. Rediscovery of *Allocebus trichotis* in northeast Madagascar. *Folia Primatol* 56:57–63.

548. Meier B, Albignac R, Peyrieras A, Rumpler Y, Wright P. 1987. A new species of *Hapalemur* primates from south east Madagascar. *Folia Primatol* 48:211–215.

549. Mellen JD, Littlewood AP, Barrow BC, Stevens VJ. 1981. Individual and social behavior in a captive troop of mandrills *(Mandrillus sphinx)*. *Primates* 22(2):206–220.

550. Melnick DJ. 1994. Pers com. Columbia Univ, New York, Oct.

551. Melnick DJ, Pearl MC. 1987. Cercopithecines in multimale groups: genetic diversity and population structure. In Smuts BB, Cheney DL, Seyfarth RM, Wrangham RW, Struhsaker TT (eds), *Primate Societies*, 121–134. Univ of Chicago Press, Chicago.

552. Menard N, Vallet D. 1993. Dynamics of fission in a wild Barbary macaque group *(Macaca sylvanus)*. *IJP* 14(3):479–500.

553. Meyers D. 1993. The effects of resource seasonality on behavior and reproduction in the golden crowned sifaka in three Malagasy forests. PhD diss, Duke Univ, Durham, NC.

554. Meyers DM, Wright PC. 1993. Resource tracking: food availability and *Propithecus* seasonal reproduction. In Kappeler PM, Ganzhorn JU (eds),

Bibliography *continued*

Lemur Social Systems and Their Ecological Basis, 179–192. Plenum, New York.

555. Michael RP, Zumpe D. 1993. A review of hormonal factors influencing the sexual and aggressive behavior of macaques. *AJP* 30:213–241.

556. Miles HL, Harper SE. 1994. Chantek: the language ability of an enculturated orangutan *(Pongo pygmaeus)*. In Ogden JJ, Perkins LA, Sheeran L (eds), *Proceedings of the International Conference on Orangutans: The Neglected Ape*, 209–219. Zoological Society of San Diego, San Diego.

557. Miller LE. 1994. Life's ups and downs: activity budgets and feeding strategies of adult female wedge-capped capuchins *(Cebus olivaceus)*. Paper presented at IPS Congress 15, Kuta, Bali, Indonesia. [Abstract available: 1994. *15th Congress IPS Handbook and Abstracts*: 84.]

558. Milton K. 1978. Behavioral adaptations to leaf eating by the mantled howler monkey *(Alouatta palliata)*. In Montgomery GG (ed), *The Ecology of Arboreal Folivores*, 535–550. Smithsonian Institution Press, Washington, DC.

559. Milton K. 1984. Habitat, diet, and activity patterns of free ranging woolly spider monkey *(Brachyteles arachnoides* E. Geoffroy 1806). *IJP* 5(5):491–514.

560. Milton K. 1993. Diet and social organization of a free ranging spider monkey population: the development of a species-typical behavior in the absence of adults. In Pereira ME, Fairbanks LA (eds), *Juvenile Primates: Life History, Development, and Behavior*, 173–181. Oxford Univ Press, New York.

561. Mitani JC. 1984. The behavioral regulation of monogamy in gibbons *(Hylobates muelleri)*. *Behav Ecol Sociobiol* 15:225–229.

562. Mitani JC. 1987. Territoriality and monogamy among agile gibbons, *Hylobates agilis. Behav Ecol Sociobiol* 20:265–269.

563. Mitani JC. 1988. Male gibbon *(Hylobates agilis)* singing behavior: natural history, song variation, and function. *Ethology* 79:177–194.

564. Mitani JC. 1990. Demography of agile gibbons *(Hylobates agilis)*. *IJP* 11(5):411–425.

565. Mitani JC. 1990. Experimental field studies of Asian ape social systems. *IJP* 11(2):103–126.

566. Mitani M. 1991. Niche overlap and polyspecific association among sympatric *Cercopithecus* in the Campo Animal Reserve, southwest Cameroon. *Primates* 32(2):137–151.

567. Mitani JC. 1992. Singing behavior of male gibbons: field observations and experiments. In Nishida T, McGrew WC, Marler P, Pickford M, de Waal FBM (eds), *Topics in Primatology: Human Origins*, vol 1, 199–210. Univ of Tokyo Press, Tokyo.

568. Mitchell AH. 1993. Talk. Paper presented at New York Consortium in Evolutionary Primatology/New York Regional Primatology Colloquium, City Univ of New York, New York.

569. Mitchell CL. 1994. Migration alliances and coalitions among adult male South American squirrel monkeys *(Saimiri sciureus)*. *Behaviour* 130(3–4):169–190.

570. Mitchell CL, Boinski S, van Schaik CP. 1991. Competitive regimes and female bonding in two species of squirrel monkeys *(Saimiri oerstedii* and *S. sciureus)*. *Behav Ecol Sociobiol* 28:55–60.

571. Mitchell G, Towers S, Soteriou S, et al. 1988. Sex differences in behavior of endangered mangabeys. *Primates* 29(1):129–134.

572. Mittermeier RA, Coimbra-Filho AF. 1981. Systematics: species and subspecies. In Coimbra-Filho AF, Mittermeier RA (eds), *Ecology and Behavior of Neotropical Primates*, vol 1, 29–109. Academia Brasileira de Ciências, Rio de Janeiro.

573. Mittermeier RA. 1987. Effects of hunting on rain forest primates. In Marsh CW, Mittermeier RA (eds), *Primate Conservation in the Tropical Rain Forest*, 109–146. AR Liss, New York.

574. Mittermeier RA. 1995. Pers com. Conservation International, Washington, DC, Dec.

575. Mittermeier RA, Konstant WR, Ginsberg H, van Roosmalen MGM, da Silva EC Jr. 1983. Further evidence of insect consumption in the bearded saki monkey. *Primates* 24(4):602–605.

576. Mittermeier RA, Konstant WR, Mast RB. 1994. Use of Neotropical and Malagasy primate species in biomedical research. *AJP* 34:73–80.

577. Mittermeier RA, Konstant WR, Nicoll ME, Langrand O, comps. 1992. *Lemurs of Madagascar: An Action Plan for Their Conservation: 1993–1999*. IUCN, Gland, Switzerland.

578. Mittermeier RA, Milton K. 1977. A brief survey of the primates of Coiba Island, Panama. *Primates* 18:931–936.

579. Mittermeier RA, Schwarz M, Ayres JM. 1992. A new species of marmoset, genus *Callithrix* Erxleben, 1777 (Callitrichidae, Primates) from the Rio Maues region, State of Amazonas, central Brazilian Amazonia. *Goeldiana Zool* 14:1–17.

580. Mittermeier RA, Tattersall I, Konstant WR, Meyers DM, Mast RB, eds. 1994. *Lemurs of Madagascar*. Conservation International, Washington, DC.

581. Molnar S. 1983. *Human Variation: Races, Types, Ethnic Groups*. Prentice-Hall, Englewood Cliffs, NJ.

582. Moore J. 1984. Female transfer in primates. *IJP* 5(6):537–581.

583. Moos R, Rock J, Salzert W. 1985. Infanticide in gelada baboons *(Theropithecus gelada)*. *Primates* 26(4):497–500.

584. Moreno-Black GS, Bent EF. 1982. Secondary compounds in the diet of *Colobus angolensis*. *Afr J Ecol* 20(1):29–36.

585. Morland HS. 1993. Seasonal behavioral variation and its relationship to thermoregulation in ruffed lemurs *(Varecia variegata variegata)*. In Kappeler PM, Ganzhorn JU (eds), *Lemur Social Systems and Their Ecological Basis*, 193–204. Plenum, New York.

586. Morrell V. 1995. Will primate genetics split one gorilla into two? *Science* 265(5179):1661.

587. Moynihan M. 1976. *New World Primates: Adaptive Radiation and the Evolution of Social Behavior, Languages, and Intelligence*. Princeton Univ Press, Princeton, NJ.

588. Mukhergee RP. 1978. Further observations on the golden langur *(Presbytis geei* Khajuria, 1959) with a note to capped langur *(Presbytis pileatus* Blyth, 1843) of Assam. *Primates* 19(4):737–747.

589. Mukhergee RP. 1986. The hoolock gibbon, *Hylobates hoolock*, in Tripura, India. In Else JG, Lee PC (eds), *Primate Ecology and Conservation*, vol 2, 115–124. Cambridge Univ Press, Cambridge.

590. Mulavwa M. 1991. Notes on the contact call of mona monkeys *(Cercopithecus wolfi)* in the Malabi Forest: frequency of emission and daily activities. In Ehara A, Kimura T, Takenaka O, Iwamoto M (eds), *Primatology Today*. Elsevier Science, Amsterdam.

591. Muleris M, Gautier JP, Lombard M, Dutrillaux B. 1985. Cytogenetic study of *Cercopithecus wolfi, C. erythrotis*, and of a hybrid *C. ascanius* × *C. pogonias grayi. Ann Genet* 28(2):75–80.

592. Musser GG. 1995. Pers com. American Museum of Natural History, New York, May.

593. Musser GG, Dagosto M. 1987. The identity of *Tarsius pumilus*: a pygmy species endemic to the montane mossy forests of central Sulawesi. *Am Mus Novit* 2867:1–53.

594. Nadler RD. 1994. Proximate and ultimate influences on the sexual behavior of orangutans *(Pongo pygmaeus)*. In Ogden JJ, Perkins LA, Sheeran L (eds), *Proceedings of the International Conference on Orangutans: The Neglected Ape*, 144–153. Zoological Society of San Diego, San Diego.

595. Nadler T. 1995. Pers com. Cuc Phuong Primate Rescue Center, Cuc Phuong National Park, Viet Nam, Oct.

596. Nagel U. 1973. A comparison of anubis baboons, hamadryas baboons, and their hybrids at a species border in Ethiopia. *Folia Primatol* 19:104–165.

597. Nakagawa N. 1989. Activity budget and diet of patas monkeys in Kala Maloue National Park, Cameroon: a preliminary report. *Primates* 30(1):27–34.

598. Napier JR, Napier PH. 1967. *A Handbook of the Living Primates: Morphology, Ecology, and Behaviour of Nonhuman Primates*. Academic Press, London.

599. Napier JR, Napier PH. 1985. *The Natural History of Primates*. MIT Press, Cambridge, MA.

600. Napier PH. 1976. *Catalogue of Primates in the British Museum (Natural History), Part 1: Family Callitrichidae and Cebidae*. British Museum (Natural History), London.

601. Napier PH. 1981. *Catalogue of Primates in the British Museum (Natural History) and Elsewhere in the British Isles, Part 2: Family Cercopithecidae, Subfamily Cercopithecinae*. British Museum (Natural History), London.

602. Napier PH. 1985. *Catalogue of Primates in the British Museum (Natural History) and Elsewhere in the British Isles, Part 3: Family Cercopithecidae, Subfamily Colobinae*. British Museum (Natural History), London.

603. Nash LT. 1986. Influence of moonlight level on traveling and calling activity in two sympatric species of galago in Kenya. In Taub DM, King FA (eds), *Current Perspectives in Primate Social Dynamics*, 357–377. Van Nostrand Reinhold, New York.

604. Nash LT. 1986. Dietary, behavioral, and morphological aspects of gummivory in primates. *Yearbk Phys Anthropol* 29:113–137.

605. Nash LT. 1986. Social organization of two sympatric galagos at Gedi, Kenya. In Else JG, Lee PC (eds), *Primate Ecology and Conservation*, vol 2, 125–132. Cambridge Univ Press, Cambridge.

606. Nash LT. 1993. Juveniles of nongregarious primates. In Pereira ME, Fairbanks LA (eds), *Juvenile Primates: Life History, Development, and Behavior*, 119–137. Oxford Univ Press, New York.

607. Nash LT. 1996. Pers com. Arizona State Univ, Tempe, Jan.

608. Nash LT, Bearder SK, Olson TR. 1989. Synopsis of galago species characteristics. *IJP* 10(1):57–80.

609. Nash LT, Fritz PG. 1982. Captive breeding and resocialization of the chimpanzee at the Primate Foundation of Arizona (Abstr 0211). In King FA, Taub D (eds), *Selected Proceedings of the 9th Congress of the International Primatology Society*. Van Nostrand Reinhold, New York.

610. Nash LT, Harcourt CS. 1986. Social organization of galagos in Kenyan coastal forests, 1: *Galago zanzibaricus*. *AJP* 10:339–355.

611. Nash LT, Harcourt CS. 1986. Social organization of galagos in Kenyan coastal forests, 2: *Galago garnettii*. *AJP* 10:357–369.

612. Nash LT, Whitten PL. 1989. Preliminary observation on the roles of *Acacia* gum chemistry in *Acacia* utilization by *Galago senegalensis* in Kenya. *AJP* 17:27–39.

613. Ndunda M, Maruhashi T, Yumoto T, Yamagiwa J. 1988. Conservation of eastern lowland gorillas in

Masisi Region, Zaire. *Primate Conserv* 9:111–113.

614. Noë R. 1994. A model of coalition formation among male baboons with fighting ability as the crucial parameter. *Anim Behav* 47(1):211–213.

615. Neville MK, Glander KE, Braza F, Rylands AB. 1988. The howling monkeys, genus *Alouatta*. In Mittermeier RA, Rylands AB, Coimbra-Filho AF, de Fonseca GAB (eds), *Ecology and Behavior of Neotropical Primates*, vol 2, 349–453. World Wildlife Fund, Washington, DC.

616. Niemitz C. 1979. Outline of the behavior of *Tarsius bancanus*. In Doyle GA, Martin RD (eds), *The Study of Prosimian Behavior*, 631. Academic Press, New York.

617. Niemitz C. 1984. Synecological relationships and feeding behavior of the genus *Tarsius*. In Niemitz C (ed), *Biology of Tarsiers*, chap 3. G Fischer, Stuttgart, Germany.

618. Niemitz C, Nietsch A, Warter S, Rumpler Y. 1991. *Tarsius dianae*: a new primate species from central Sulawesi (Indonesia). *Folia Primatol* 56:105–116.

619. Nietsch A, Niemitz C. 1991. Use of habitat and space in free ranging *Tarsius spectrum*. *Primate Rep* 31:27–28.

620. Nishida T, Hiraiwa-Hasegawa M. 1987. Chimpanzees and bonobos: cooperative relationships among males. In Smuts BB, Cheney DL, Seyfarth RM, Wrangham RW, Struhsaker TT (eds), *Primate Societies*, 165–178. Univ of Chicago Press, Chicago.

621. Nishimura A, de Fonseca GAB, Mittermeier RA, Young AL, Strier KB, Valle CMC. 1988. The muriqui, genus *Brachyteles*. In Mittermeier RA, Rylands AB, Coimbra-Filho AF, de Fonseca GAB (eds), *Ecology and Behavior of Neotropical Primates*, vol 2, 577–610. World Wildlife Fund, Washington, DC.

622. Noë R. 1992. Alliance formation among male baboons: shopping for profitable partners. In Harcourt AH, de Waal FBM (eds), *Coalitions and Alliances in Humans and Other Animals*, 311–346. Oxford Univ Press, Oxford.

623. Noë R. 1994. Pers com. Max-Planck Institut für Verhaltensphysiologie—Seewiesen, Starnberg, Germany, Apr.

624. Norconk MA, Kinzey WG. 1994. Challenge of Neotropical frugivory: travel patterns of spider monkeys and bearded sakis. *AJP* 34:171–183.

625. Nowak RM, ed. 1991. Primates. In *Walker's Mammals of the World*, 5th ed, vol 1, 400–514. Johns Hopkins Univ Press, Baltimore.

626. O'Brien TG, Robinson JG. 1993. Stability of social relationships in female wedge-capped capuchin monkeys. In Pereira ME, Fairbanks LA (eds), *Juvenile Primates: Life History, Development, and Behavior*, 197–210. Oxford Univ Press, New York.

627. O'Keefe RT, Lifshitz K. 1985. A behavioral profile for stumptailed macaques *(Macaca arctoides)*. *Primates* 26(2):143–160.

628. Oates JF. 1977. The guereza and its food. In Clutton-Brock TH (ed), *Primate Ecology: Studies of Feeding and Ranging Behaviour in Lemurs, Monkeys, and Apes*, 276–323. Academic Press, London.

629. Oates JF. 1984. The niche of the potto, *Perodicticus potto*. *IJP* 5(1):51–61.

630. Oates JF, comp. 1985. *Action Plan for African Primate Conservation: 1986–1990*. IUCN, Gland, Switzerland.

631. Oates JF. 1985. The Nigeria guenon *Cercopithecus erythrogaster*: ecological, behavioural, systematic, and historical observations. *Folia Primatol* 45:25–43.

632. Oates JF. 1988. The diet of the olive colobus monkey, *Procolobus verus*, in Sierra Leone. *IJP* 9(5):457–478.

633. Oates JF. 1991. Africa's primates in 1992: conservation issues. *AJP* 34:61–71.

634. Oates JF. 1994. Pers com. Hunter College, City Univ of New York, New York, Jan.

635. Oates JF. 1994. The natural history of African colobines. In Davies AG, Oates JF (eds), *Colobine Monkeys: Their Ecology, Behaviour, and Evolution*, 75–128. Cambridge Univ Press, Cambridge.

636. Oates JF, comp. 1996. *A New Action Plan for African Primate Conservation*. IUCN, Gland, Switzerland.

637. Oates JF, Anadu PA, Gadsby EL, Werre JL. 1992. Sclater's guenon—a rare Nigerian monkey threatened by deforestation. *Res Explor* 8(4):476–479.

638. Oates JF, Anadu PA. 1989. A field observation of Sclater's guenon. *Folia Primatol* 52:93–96.

639. Oates JF, Davies AG, Delson E. 1994. The diversity of living colobines—the natural history of African colobines. In Davies AG, Oates JF (eds), *Colobine Monkeys: Their Ecology, Behaviour, and Evolution*, 45–75. Cambridge Univ Press, Cambridge.

640. Oates JF, Trocco TF. 1983. Taxonomy and phylogeny of black-and-white colobus monkeys, inferences from an analysis of loud call variation. *Folia Primatol* 40:83–113.

641. Oates JF, Whitesides GH. 1990. Association between olive colobus *(Procolobus verus)*, diana guenons *(Cercopithecus diana)*, and other forest monkeys in Sierra Leone. *AJP* 21:129–146.

642. Ogawa H. 1995. Recognition of social relationships in bridging behavior among Tibetan macaques *(Macaca thibetana)*. *AJP* 35:305–310.

643. Oi T. 1990. Patterns of dominance and affiliation in wild pig-tailed macaques *(Macaca nemestrina)* in west Sumatra. *IJP* 11(4):339–356.

644. Okayasu N. 1991. Vocal communication and its sociological interpretation of wild bonobos in Wamba, Zaire. In Ehara A, Kimura T, Takenaka O, Iwamonto M (eds), *Primatology Today*. Elsevier Science, Amsterdam.

645. Overdorff DJ. 1993. Ecological and reproductive correlates to range use in red-bellied lemurs *(Eulemur rubriventer)* and rufous lemurs *(Eulemur fulvus rufus)*. In Kappeler PM, Ganzhorn JU (eds), *Lemur Social Systems and Their Ecological Basis*, 167–178. Plenum, New York.

646. Overdorff DJ. 1993. Similarities, differences, and seasonal patterns in the diets of *E. rubriventer* and *E. fulvus rufus* in the Ranomafana National Park, Madagascar. *IJP* 14(5):721–752.

647. Pages E. 1980. Ethoecology of *Microcebus coquereli*. In Charles-Dominique P, Cooper HM, Hladik A, et al (eds), *Nocturnal Malagasy Primates: Ecology, Physiology, and Behavior*, 97–115. Academic Press, New York.

648. Pagel M. 1994. The evolution of conspicuous oestrous advertisement in Old World monkeys. *Anim Behav* 47:1333–1334.

649. Pariente C. 1979. Influence of light on the activity rhythms of two Malagasy lemurs: *Phaner furcifer* and *Lepilemur mustelinus leucopus*. In Martin RD, Doyle GA, Walker AC (eds), *Prosimian Biology*, 183–200. Univ of Pittsburgh Press, Pittsburgh.

650. Parker SP, ed. 1990. *Grzimek's Encyclopedia of Mammals*. Vol 2. McGraw Hill, New York.

651. Paul A. 1989. Determinants of male mating success in a large group of Barbary macaques *(Macaca sylvanus)* at Affenberg Salem. *Primates* 30(4):461–476.

652. Paul A, Kuester J, Arnemann J. 1991. The role of kinship in the allomothering system of the Barbary macaque *(Macaca sylvanus)*. *Primate Rep* 31:30.

653. Paul A, Kuester J, Podzuweit D. 1993. Reproductive senescence and terminal investment in female Barbary macaques *(Macaca sylvanus)* at Salem. *IJP* 14(1):105–121.

654. Payne J, Francis CM. 1985. *A Field Guide to Mammals of Borneo*. Sabah Society, Sabah, Malaysia.

655. Pereira ME. 1993. Agonistic interaction, dominance, relation, and ontogenetic trajectories in ring-tail lemurs. In Pereira ME, Fairbanks LA (eds), *Juvenile Primates: Life History, Development, and Behavior*, 285–308. Oxford Univ Press, New York.

656. Peres C. 1993. The ecology of buffy saki monkeys *(Pithecia albicans*, Gray 1860): a canopy seed predator. *AJP* 31:129–140.

657. Peres C. 1994. Diet and feeding ecology of gray woolly monkey *(Lagothrix lagotricha cana)* in central Amazonia: comparison with other atelines. *IJP* 15(3):333–372.

658. Peres CA. 1988. Primate community structure in western Brazilian Amazonia. *Primate Conserv* 9:83–86.

659. Peres CA. 1990. A harpy eagle successfully captures an adult male red howler monkey. *Wilson Bull* 102(3):560–561.

660. Persson V. 1993. Pers com. Museum de Historia Natural Capao da Imbuia, Curitiba, Brazil, Aug.

661. Petit O, Thierry B. 1991. Aggressive and peaceful interventions on conflicts in Tonkean macaque. *Primate Rep* 31:30–31.

662. Petter JJ, Charles-Dominique P. 1979. Vocal communications in prosimians. In Doyle GA, Martin RD (eds), *The Study of Prosimian Behavior*, 247–306. Academic Press, New York.

663. Petter JJ, Petter-Rousseaux A. 1979. Classification of the prosimians. In Doyle GA, Martin RD (eds), *The Study of Prosimian Behavior*, 1–44. Academic Press, New York.

664. Petter-Rousseaux A. 1980. Seasonal rhythm in nocturnal prosimians reproduction, and body weight variation in five sympatric nocturnal prosimians in simulated light and climatic conditions. In Charles-Dominique P, Cooper HM, Hladik A, et al (eds), *Nocturnal Malagasy Primates: Ecology, Physiology, and Behavior*, 137–152. Academic Press, New York.

665. Phillips KA, Bernstein IS, Dettermer EL, Devermann H, Power M. 1994. Sexual behavior in brown capuchins *(Cebus apella)*. *IJP* 15(6):907–917.

666. Pilbeam D. 1988. Primates. In Harrison GA, Tanner JM, Pilbeam DR, Baker PT (eds), *Human Biology: An Introduction to Human Evolution, Variation, Growth, and Adaptability*, 3d ed, 27–71. Oxford Univ Press, Oxford.

667. Pitts R. 1988. Classical Heinrothian intimidation displays exhibited by *Galago demidoff demidoff*: a paradigm for pinpointing galagine species diversity. *IJP* 9(6):529–557.

668. Poirier FE. 1970. Dominance structure of the Nilgiri langur *(Presbytis johnii)* of south India. *Folia Primatol* 12:161–186.

669. Pola YV, Snowdon CT. 1975. The vocalizations of pygmy marmosets *(Cebuella pymnaea)*. *Anim Behav* 23:826–842.

670. Pollock JI. 1977. The ecology and sociology of feeding in *Indri indri*. In Clutton-Brock TH (ed), *Primate Ecology: Studies of Feeding and Ranging Behaviour in Lemurs, Monkeys, and Apes*, 38–71. Academic Press, London.

671. Pook AG. 1978. A comparison between the reproduction and parental behaviour of the Goeldi's monkey *(Callimico goeldii)* and the true marmosets (Callitrichidae). In Rothe H, Wolters HJ, Hearn JP (eds), *Biology and Behaviour of Marmosets*, 1–16. Eigenverlag Rothe, Göttingen, Germany.

672. Pook AG, Pook G. 1981. A field study of the socio-ecology of the Goeldi's monkey, *Callimico goeldii*, in northern Bolivia. *Folia Primatol* 35:288–317.

673. Porter LM. 1996. Pers com. State Univ of New York at Stony Brook.

674. Powell B. 1994. Pers com. River State Univ of Science and Technology, Port Harcourt, Nigeria, Jan.

675. Pratcs JC, Gayer SMP, Kunz LF Jr, Buss G. 1987. Feeding habits of the brown howler monkey (*Alouatta fusca clamitans*) in Itapua State Park (30°20´S; 50°55´W); RS; Brasil. *IJP* 8(5):534.

676. Price EC, Feistner ATC. 1993. Food sharing in lion tamarins: tests of three hypotheses. *AJP* 31:211–221.

677. Adiputra IMW. 1994. Feeding behavior of Javan leaf monkey (*Presbytis comata comata*) in Lake Patengan, West Java. Paper presented at IPS Congress 15, Kuta, Bali, Indonesia. [Abstract available: 1994. *15th Congress IPS Handbook and Abstracts*: 82.]

678. Queiroz HL. 1992. A new species of capuchin monkey, genus *Cebus* Erxleben, 1777 (Cebidae: Primates) from eastern Brazilian Amazonia. *Goeldiana Zool* 15:1–11.

679. Quris R. 1980. Emission vocale de forte intensite chez *Cercocebus galeritus agilis*: structure, caracteristiques, specifiques, et individuelles, modes d'emission. *Mammalia* 44(1):49.

680. Radetsky P. 1995. Gut thinking. *Discover* 16(2):76–81.

681. Raemaekers J. 1979. Ecology of sympatric gibbons. *Folia Primatol* 31:227–245.

682. Raemaekers J. 1984. Large versus small gibbons: relative roles of bioenergetics and competition in their ecological segregation in sympatry. In Preuschoft H, Chivers DJ, Brockelman WY, Creel N (eds), *The Lesser Apes: Evolutionary and Behavioural Biology*, 209–218. Edinburgh Univ Press, Edinburgh.

683. Rajpurohit LS, Sommer V. 1993. Juvenile male emigration from natal one-male troops in Hanuman langurs. In Pereira ME, Fairbanks LA (eds), *Juvenile Primates: Life History, Development, and Behavior*, 86–103. Oxford Univ Press, New York.

684. Ramirez M. 1988. The woolly monkeys, genus *Lagothrix*. In Mittermeier RA, Rylands AB, Coimbra-Filho AF, de Fonseca GAB (eds), *Ecology and Behavior of Neotropical Primates*, vol 2, 539–575. World Wildlife Fund, Washington, DC.

685. Ratsirarson J, Anderson J, Warter S, Rumpler Y. 1987. Notes on the distribution of *Lepilemur septentrionalis and L. mustelinus* in northern Madagascar. *Primates* 28(1):119–122.

686. Ray JC, Sapolsky RM. 1992. Styles of male social behavior and their endocrine correlates among high ranking wild baboons. *AJP* 28(4):231–250.

687. Remis M. 1995. Gorillas in the trees: the implications of body size and social context. Paper presented at New York Consortium in Evolutionary Primatology/New York Regional Primatology Colloquium, City Univ of New York, New York.

688. Richard A. 1970. A comparative study of the activity patterns and behaviour of *Alouatta villosa* and *Ateles geoffroyi*. *Folia Primatol* 12:241–263.

689. Richard A. 1979. Patterns of mating in *Propithecus verreauxi verreauxi*. In Martin RD, Doyle GA, Walker AC (eds), *Prosimian Biology*, 49–74. Univ of Pittsburgh Press, Pittsburgh.

690. Richard A. 1985. Lemurs; Monkeys in clover. In Macdonald D (ed), *Primates*, 24–28, 30–31; 90–91. Torstar Books, New York.

691. Richard AF. 1985. *Primates in Nature*. WH Freeman, New York.

692. Richard AF. 1987. Malagasy prosimians: female dominance. In Smuts BB, Cheney DL, Seyfarth RM, Wrangham RW, Struhsaker TT (eds), *Primate Societies*, 25–33. Univ of Chicago Press, Chicago.

693. Richard AF, Goldstein SJ, Dewar RE. 1989. Weed macaques: the evolutionary implications of macaque feeding ecology. *IJP* 10(6):569–591.

694. Richardson S. 1994. A monopoly on maternity. *Discover* 15(2):28–29.

695. Roberts M. 1994. Growth, development, and parental care in the western tarsier (*Tarsius bancanus*) in captivity: evidence for a slow life history and nonmonogamous mating system. *IJP* 15(1):1–28.

696. Robinson JG. 1979. An analysis of the organization of vocal communication in the titi monkey *Callicebus moloch*. *Z Tierpsychol* 49:381–405.

697. Robinson JG. 1982. Vocal systems regulating within group spacing. In Snowdon CT, Brown CH, Petersen MR (eds), *Primate Communication*, 94–116. Cambridge Univ Press, Cambridge.

698. Robinson JG, Janson CH. 1987. Capuchins, squirrel monkeys, and atelines: socioecological convergence with Old World primates. In Smuts BB, Cheney DL, Seyfarth RM, Wrangham RW, Struhsaker TT (eds), *Primate Societies*, 69–82. Univ of Chicago Press, Chicago.

699. Robinson JG, Wright PC, Kinzey WG. 1987. Monogamous cebids and their relatives: intergroup calls and spacing. In Smuts BB, Cheney DL, Seyfarth RM, Wrangham RW, Struhsaker TT (eds), *Primate Societies*, 44–53. Univ of Chicago Press, Chicago.

700. Rodgers LJ, Ward JP. 1994. Head cocking and visual mechanisms in primates. Paper presented at IPS Congress 15, Kuta, Bali, Indonesia. [Abstract available: 1994. *15th Congress IPS Handbook and Abstracts*: 165.]

701. Rodgers ME, Maisels F, Williamson EA, Fernandez M, Tutin CEG. 1990. Gorilla diet in the Lopé Reserve, Gabon: a nutritional analysis. *Oecologia* 84:326–339.

702. Rodman PS. 1977. Feeding behaviour of orangutans of the Kutai Nature Reserve, east Kalimantan. In Clutton-Brock TH (ed), *Primate Ecology: Studies of Feeding and Ranging Behaviour in Lemurs, Monkeys, and Apes*, 384–414. Academic Press, London.

703. Rodriguez GAC, Boher SB. 1988. Notes on the biology of *Cebus nigrivittatus* and *Alouatta seniculus*. *Primate Conserv* 9:61–65.

704. Roonwal ML, Mohnot SM. 1977. *Primates of South Asia: Ecology, Sociology, and Behavior*. Harvard Univ Press, Cambridge.

705. Rose LM. 1994. Sex differences in diet and foraging behavior in white-faced capuchins (*Cebus capucinus*). *IJP* 15(1):95–114.

706. Rose MD. 1978. Feeding and associated positional behavior of black and white colobus monkeys (*Colobus guereza*). In Montgomery GG (ed), *The Ecology of Arboreal Folivores*, 253–264. Smithsonian Institution Press, Washington, DC.

707. Rosenberger AL. 1992. Evolution of feeding niches in New World monkeys. *AJPA* 88:525–562.

708. Ross C. 1991. Life history pattern of New World monkeys. *IJP* 12(5):481–502.

709. Ross C. 1992. Life history patterns and ecology of macaque species. *Primates* 33(2):207–215.

710. Ross C. 1992. Basal metabolic rate, body weight, and diet in primates: an evaluation of the evidence. *Folia Primatol* 58:7–23.

711a. Rowe NB. 1989. Pers obs. Sawai Madhopur, Rajasthan, India, Dec.

711b. Rowe NB. 1993. Pers obs. Caratinga Biological Station, Caratinga, Brazil, Aug.

712. Rowell TE. 1966. Forest living baboons in Uganda. *J Zool Lond* 149:344–364.

713. Rowell TE. 1970. Reproductive cycles of two *Cercopithecus* monkeys. *J Reprod Fertil* 22:321–338.

714. Rowell TE. 1988. The social system of guenons, compared with baboons, macaques, and mangabeys. In Gautier-Hion A, Bourlière F, Gautier JP, Kingdon J (eds), *A Primate Radiation: Evolutionary Biology of the African Guenons*, 439–451. Cambridge Univ Press, Cambridge.

715. Russon AE, Bard KA, Parker S, eds. 1996. *Reaching into Thought: The Minds of the Great Apes*. Cambridge Univ Press, Cambridge.

716. Rubenstein DI. 1993. On the evolution of juvenile life styles in mammals. In Pereira ME, Fairbanks LA (eds), *Juvenile Primates: Life History, Development, and Behavior*, 38–56. Oxford Univ Press, New York.

717. Rudran R. 1973. Adult male replacement in one-male troops of purple-faced langurs (*Presbytis senex senex*) and its effect on population structure. *Folia Primatol* 19:166–192.

718. Rudran R. 1973. The reproductive cycles of two subspecies of purple-faced langurs (*Presbytis senex*) with relation to environmental factors. *Folia Primatol* 19:41–60.

719. Ruhiyat Y. 1983. Socio-ecological study of *Presbytis aygula* in west Java. *Primates* 24(3):344–359.

720. Rylands AB, Mittermeier RA. 1982. Parks reserves and primate conservation in Brazilian Amazonia. *Oryx* 17(2):78–87.

721. Rylands AB. 1993. The ecology of the lion tamarins, *Leontopithecus*: some intrageneric differences and comparisons with other callitrichids. In Rylands AB (ed), *Marmosets and Tamarins: Systematics, Behaviour, Ecology*, 296–314. Oxford Univ Press, Oxford.

722. Rylands AB, de Faria DS. 1993. Habitats, feeding ecology, and home range size in the genus *Callithrix*. In Rylands AB (ed), *Marmosets and Tamarins: Systematics, Behaviour, Ecology*, 262–272. Oxford Univ Press, Oxford.

723. Rylands AB, Mittermeier RA, Rodriguez-Luna E. 1995. A species list for the New World primates (Platyrrhini): distribution by country, endemism, and conservation status according to the Mace-Land system. *Neotrop Primates* 3(suppl):113–160.

724. Sabater Pi JS, Bermejo M, Illera G, Vea JJ. 1993. Behavior of bonobos (*Pan paniscus*) following their capture of monkeys in Zaire. *IJP* 14(5):797–804.

725. Sampaio MI de C, Schneider MPC, Schneider H. 1993. Contribution of genetic distances studies to the taxonomy of *Ateles*, particularly *Ateles paniscus paniscus* and *Ateles paniscus chamek*. *IJP* 14(6):895–904.

726. Sauther ML, Sussman RW. 1993. A new interpretation of the social organization and mating system of the ring-tailed lemur, *Lemur catta*. In Kappeler PM, Ganzhorn JU (eds), *Lemur Social Systems and Their Ecological Basis*, 111–122. Plenum, New York.

727. Sawaguchi T. 1992. The size of neocortex in relation to ecology and social structure in monkeys and apes. *Folia Primatol* 58:131–145.

728. Scanlon E, Chalmers NR, Monteiro da Cruz MAO. 1989. Home range use and the exploitation of gum in the marmoset, *Callithrix jacchus*. *IJP* 10(2):123–126.

729. Schaffner CM, Caine NG. 1992. Who needs to reconcile? Post conflict behavior in red-bellied tamarins [abstract]. *AJP* 27(1):56.

730. Schaller GB. 1964. *The Year of the Gorilla*. Univ of Chicago Press, Chicago.

731. Schilling A. 1979. A study of the marking behaviour in *Lemur catta*. In Martin RD, Doyle GA,

Walker AC (eds), *Prosimian Biology*, 347–364. Univ of Pittsburgh Press, Pittsburgh.

732. Schilling A. 1980. Seasonal variation in the fecal marking of *Cheirogaleus medius* in stimulated climatic condition. In Charles-Dominique P, Cooper HM, Hladik A, et al (eds), *Nocturnal Malagasy Primates: Ecology, Physiology, and Behavior*, 181–190. Academic Press, New York.

733. Schilling D. 1984. Song bouts and dueting in the *concolor* gibbons. In Preuschoft H, Chivers DJ, Brockelman WY, Creel N (eds), *The Lesser Apes: Evolutionary and Behavioural Biology*, 390–403. Edinburgh Univ Press, Edinburgh.

734. Schlichte HJ. 1978. A preliminary report on the habitat utilization of a group of howler monkeys *(Alouatta villosa pigra)* in the National Park of Tikal, Guatemala. In Montgomery GG (ed), *The Ecology of Arboreal Folivores*, 551–560. Smithsonian Institution Press, Washington, DC.

735. Schmid J, Kappeler PM. 1994. Sympatric mouse lemurs *(Microcebus* spp.) in western Madagascar. *Folia Primatol* 63:162–170.

736. Schneider H, Sampaio MIC, Schneider MPC, et al. 1991. Coat color and biochemical variation in Amazonian wild populations of *Alouatta belzebul*. *AJPA* 85:85–93.

737. Schultze H. 1995. Pers com. Ruhr-Universität Bochum, Germany, Nov.

738. Schultze H. 1995. Species review: threat and conservation of nocturnal prosimians. Draft ms.

739. Schultze H. 1995. An identification key for species, subspecies, and local populations of the Lorisinae *(Loris, Nycticebus, Arctocebus, Perodicticus)*. Draft ms.

740. Schürmann CL. 1981. The courtship and mating behavior of wild orangutans in Sumatra. In Chiarelli AB, Corruccini RS (eds), *Primate Behavior and Sociobiology*, 130–135. Springer-Verlag, Berlin.

741. Scott LM. 1984. Reproductive behavior of adolescent female baboons *(Papio anubis)* in Kenya. In Small MF (ed), *Female Primates*, 77–102. AR Liss, New York.

742. Seal US, Ballou JD, Padua CV. 1990. *Leontopithecus* population viability analysis workshop report. Captive Breeding Specialist Group, Species Survival Commission, IUCN, Belo Horizonte, Brazil.

743. Sekulic R. 1982. Daily and seasonal patterns of roaring and spacing in four red howler *Alouatta seniculus* troops. *Folia Primatol* 39:22–48.

744. Sekulic R. 1983. The effect of female call on male howling in red howler monkeys *(Alouatta seniculus)*. *IJP* 4(3):291–305.

745. Seth PK, Seth S. 1986. Ecology and behavior of rhesus monkeys in India. In Else JG, Lee PC (eds), *Primate Ecology and Conservation*, vol 2, 89–104. Cambridge Univ Press, Cambridge.

746. Shaifuddin A, Erwin J, Davidson M, Supriatna J. 1993. Population studies of Sulawesi macaques: geographic distribution of Tonkean macaques *(Macaca tonkeana)* and the booted macaque *(M. ochreata)* [abstract]. *AJP* 30(4):348.

747. Shapiro G, Galdikas BMF. 1994. Attentiveness in orangutans within the sign learning context. In Ogden JJ, Perkins LA, Sheeran L (eds), *Proceedings of the International Conference on Orangutans: The Neglected Ape*, 191–208. Zoological Society of San Diego, San Diego.

748. Sheeran L, Sheeran J, Das J. 1991. The black-crested gibbon of Yunnan province, China [abstract]. *AJP* 24(2):135.

749. Sheeran LK. 1993. Behavior of wild black gibbons *Hylobates concolor jingdongensis* in Yunnan province PRC [abstract]. *AJP* 30(4):348.

750. Shepherd RE, Schaffner CM, Santos CV, French JA. 1992. Sociosexual behavior and the identifica-tion of open mouth display in Wied's black tufted-ear marmoset *(Callithrix kuhlii)*. *AJP* 27(1):58.

751. Shively C, Clark S, King N, Schapiro S, Mitchell G. 1982. Patterns of sexual behavior in male macaques. *AJP* 2:373–384.

752. Sigg H, Stolba A. 1981. Home range and daily march in a hamadryas baboon troop. *Folia Primatol* 36:40–75.

753. Sigg H, Stolba A, Abegglen JJ, Dasser V. 1982. Life history of hamadryas baboons: physical development, infant mortality, reproductive parameters, and family relationships. *Primates* 23(4):473–487.

754. Silk JB. 1980. Kidnapping and female competition among captive bonnet macaques. *Primates* 21(1):100–110.

755. Silk JB. 1994. Social relationships of male bonnet macaques: male bonding in a matrilineal society. *Behaviour* 130(3–4):271–292.

756. Silk JB, Samuels A. 1984. Triadic interactions among *M. radiata* passports and buffers. *AJP* 6:373–376.

757. Silva J de S Jr, Noronha M de A. 1995. A new record for *Callithrix mauesi* Mittermeier, Schwarz and Ayres, 1992. *Neotrop Primates* 3(3):79–80.

758. Simmen B. 1992. Competitive utilization of *Bagassa* fruits by sympatric howler and spider monkeys. *Folia Primatol* 58:155–161.

759. Simons EL. 1988. A new species of *Propithecus* (Primates) from northeast Madagascar. *Folia Primatol* 50:143–151.

760. Singh M, Pirta RS. 1983. Field experiments and observations on rhesus monkeys and bonnet monkeys: a case for primate sociobiology. In Seth PK (ed), *Perspectives in Primate Biology*, 81–88. Today and Tomorrow's Printers and Publishing, New Delhi.

761. Skinner C. 1985. A field study of Geoffroy's tamarin *(Saguinus geoffroyi)* in Panama. *AJP* 9(1):15–25.

762. Smith CC. 1977. Feeding behaviour and social organization in howling monkeys. In Clutton-Brock TH (ed), *Primate Ecology: Studies of Feeding and Ranging Behaviour in Lemurs, Monkeys, and Apes*, 97–126. Academic Press, London.

763. Snowdon CT. 1993. A vocal taxonomy of callitrichids. In Rylands AB (ed), *Marmosets and Tamarins: Systematics, Behaviour, Ecology*, 78–94. Oxford Univ Press, Oxford.

764. Soini P. 1993. The ecology of the pygmy marmoset, *Cebuella pygmaea*: some comparisons with two sympatric tamarins. In Rylands AB (ed), *Marmosets and Tamarins: Systematics, Behaviour, Ecology*, 257–261. Oxford Univ Press, Oxford.

765. Soini P. 1982. Ecology and population dynamics of the pygmy marmoset, *Cebuella pygmaea*. *Folia Primatol* 39:1–21.

766. Spoegler C. 1995. Pers com. State Univ of New York at Stony Brook, Oct.

767. Srikosamatara S. 1984. Ecology of pileated gibbon in South-east Thailand. In Preuschoft H, Chivers DJ, Brockelman WY, Creel N (eds), *The Lesser Apes: Evolutionary and Behavioural Biology*, 242–257. Edinburgh Univ Press, Edinburgh.

768. Srikosamatara S, Brockelman WY. 1983. Patterns of territorial vocalization in the pileated gibbon. In Seth PK (ed), *Perspectives in Primate Biology*. Today and Tomorrow's Printers and Publishing, New Delhi.

769. Stammbach E. 1987. Desert, forest, and montane baboons: multilevel societies. In Smuts BB, Cheney DL, Seyfarth RM, Wrangham RW, Struhsaker TT (eds), *Primate Societies*, 112–120. Univ of Chicago Press, Chicago.

770. Stanford CB, Wallis J, Mpongo E, Goodall J. 1994. Hunting decisions in wild chimpanzees [abstract]. *AJP* 33:240.

771. Stanford CB. 1988. Ecology of the capped langur and Phayre's leaf monkey in Bangladesh. *Primate Conserv* 9:125–128.

772. Stanford CB. 1991. *The Capped Langur in Bangladesh: Behavioral Ecology and Reproductive Tactics*. Karger, New York.

773. Stanford CB. 1991. The diet of the capped langur *(Presbytis pileata)* in moist deciduous forest in Bangladesh. *IJP* 12(3):199–215.

774. Stanford CB. 1991. Social dynamics of intergroup encounters in the capped langur *(Presbytis pileata)*. *AJP* 25:35–47.

775. Stanford CB. 1992. Costs and benefits of allo-mothering in wild capped langurs *(Presbytis pileata)*. *Behav Ecol Sociobiol* 30:29–34.

776. Starin ED. 1978. Food transfer by wild titi monkeys: *Callicebus torquatus torquatus*. *Folia Primatol* 30:145–151.

777. Starin ED. 1993. The kindness of strangers. *Nat Hist* 102(10):44–48.

778. Starin ED. 1994. Philopatry and affiliation among red colobus. *Behaviour* 130(3–4):257–268.

779. Starz TE, Funy J, Tzakis A, et al. 1993. Baboon-to-human liver transplant. *Lancet* 341(8837):65–71.

780. Staski E, Marks J. 1992. *Evolutionary Anthropology: An Introduction to Physical Anthropology and Archeology*. Harcourt Brace Jovanovich College Publishers, Fort Worth, TX.

781. Steenbeek R. 1994. Constraints for female migration in Thomas langurs *(Presbytis thomasi)*: a natural experiment. Paper presented at IPS Congress 15, Kuta, Bali, Indonesia. [Abstract available: 1994. *15th Congress IPS Handbook and Abstracts*: 255.]

782. Steklis HD, Fox R. 1988. Menstrual-cycle phase and sexual behavior in semi-free-ranging stump-tailed macaques, *Macaca arctoides*. *IJP* 9(5):443–456.

783. Sterling E, Feistner ATC. 1994. Recent developments in the conservation of the aye-aye, *Daubentonia madagascariensis*. Paper presented at IPS Congress 15, Kuta, Bali, Indonesia. [Abstract available: 1994. *15th Congress IPS Handbook and Abstracts*: 134.]

784. Sterling EJ. 1992. Feeding behavior of the aye-aye *(Daubentonia madagascariensis)* on Nosy Mangabe, Madagascar. Paper presented at IPS Congress 14, Strasbourg, France. [Abstract available: 1992. *14th Congress IPS Handbook and Abstracts*: 20.]

785. Sterling EJ. 1993. Patterns of range use and social organization in aye-ayes, *Daubentonia madagascariensis* on Nosy Mangabe. In Kappeler PM, Ganzhorn JU (eds), *Lemur Social Systems and Their Ecological Basis*, 1–10. Plenum, New York.

786. Stevenson MF. 1978. Preliminary report on the behaviour of the silvery marmoset, *Callithrix argentata melanura*. In Rothe H, Wolters HJ, Hearn JP (eds), *Biology and Behaviour of Marmosets*, 197–202. Eigenverlag Rothe, Göttingen, Germany.

787. Strier KB. 1992. Presentation given at State Univ of New York at Stony Brook.

788. Strier KB. 1993. Growing up in a patrifocal society: sex differences in the spatial relations of immature muriquis. In Pereira ME, Fairbanks LA (eds), *Juvenile Primates: Life History, Development, and Behavior*, 138–147. Oxford Univ Press, New York.

789. Strier KB. 1990. *Faces in the Forest: The Endangered Muriqui Monkeys of Brazil*. Oxford Univ Press, New York.

790. Strier KB. 1992. Atelinae adaptations: behavioral strategies and ecological constraints. *AJPA* 88:515–524.

791. Strier KB. 1993. Sex ratios, grouping patterns, and competition in wild muriqui monkeys. *AJPA* 16(suppl):191.

792. Strier KB. 1994. Brotherhoods among atelins: kinship, affiliation, and competition. *Behaviour* 130(3–4):151–166.

793. Struhsaker TT. 1967. Ecology of vervet monkeys *(Cercopithecus aethiops)* in the Masai-Amboseli Game Reserve. *Ecology* 48(6):891–904.

794. Struhsaker TT. 1970. Phylogenetic implications of some vocalizations of *Cercopithecus* monkeys. In Napier JR, Napier PH (eds), *Old World Monkeys: Evolution, Systematics, and Behavior*, 365–444. Academic Press, New York.

795. Struhsaker TT. 1975. *The Red Colobus Monkey.* Univ of Chicago Press, Chicago.

796. Struhsaker TT. 1977. Infanticide and social organization in the redtail monkey *(Cercopithecus ascanius schmidti)* in the Kibale Forest, Uganda. *Z Tierpsychol* 45:75–84.

797. Struhsaker TT. 1981. Vocalizations, phylogeny, and paleogeography of red colobus monkeys *(Colobus badius)*. *Afr J Ecol* 19(3):265–283.

798. Struhsaker TT. 1981. Polyspecific associations among tropical rain-forest primates. *Z Tierpsychol* 57:268–304.

799. Struhsaker TT. 1988. Male tenure, multi-male influxes, and reproductive success in redtailed monkeys *(Cercopithecus ascanius)*. In Gautier-Hion A, Bourlière F, Gautier JP, Kingdon J (eds), *A Primate Radiation: Evolutionary Biology of the African Guenons*, 340–363. Cambridge Univ Press, Cambridge.

800. Struhsaker TT, Butynski TM, Lwanga JS. 1988. Hybridization between redtail *(Cercopithecus ascanius schmidti)* and blue *(C. mitis stuhlmanni)* monkeys in the Kibale Forest, Uganda. In Gautier-Hion A, Bourlière F, Gautier JP, Kingdon J (eds), *A Primate Radiation: Evolutionary Biology of the African Guenons*, 477–498. Cambridge Univ Press, Cambridge.

801. Struhsaker TT, Leakey M. 1990. Prey selectivity by crowned hawk eagles on monkeys in the Kibale Forest, Uganda. *Behav Ecol Sociobiol* 26(6):435–443.

802. Struhsaker TT, Leland L. 1980. Observations on two rare and endangered populations of red colobus monkeys in East Africa: *Colobus badius gordonorum* and *Colobus badius kirki*. *Afr J Ecol* 18:191–216.

803. Struhsaker TT, Leland L. 1985. Infanticide in a patrilineal society of red colobus monkeys. *Z Tierpsychol* 69(2):89–132.

804. Struhsaker TT, Leland L. 1987. Colobines: infanticide by adult males. In Smuts BB, Cheney DL, Seyfarth RM, Wrangham RW, Struhsaker TT (eds), *Primate Societies*, 83–97. Univ of Chicago Press, Chicago.

805. Struhsaker TT, Oates JF. 1975. Comparison of the behavior and ecology of red colobus and black-and-white colobus monkeys in Uganda: a summary. In Tuttle RH (ed), *Socioecology and Psychology of Primates*, 103–124. Mouton, The Hague.

806. Struhsaker TT, Pope TR. 1991. Mating system and reproductive success: a comparison of two African forest monkeys *(Colobus badius* and *Cercopithecus ascanius)* behaviour. *Behaviour* 117(3–4):182–205.

807. Strum S. 1987. *Almost Human.* Random House, New York.

808. Su Y, Ren R, Zhu Z, Hu Z. 1994. Preliminary survey of the home range and ranging behavior of golden monkeys *(Rhinopithecus roxellana)* in Shennongjia National Natural Reserve. Paper pre-

sented at IPS Congress 15, Kuta, Bali, Indonesia. [Abstract available: 1994. *15th Congress IPS Handbook and Abstracts:* 280.]

809. Subba PB, Santiapillai C. 1989. Golden langur in the Royal Manas National Park of Bhutan. *Primate Conserv* 10:31–32.

810. Sugardjito J. 1983. Selecting nest-sites of Sumatran orang-utans, *Pongo pygmaeus abelii* in Gunung Leuser National Park, Indonesia. *Primates* 24(4):467–474.

811. Sussman RW. 1977. Feeding behaviour of *Lemur catta* and *Lemur fulvus*. In Clutton-Brock TH (ed), *Primate Ecology: Studies of Feeding and Ranging Behaviour in Lemurs, Monkeys, and Apes*, 1–37. Academic Press, London.

812. Swayamprabha MS, Kadam KM. 1980. Mother infant relationship in slender loris *(Loris tardigradus lydekkerianus)*. *Primates* 21(4):561–566.

813. Symington MM. 1987. Predation and party size on the black spider monkey *Ateles paniscus chamek*. *IJP* 8(5):534.

814. Symington MM. 1988. Food competition and foraging party size in the black spider monkey *(Ateles paniscus chamek)*. *Behaviour* 105(1–2):117–132.

815. Symington MM. 1988. Demography, ranging patterns, and activity budgets of black spider monkeys *(Ateles paniscus chamek)* in the Manu National Park, Peru. *AJP* 15:45–67.

816. Tan CL. 1994. Survey of *Nycticebus pygmaeus* in southern Vietnam. Paper presented at IPS Congress 15, Kuta, Bali, Indonesia. [Abstract available: 1994. *15th Congress IPS Handbook and Abstracts:* 136.]

817. Tardif SD, Garber PA. 1994. Social and reproductive patterns in Neotropical primates: relation to ecology, body size, and infant care. *AJP* 34:111–114.

818. Tardif SD, Harrison ML, Simek MA. 1993. Communal infant care in marmosets and tamarins: relation to energetics, ecology, and social organization. In Rylands AB (ed), *Marmosets and Tamarins: Systematics, Behaviour, Ecology*, 220–234. Oxford Univ Press, Oxford.

819. Tardif SD, Richter CB, Carson RL. 1984. Effects of sibling rearing experience on future reproductive success in two species of Callitrichidae. *AJPA* 6:377–380.

820. Tattersall I. 1982. *The Primates of Madagascar.* Columbia Univ Press, New York.

821. Taub DM. 1982. Sexual behavior of wild Barbary macaque male *(Macaca sylvanus)*. *AJP* 2:109–113.

822. Tenaza R. 1985. Songs of hybrid gibbons *Hylobates lar* and *H. muelleri*. *AJP* 8:249–253.

823. Tenaza R, Tilson RL. 1985. Human predation and Kloss's gibbon *(Hylobates klossii)* sleeping trees in Siberut, Indonesia. *AJP* 8:299–309.

824. Tenaza RR. 1989. Intergroup calls of male pig-tailed langurs *(Simias concolor)*. *Primates* 30(2):199–206.

825. Tenaza RR, Fuentes A. 1995. Monandrous social organization of pigtailed langurs *(Simias concolor)* in the Pagai Islands, Indonesia. *IJP* 16(2):295–310.

826. Terborgh J. 1983. *Five New World Primates: A Study in Comparative Ecology.* Princeton Univ Press, Princeton, NJ.

827. Terborgh J. 1985. Cooperation is better than conflict. In Macdonald D (ed), *Primates*, 52–53. Torstar Books, New York.

828. Terborgh J. 1990. Mixed flock and polyspecific associations: costs and benefits to mixed groups of birds and monkeys. *AJP* 21:87–100.

829. Terborgh J, Goldizen AW. 1985. On the mating system of the cooperatively breeding saddle-

backed tamarin *(Saguinus fuscicollis)*. *Behav Ecol Sociobiol* 16:293–299.

830. Thalmann U, Geissmann T, Simona A, Mutschler T. 1993. The indris of Anjanaharibe-Sud, northeastern Madagascar. *IJP* 14(3):357–381.

831. Thierry B. 1986. Affiliative interference in mounts in a group of Tonkean macaques *(Macaca tonkeana)*. *AJP* 11:89–97.

832. Thierry B, Demaria C, Preuschoft S, Desportes C. 1989. Structural convergence between silent bared-teeth display and relaxed open-mouth display in the Tonkean macaque *(Macaca tonkeana)*. *Folia Primatol* 52:178–184.

833. Thierry B, Gauthier C, Peignot P. 1990. Social grooming in Tonkean macaques *(Macaca tonkeana)*. *IJP* 11(4):357–374.

834. Thomas O. 1902. On some mammals from Coiba Island, off the west coast of Panama. *Novit Zool* 9:135–137.

835. Thomas SC. 1991. Population densities and patterns of habitat use among anthropoid primates of the Ituri forest, Zaire. *Biotropica* 23(1):68–83.

836. Thorington RW Jr, Ruiz JC, Eisenberg JF. 1984. A study of a black howling monkey *(Alouatta caraya)* population in northern Argentina. *AJP* 6:357–366.

837. Tien DV. 1983. On the north Indochinese gibbons *(Hylobates concolor)* (Primates: *Hylobates)* in North Vietnam. *J Hum Evol* 12:367–372.

838. Tilson RL. 1977. Social organization of simakobu monkey *(Nasalis concolor)* in Siberut Island, Indonesia. *J Mammal* 58(2):202–212.

839. Tilson RL. 1981. Family formation strategies of Kloss's gibbon. *Folia Primatol* 35:259–285.

840. Tilson RL, Tenaza RR. 1982. Interspecific spacing between gibbons *(Hylobates klossii)* and langurs *(Presbytis potenziani)* on Siberut island, Indonesia. *AJP* 2:355–362.

841. Uehara S. 1990. Utilization patterns of a marsh grassland within the tropical rain forest by the bonobos *(Pan paniscus)* of Yalosidi, Republic of Zaire. *Primates* 31(3):311–322.

842. Uehara S, Nishida T, Hamai M, et al. 1992. Characteristics of predation by the chimpanzees in the Mahale Mountains National Park, Tanzania. In Nishida T, McGrew WC, Marler P, Pickford M, de Waal FBM (eds), *Symposium Proceedings of the 13th Congress of the International Primatological Society.* Karger, Basel, Switzerland.

843. Ueno Y. 1991. Urine washing in tufted capuchins *(Cebus apella)*: discrimination between groups by urine odor. In Ehara A, Kimura T, Takenaka O, Iwamonto M (eds), *Primatology Today.* Elsevier Science, Amsterdam.

844. Ungar PS. 1995. Fruit preferences of four sympatric primate species at Ketambe, northern Sumatra. *IJP* 16(2):221–244.

845. Utami SS. 1994. Meat eating behavior of adult female orangutans. Paper presented at IPS Congress 15, Kuta, Bali, Indonesia. [Abstract available: 1994. *15th Congress IPS Handbook and Abstracts:* 336.]

846. Valenzuela N. 1992. Early development of three wild infant *Cebus apella* at La Macarena, Colombia. *Field Stud New World Monkeys La Macarena Colomb* 6:15–23.

847. Valenzuela N. 1993. Social contacts between infants and other group members in a wild *Cebus apella* group at La Macarena, Colombia. *Field Stud New World Monkeys La Macarena Colomb* 8:1–9.

848. Van Horn RN, Eaton GG. 1979. Reproductive physiology and behavior in prosimians. In Doyle GA, Martin RD (eds), *The Study of Prosimian Behavior*, 79–122. Academic Press, New York.

849. van Noordwijk MA, Hemelrijk CK, Herremans L, Sterck EHM. 1993. Spatial position and behavioral

sex differences in juvenile long-tailed macaques. In Pereira ME, Fairbanks LA (eds), *Juvenile Primates: Life History, Development, and Behavior*, 77–85. Oxford Univ Press, New York.

850. van Roosmalen MGM, Klein LL. 1988. The spider monkey, genus *Ateles*. In Mittermeier RA, Rylands AB, Coimbra-Filho AF, de Fonseca GAB (eds), *Ecology and Behavior of Neotropical Primates*, vol 2, 455–538. World Wildlife Fund, Washington, DC.

851. van Roosmalen MGM, Mittermeier RA, Milton K. 1981. The bearded sakis, genus *Chiropotes*. In Coimbra-Filho AF, Mittermeier RA (eds), *Ecology and Behavior of Neotropical Primates*, vol 1, 419–442. Academia Brasileira de Ciências, Rio de Janeiro.

852. van Schaik C. 1994. Tool use in wild Sumatran orangutans *(Pongo pygmaeus)*. Paper presented at IPS Congress 15, Kuta, Bali, Indonesia. [Abstract available: 1994. *15th Congress IPS Handbook and Abstracts:* 339.]

853. van Schaik CP, Assink PR, Salafsky N. 1992. Territorial behavior in Southeast Asian langurs: resource defense or mate defense? *AJP* 26:233–242.

854. van Schaik CP, Dunbar RIM. 1990. The evolution of monogamy in large primates: a new hypothesis and some crucial tests. *Behaviour* 115(1–2):30–59.

855. van Schaik CP, Horstermann M. 1994. Predation risk and the number of adult males in a primate group: a comparative test. *Behav Ecol Sociobiol* 35:261–272.

856. Vasey N. 1995. Ranging behavior, community size, and community composition of red ruffed lemurs, *Varecia variegata rubra*, at Andranobe, Masoala Peninsula, Madagascar. Paper presented at the International Conference on the Biology and Conservation of Prosimians, The North of England Zoological Society, Chester, UK. [Abstract available: 1996. *Folia Primatol.* In press.]

857. Vivo MD. 1991. *Taxonomia de Callithrix Erxleben, 1777 (Callitrichidae, Primates)*. Fundação Biodiversitas, Belo Horizonte, Brazil.

858. Walker SE. 1993. Habitat use by *Pithecia pithecia* and *Chiropotes satanas*. *AJPA* 16(suppl):202–203.

859. Walker SE, Jolly CJ, Oates JF. 1988. Electrophoretic evidence for the evolutionary position of *Cercopithecus erythrogaster* and *C. erythrotis*. *Folia Primatol* 51:220–226.

860. Wallis SJ. 1981. The behavioural repertoire of the grey-cheeked mangabey *Cercocebus albigena johnstoni*. *Primates* 22(4):523–532.

861. Wallis SJ. 1983. Sexual behavior and reproduction of *Cercocebus albigena johnstoni* in Kibale Forest, western Uganda, 1. *IJP* 4(2):153–166.

862. Wang Y, Jiang X, Li D. 1994. Classification of existing subspecies of golden snub-nosed monkey, *Rhinopithecus roxellana* (Colobinae, Primates). Paper presented at IPS Congress 15, Kuta, Bali, Indonesia. [Abstract available: 1994. *15th Congress IPS Handbook and Abstracts:* 266.]

863. Waser PM. 1980. Polyspecific associations of *Cercocebus albigena*: geographic variation and ecological correlates. *Folia Primatol* 33:57–76.

864. Waser PM. 1982. The evolution of male loud calls among mangabeys and baboons. In Snowdon CT, Brown CH, Petersen MR (eds), *Primate Communication*, 117–143. Cambridge Univ Press, Cambridge.

865. Waser SK. 1984. Ecological difference and behavioral contrasts between two mangabey species: adaptations for foraging. In Rodman PS, Cant JGH (eds) *Adaptations for Foraging in Nonhuman Primates: Contributions to an Organismal Biology of Prosimians, Monkeys, and Apes*, 195–216. Columbia Univ Press, New York.

866. Wasser SK, Norton G. 1993. Baboons adjust secondary sex ratio in response to predictors of sex-specific offspring survival. *Behav Ecol Sociobiol* 32:273–281.

867. Watanabe K. 1981. Variations in group composition and population density of the two sympatric Mentawaion leaf-monkeys. *Primates* 22(2):145–160.

868. Waterman PG, Ross JAM, Bennett EL, Davies AG. 1988. A comparison of the floristics and leaf chemistry of the tree flora in two Malaysian rain forests and the influence of leaf chemistry on populations of colobine monkeys in the Old World. *Biol J Linn Soc* 34:1–32.

869. Watts DP, Pusey AE. 1993. Behavior of juvenile and adolescent great apes. In Pereira ME, Fairbanks LA (eds), *Juvenile Primates: Life History, Development, and Behavior*, 148–172. Oxford Univ Press, New York.

870. Watts ES. 1990. Evolutionary trends in primate growth and development. In DeRousseau CJ (ed), *Primate Life History and Evolution*, 89–104. Wiley-Liss, New York.

871. Weisenseel KA, Izard MK, Nash LT, Ange RL, Poorman-Allen P. 1995. A comparison of reproduction between two species of *Nycticebus*. Poster presented at the International Conference on the Biology and Conservation of Prosimians, The North of England Zoological Society, Chester, UK. [Abstract available: 1996. *Folia Primatol.* In press.]

872. Weitzel V. 1983. A preliminary study of the dental and cranial morphology of *Presbytis* and *Trachypithecus*. Master's thesis, Australian National Univ, Canberra, Australia.

873. Weitzel V. 1995. Pers com. Australian National Univ, Canberra, Australia, Dec.

874. Weitzel V, Yang CM, Groves CP. 1988. Catalogue of primates in the Zoological Reference Collection, Department of Zoology, National Univ of Singapore. *Raffles Bull Zool* 36(1):1–166.

875. Welker C, Schafer-Witt C. 1988. Preliminary observations on behavioral differences among thick-tailed bushbabies. *IJP* 9(6):507–518.

876. Westergaard GC. 1988. Lion-tailed macaques *(Macaca silenus)* manufacture and use tools. *J Comp Psychol* 102:152–159.

877. White F. 1986. Census and preliminary observations of the ecology of the black-faced spider monkey *(Ateles paniscus chamek)* in Manu National Park, Peru. *AJP* 11:125–132.

878. Whitehead JM. 1990. Comparison of the loud calls of howling monkeys [abstract]. *AJP* 20(3):244.

879. Whitington CL. 1992. Interactions between lar gibbons and pig-tailed macaques at fruit sources. *AJP* 26:61–64.

880. Whitten AJ. 1982. The ecology of singing in Kloss gibbons *(Hylobates klossii)* on Siberut Island, Indonesia. *IJP* 3(1):33–51.

881. Whitten AJ. 1984. Ecological comparisons between Kloss gibbons and other small gibbons. In Preuschoft H, Chivers DJ, Brockelman WY, Creel N (eds), *The Lesser Apes: Evolutionary and Behavioural Biology*, 219–227. Edinburgh Univ Press, Edinburgh.

882. Whitten AJ. 1984. Trilling handicap in Kloss gibbons. In Preuschoft H, Chivers DJ, Brockelman WY, Creel N (eds), *The Lesser Apes: Evolutionary and Behavioural Biology*, 416–419. Edinburgh Univ Press, Edinburgh.

883. Whitten AJ. 1987. The *Presbytis* of Sumatra. *Primate Conserv* 8:46–47.

884. Whitten AJ, Whitten JEJ. 1982. Preliminary observations of the Mentawai macaque in Siberut Island, Indonesia. *IJP* 3(4):445–459.

885. Whitten PL. 1984. Competition among female vervet monkeys. In Small MF (ed), *Female Primates*, 127–140. AR Liss, New York.

886. Whitten T. 1982. *The Gibbons of Siberut*. J Dent, London.

887. Wickings EJ, Ambrose L, Bearder SK. 1995. Sympatric populations of *Galago demidoff* and *G. thomasi* in the Haut Ogooué region of Gabon. Poster presented at the International Conference on the Biology and Conservation of Prosimians, The North of England Zoological Society, Chester, UK. [Abstract available: 1996. *Folia Primatol.* In press.]

888. Wilkie DS, Sidle JG, Boundzanga GC. 1992. Mechanized logging, market hunting, and a bank loan in the Congo. *Conserv Biol* 6(4):570–580.

889. Willard MJ, Dana K, Stark L, Owen J, Zazula J, Cocoran P. 1982. Training a capuchin *(Cebus apella)* to perform as an aide for a quadriplegics. *Primates* 23(4):520–532.

890. Williams L, Gibson S, McDaniel M, Bazzel J, Barnes S, Abee C. 1994. Allomaternal interactions in the Bolivian squirrel monkey, *Saimiri boliviensis*. *AJP* 34:145–156.

891. Williams L, Vitulli W, McElhinney T, Weibe RH, Abee CR. 1986. Male behavior through the breeding season in *Saimiri boliviensis boliviensis*. *AJP* 11:27–35.

892. Wilson JM, Stewart PD, Ramangason GS, Denning AM, Hutchings MS. 1989. Ecology and conservation of the crowned lemur, *Lemur coronatus*, at Ankarana, N. Madagascar. *Folia Primatol* 52:21–26.

893. Winter M. 1978. Some aspects of the ontogeny of vocalizations of hand-reared common marmosets. In Rothe H, Wolters HJ, Hearn JP (eds), *Biology and Behaviour of Marmosets*, 127–140. Eigenverlag Rothe, Göttingen, Germany.

894. Wirth R, Adler HJ, Thang NQ. 1991. Douc langurs: how many species are there? *ZooNooz* 64(6):12–13.

895. Wolfe KE, Fleagle JG. 1977. Adult male replacement in a group of silvered leaf monkeys *(Presbytis cristata)* at Kuala Selangor. *Primates* 18(4):949–955.

896. Wolfe LD. 1984. Japanese macaque female sexual behavior: a comparison of Arashiyama east and west. In Small MF (ed), *Female Primates*, 141–158. AR Liss, New York.

897. Wolfheim JH, Rowell TE. 1972. Communications among captive talapoin monkeys *(Miopithecus talapoin)*. *Folia Primatol* 18:224–255.

898. Wolfheim JH, ed. 1983. *Primates of the World: Distribution, Abundance, and Conservation*. Univ of Washington Press, Seattle.

899. Wrangham RW. 1977. Feeding behaviour of chimpanzees in Gombe National Park, Tanzania. In Clutton-Brock TH (ed), *Primate Ecology: Studies of Feeding and Ranging Behaviour in Lemurs, Monkeys, and Apes*, 504–538. Academic Press, London.

900. Wrangham RW. 1985. Chimpanzees. In Macdonald D (ed), *Primates*, 126–131. Torstar Books, New York.

901. Wrangham RW. 1994. Overview: ecology, diversity, and culture. In Wrangham RW, McGrew WC, de Waal FBM, Heltne PG (eds), *Chimpanzee Culture*, 21–24. Harvard Univ Press, Cambridge.

902. Wrangham RW, Clark AD, Isabirye-Basuta G. 1992. Female social relationships and social organization of Kibale Forest chimpanzees. In Nishida T, McGrew WC, Marler P, Pickford M, de Waal FBM (eds), *Topics in Primatology: Human Origins*, vol 1, 81–98. Univ of Tokyo Press, Tokyo.

903. Wrangham RW, Gittleman JL, Chapman CA. 1993. Constraints on group size in primates and carnivores: population density and day range as

assays of exploitation competition. *Behav Ecol Sociobiol* 32:199–209.

904. Wrangham RW, Goodall J. 1989. Chimpanzee use of medicinal plants. In Heltne PG, Marquardt LA (eds), *Understanding Chimpanzees*, 22–37. Harvard Univ Press, Cambridge.

905. Wright PC. 1986. Diet, ranging behavior, and activity patterns of the gentle lemur *(Hapalemur griseus)* in Madagascar. *AJPA* 69(1–4):283.

906. Wright PC. 1988. Social behavior of *Propithecus diadema edwardsi* in Madagascar. *AJPA* 75(1–4):289.

907. Wright PC. 1981. The night monkeys, genus *Aotus.* In Coimbra-Filho AF, Mittermeier RA (eds), *Ecology and Behavior of Neotropical Primates*, vol 1, 211–240. Academia Brasileira de Ciências, Rio de Janeiro.

908. Wright PC. 1984. Biparental care in *A. trivirgatus* and *C. moloch.* In Small MF (ed), *Female Primates*, 59–76. AR Liss, New York.

909. Wright PC. 1990. Patterns of paternal care in primates. *IJP* 11(2):89–102.

910. Wright PC. 1993. The evolution of female dominance and biparental care among non-human primates. In Miller B (ed), *Sex and Gender Hierarchies*, 127–145. Univ of Cambridge Press, Cambridge.

911. Wright PC. 1994. The behavior and ecology of the owl monkey. In Baer J, Weller RE, Kakoma I (eds), Aotus: *The Owl Monkey*, 97–112. Academic Press, New York.

912. Wright PC. 1996. Pers com. State Univ of New York at Stony Brook, Jan.

913. Wright PC. 1996. Demography and life history of free-ranging *Propithecus diadema edwardsi* in Ranomafana National Park, Madagascar. *IJP* 16(5):835–854.

914. Wright PC, Izard MK, Simons EL. 1986. Reproductive cycles in *Tarsius bancanus. AJP* 11:207–215.

915. Wright PC, Randrimanantena M. 1989. Comparative ecology of three sympatric bamboo lemurs in Madagascar. *AJPA* 78(2):327.

916. Wright PC, Toyama LM, Simons EL. 1986. Courtship and copulation in *Tarsius bancanus. Folia Primatol* 46:142–148.

917. Wu HY, Lin YS. 1992. Life history variables of wild troops of Formosan macaques *(Macaca cyclopis)* in Kenting, Taiwan. *Primates* 33(1):85–97.

918. Wu HY, Lin YS. 1994. Study on population ecology of Formosan macaque *(Macaca cyclopis),* Yushan National Park, Taiwan. Paper presented at IPS Congress 15, Kuta, Bali, Indonesia. [Abstract available: 1994. *15th Congress IPS Handbook and Abstracts:* 152.]

919. Yamagiwa J, Mwanza N, Yumoto T, Maruhashi T. 1991. Ant eating by eastern lowland gorillas. *Primates* 32(2):247–253.

920. Yamamoto ME. 1993. From dependence to sexual maturity: the behavioural ontogeny of Callitrichidae. In Rylands AB (ed), *Marmosets and Tamarins: Systematics, Behaviour, Ecology,* 235–256. Oxford Univ Press, Oxford.

921. Yang DH, Zhang J, Li C. 1987. Preliminary survey on the population and distribution of gibbons in Yunnan Province. *Primates* 28(4):547–549.

922. Yang DH, Zu PK. 1990. A preliminary study on the food of *Hylobates concolor concolor. Primate Rep* 26:81–87.

923. Yangur A. 1994. Survey of the banded leaf monkey *(Presbytis femoralis rhionis)* on Bintang Island, Indonesia. *Asian Primates* 3(3–4):1–2.

924. Yeager CP. 1989. Feeding ecology of the proboscis monkey *(Nasalis larvatus). IJP* 10(6):497–529.

925. Zeeve SR. 1991. Behavior and ecology of primates in the Lomako Forest, Zaire. PhD thesis, State Univ of New York at Stony Brook.

926. Zhang SY. 1995. Activity and ranging in relation to fruit utilization by brown capuchin monkeys *(Cebus apella)* in French Guiana. *IJP* 16(3):489–508.

927. Zhang SY. 1994. Sleeping habits of the brown capuchin monkey *(Cebus apella)* in French Guiana. Paper presented at IPS Congress 15, Kuta, Bali, Indonesia. [Abstract available: 1994. *15th Congress IPS Handbook and Abstracts:* 89.]

928. Zhao Q, Deng Z. 1988. Ranging behavior of *Macaca thibetana* at Mt. Emei, China. *IJP* 9(1):37–48.

929. Zhao Q, Deng Z. 1988. *Macaca thibetana* at Mt. Emei, China: a cross-sectional study of growth and development. *AJP* 16:251–260.

930. Zhao Q, He S, Wu B, Nash LT. 1988. Excrement distribution and habitat use in *Rhinopithecus bieti* in winter. *AJP* 16:275–284.

931. Zhao QK. 1993. Sexual behavior of Tibetan macaques at Mt. Emei, China. *Primates* 34(4):431–444.

932. Zhu Z, Ren R, Su Y, Hu Z. 1994. Field observations of the reproductive ecology and behavior of golden monkeys *(Rhinopithecus roxellana)* in Shennongjia National Natural Reserve. Paper presented at IPS Congress 15, Kuta, Bali, Indonesia. [Abstract available: 1994. *15th Congress IPS Handbook and Abstracts:* 279.]

933. Zimmermann E, Bearder SK, Doyle GA, Anderson AB. 1988. Variations in vocal pattern of Senegal and South African lesser bush baby and their implications for taxonomic relationships. *Folia Primatol* 51:87–105.

934. Zimmermann E, Lerch C. 1993. The complex acoustic design of an advertisement call in male mouse lemurs *(Microcebus murinus),* Prosimii, Primates and sources of its variation. *Ethology* 93:211–224.

935. Shapiro GL. 1982. Sign acquisition in a home reared/freeranging orangutan: comparison with other signing apes. *AJP* 3:121–129.

936. Bearder SK. 1995. Pers com. Oxford Brookes Univ, Oxford, UK, Sep.

937. Richard AF. 1977. Feeding behaviour of *Propithecus verreauxi.* In Clutton-Brock TH (ed), *Primate Ecology: Studies of Feeding and Ranging Behaviour in Lemurs, Monkeys, and Apes,* 72–96. Academic Press, London.

938. Niemitz C. 1984. Activity and use of space in semi-wild Bornean tarsiers with remarks on wild spectral tarsiers. In Niemitz C (ed), *Biology of Tarsiers,* 85–115. G Fischer, Stuttgart, Germany.

939. Glander KE, Powzyk JA. 1995. Morphometrics of wild *Indri indri* and *Propithecus diadema diadema.* Paper presented at the International Conference on the Biology and Conservation of Prosimians, The North of England Zoological Society, Chester, UK. [Abstract available: 1996. *Folia Primatol.* In press.]

940. Wright PC. 1986. Ecological correlates of monogamy in *Aotus* and *Callicebus.* In Else JG, Lee PC (eds), *Primate Ecology and Conservation,* vol 2, 159–168. Cambridge Univ Press, Cambridge.

941. Tremble M, Muskita Y, Supriatna J. 1993. Field observations of *Tarsius dianae* at Lore Lindu National Park, central Sulawesi, Indonesia. *Trop Biodiversity* 1(2):67–76.

942. Niemitz C. 1984. Taxonomy and distribution of the genus *Tarsius* Storr, 1780. In Niemitz C (ed), *Biology of Tarsiers,* 1–16. G Fischer, Stuttgart, Germany.

943. Robinson JG, Redford KH. 1991. The use and conservation of wildlife; Sustainable harvest of Neotropical forest animals. In Robinson JG, Redford KH (eds), *Neotropical Wildlife Use and Conservation,* 3–5; 415–929. Univ of Chicago Press, Chicago.

944. Horwich RH, Lyon J. 1990. The baboon or black howler monkey. In *A Belizean Rain Forest: The Community Baboon Sanctuary,* 24–37. Orang-utan Press, Gays Mills, WI.

945. Schwartz J. 1996. *Pseudopotto martini:* a new genus and species of extant lorisiform primate. *Anthropol Pap Amer Mus Nat Hist* 78:1–14.

946. Andrews J, Birkinshaw C. 1995. A comparison between the daytime and night-time feeding ecology of the black lemur, *Eulemur macaco,* in Lokobe Forest, Madagascar. Paper presented at the International Conference on the Biology and Conservation of Prosimians, The North of England Zoological Society, Chester, UK. [Abstract available: 1996. *Folia Primatol.* In press.]

947. Milton K. 1985. Mating patterns of woolly spider monkeys, *Brachyteles arachnoides:* implications for female choice. *Behav Ecol Sociobiol* 17:53–59.

948. White FJ, Overdorff DJ, Balko EA, Wright PC. 1995. Distribution of ruffed lemurs *(Varecia variegata)* in Ranomafana National Park, Madagascar. *Folia Primatol* 64:124–131.

949. Silva J de S Jr, Noronha M de A. 1996. On a new species of bare-eared marmoset, genus *Callithrix* Erxleben, 1777, from Central Amazonia, Brazil (Primates: Callitrichidae). *Goeldiana Zool* 18. In press.

950. Rylands AB. 1995. Pers com. Instituto de Ciências Biologicas, Univ Federal de Minas Gerais, Belo Horizonte, Minas Gerais, Brazil, Jan.

951. Savage-Rumbaugh S, Lewin R. 1994. *Kanzi: The Ape at the Brink of the Human Mind.* J Wiley, New York.

952. See 328.

953. Rylands AB, Coimbra-Filho AF, Mittermeier RA. 1993. Systematics, geographic distribution, and some notes on the conservation status of the Callitrichidae. In Rylands AB (ed), *Marmosets and Tamarins: Systematics, Behaviour, Ecology,* 11–77. Oxford Univ Press, Oxford.

954. Moynihan MA. 1964. Some behavior patterns of platyrrhine monkeys, 1: The night monkey *(Aotus trivirgatus). Smithson Misc Collect* 146(5):1–84.

955. Shekelle M, Mukti S, Masala Y. In prep. Natural history of the tarsiers of Sulawesi, part 1: north Sulawesi. *Trop Biodiversity.*

956. Fooden J. 1990. The bear macaque *Macaca arctoides:* a systematic review. *J Hum Evol* 19:607–686.

957. Fooden J. 1995. Systematic review of Southeast Asia long-tail macaques, *Macaca fascicularis* Raffles (1821). *Fieldiana Zool,* ns, 81.

958. Huffman MA. 1991. Mate selection and partner preferences in female Japanese macaques. In Fedigan LM, Asquinth PJ (eds), *The Monkeys of Arashiyama,* 101–123. State Univ of New York Press, Albany.

959. Huffman MA. 1991. History of the Arashiyama Japanese macaques. In Fedigan LM, Asquinth PJ (eds), *The Monkeys of Arashiyama,* 21–53. State Univ of New York Press, Albany.

960. Fooden J. 1975. Taxonomy and evolution of liontailed and pigtail macaques (Primates: Cercopithecidae). *Fieldiana Zool* 67:1–169.

961. Fuentes A. 1996. Pers com. Univ of California, Berkeley, Dec.

962. Mootnick AR, Sheeran LK. 1995. Agile gibbon *(Hylobates agilis).* In Beacham W (ed), *Beacham's International Threatened, Endangered, and Extinct Species.* Beacham, Washington, DC.

963. Porter LM. 1996. Influences on sportive lemur distribution in the Ranomafana National Park. *AJPA.* In press.

964. Wright PC, Heckscher K, Dunham A. 1995. Predation on rain forest prosimians in Ranomafana National Park, Madagascar. Paper presented at the International Conference on the Biology and Conservation of Prosimians, The North of England Zoological Society, Chester, UK. [Abstract available: 1996. *Folia Primatol.* In press.]

965. Sekulic R. 1982. Birth in free ranging howlers *(Alouatta seniculus)*. *Primates* 23(4):580–582.

966. Kaplan G, Rogers L. 1994. *Orangutans in Borneo.* Univ of Northern England Press, Armidale, UK.

967. Lion tail macaque *Macaca silenus:* population and habitat viability assessment workshop report. 1993. *Zoos' Print* 8(12):11–28.

968. Fooden J. 1979. Taxonomy and evolution of the *sinica* group of macaques, part 1: species and subspecies account of the *sinica* group of macaques. *Primates* 20(1):109–140.

969. Whiten A, Byrne RW, Waterman PG, Henzi SP, McCullough FM. 1990. Specifying the rules underlying selective foraging in wild mountain baboons, *P. ursinus.* In de Mello MT, Whiten A, Byrne RW (eds), *Baboons: Behaviour and Ecology, Use and Care*, 5–22. Brasília, Brazil.

970. Feistner ATC. 1990. Reproductive parameters in a semifree-ranging group of mandrills. In de Mello MT, Whiten A, Byrne RW (eds), *Baboons: Behaviour and Ecology, Use and Care*, 77–88. Brasília, Brazil.

971. Bercovitch FB. 1990. Female choice, male reproductive success, and the origin of one male groups in baboons. In de Mello MT, Whiten A, Byrne RW (eds), *Baboons: Behaviour and Ecology, Use and Care*, 61–76. Brasília, Brazil.

972. Anderson CM. 1990. Desert, mountain, and savanna baboons: a comparison with special references to the Suikerbosrand population. In de Mello MT, Whiten A, Byrne RW (eds), *Baboons: Behaviour and Ecology, Use and Care*, 89–104. Brasília, Brazil.

973. Fuentes A. 1994. The socioecology of the Mentawai Island langur *Presbytis potenziana*. PhD diss, Univ of California, Davis.

974. Pavelka MSM. 1993. *Monkeys of the Mesquite: The Social Life of the South Texas Snow Monkey.* Kendall/Hunt, Dubuque, IA.

975. Fuentes A. 1995. Feeding and ranging in the Mentawai langur *Presbytis potenziani*. Manuscript.

976. Li Z. 1993. Time budgets of *Presbytis leucocephalus*. *Acta Theriol Sin* 12(1):7–13.

977. Bleisch WV. 1995. Pers com. New York, Oct.

978. Wolfe K. 1984. Reproductive competition among co-resident male silvered leaf monkeys *(Presbytis cristata)*. PhD diss, Yale Univ, New Haven, CT.

979. Stewart KJ, Harcourt AH. 1987. Gorillas: variation in female relationships. In Smuts BB, Cheney DL, Seyfarth RM, Wrangham RW, Struhsaker TT (eds), *Primate Societies*, 155–164. Univ of Chicago Press, Chicago.

980. Doran D. 1996. Comparative positional behavior of the African apes. In McGrew W, Marchant L, Nishida T (eds), *Great Ape Societies*, chap 16. Cambridge Univ Press, Cambridge.

981. Doran D. 1993. Sex differences in adult chimpanzee positional behavior: the influence of body size on locomotion and posture. *AJPA* 91:99–115.

982. Thompson-Handler NT, Malenky RK, Reinartz GE (eds). 1995. *Action Plan for Pan paniscus: Report on Free Ranging Populations and Proposal for Their Preservation.* Zoological Society of Milwaukee County, Milwaukee.

983. Nhat P. 1994. Some data on the food of the Tonkin snub-nosed monkey *(Rhinopithecus avunculus)*. *Asian Primates* 3(3):4–5.

984. Mootnick A, Haimoff E, Nyunt-Iwin K. 1987. Conservation and captive management of hoolock gibbons in the Socialist Republic of the Union of Burma. In *American Association of Zoological Parks and Aquariums Annual Proceedings*, 398–423.

985. Mootnick AR, Sheeran LK. 1995. Hoolock gibbon *(Hylobates hoolock)*. In Beacham W (ed), *Beacham's International Threatened, Endangered, and Extinct Species.* Beacham, Washington, DC.

986. Mootnick AR, Sheeran LK. 1995. Kloss' gibbon *(Hylobates klossi)*. In Beacham W (ed), *Beacham's International Threatened, Endangered, and Extinct Species.* Beacham, Washington, DC.

987. Sheeran LK, Mootnick AR. 1995. Black gibbon *(Hylobates concolor)*. In Beacham W (ed), *Beacham's International Threatened, Endangered, and Extinct Species.* Beacham, Washington, DC.

988. Sheeran LK, Mootnick AR. 1995. Crested gibbon *(Hylobates leucogenys)*. In Beacham W (ed), *Beacham's International Threatened, Endangered, and Extinct Species.* Beacham, Washington, DC.

989. Lernould JM. 1988. Classification and geographical distributions of guenons: a review. In Gautier-Hion A, Bourlière F, Gautier JP, Kingdon J (eds), *A Primate Radiation: Evolutionary Biology of the African Guenons*, 54–78. Cambridge Univ Press, Cambridge.

990. Fooden J. 1991. Systematic review of Philippine macaques (Primates, Cercopithecidae: *Macaca fascicularis* subspp.). *Fieldiana Zool*, ns, 64:1–32.

991. Terborgh J. 1992. *Diversity and the Tropical Rain Forest.* Scientific American Library, New York.

992. Anderson S. 1983. *Simon and Schuster's Guide to Mammals.* Simon and Schuster, New York.

993. Milton K. 1984. The role of food-processing factors in primate food choice. In Rodman PS, Cant JGH (eds), *Adaptations for Foraging in Nonhuman Primates: Contributions to an Organismal Biology of Prosimians, Monkeys, and Apes*, 249–279. Columbia Univ Press, New York.

994. de Waal FBM. 1982. *Chimpanzee Politics: Power and Sex among Apes.* Harper and Row, New York.

995. Mootnick AR, Sheeran LK. 1995. Lar gibbon *(Hylobates lar)*. In Beacham W (ed), *Beacham's International Threatened, Endangered, and Extinct Species.* Beacham, Washington, DC.

996. Mootnick AR, Sheeran LK. 1995. Moloch gibbon *(Hylobates moloch)*. In Beacham W (ed), *Beacham's International Threatened, Endangered, and Extinct Species.* Beacham, Washington, DC.

997. Li Z. 1993. Ecology and behavior of the leaf monkey in China. In Ye S (ed), *Biology of the Leaf Monkey* (Presbytis), 15–87. Yunnan Science and Technology Press, Kunming, PRC.

998. Mootnick AR, Sheeran LK. 1995. Mueller's gibbon *(Hylobates muelleri)*. In Beacham W (ed), *Beacham's International Threatened, Endangered, and Extinct Species.* Beacham, Washington, DC.

999. Mootnick AR, Sheeran LK. 1995. Pileated gibbon *(Hylobates pileatus)*. In Beacham W (ed), *Beacham's International Threatened, Endangered, and Extinct Species.* Beacham, Washington, DC.

1000. Mootnick AR, Sheeran LK. 1995. Siamang *(Hylobates syndactylus)*. In Beacham W (ed), *Beacham's International Threatened, Endangered, and Extinct Species.* Beacham, Washington, DC.

1001. McGrew W. 1994. Tools compared: the material of culture. In Wrangham RW, McGrew WC, de Waal FBM, Heltne PG (eds), *Chimpanzee Culture*, 25–41. Harvard Univ Press, Cambridge.

1002. Diamond JM. 1993. *The Third Chimpanzee: The Evolution and Future of the Human Animal*, Harper Perennial, New York.

1003. Garber PA. 1996. Pers com. Univ of Illinois, Urbana.

1004. Boesch C, Boesch H. 1990. Tool use and tool making in wild chimps. *Folia Primatol* 54:86–99.

1005. van Schaik CP, Fox EA, Sitompul A. In press. Tool use in wild Sumatran orangutans *(Pongo pygmaeus)*. *Naturwiss.*

1006. Müller KH. In press. Diet and feeding ecology of masked titis *(Callicebus personatus)*. In Norconk MA, Rosenberger A, Garber PA (eds), *Adaptive Radiations of Neotropical Primates.* Plenum, New York.

1007. van Schaik CP. 1996. Pers com. Duke Univ, Durham, NC, Mar.

1008. Canh LX. 1994. Pers com. Institute for Ecology and Biological Resources, Hanoi, Viet Nam, Aug.

1009. Nhat P. 1993. First result on the diet of the red-shanked douc langur *(Pygathrix nemaeus)*. *Asian Primates* 3(3–4):4.

1010. Huffman MA, Wrangham RW. 1994. Diversity of medicinal plant use by chimpanzees in the wild. In Wrangham RW, McGrew WC, de Waal FBM, Heltne PG (eds), *Chimpanzee Cultures*, 129–14148. Harvard Univ Press, Cambridge.

1011. Wrangham RW, de Waal FBM, McGrew WC. 1994. The challenge of behavioral diversity. In Wrangham RW, McGrew WC, de Waal FBM, Heltne PG (eds), *Chimpanzee Cultures*, 1–17. Harvard Univ Press, Cambridge.

1012. Shideler SE, Savage A, Ortuno AM, Moorman EA, Lasley BL. 1994. Monitoring female reproduction function by measurement of fecal estrogen and progesterone metabolites in the white-faced saki *(Pithecia pithecia)*. *AJP* 32:95–108.

1013. Kinzey WG, Norconk MA. 1990. Hardness as a basis of fruit choice in two sympatric primates. *AJPA* 81:5–15.

1014. McGreal S. 1996. Pers com. International Primate Protection League, Summerville, SC, Mar.

1015. Singer P, Cavalieri P. 1993. *The Great Ape Project.* St Martin's Press, New York.

1016. Winkler LA. 1989. Morphology and relationships of the orangutan fatty cheek pads. *AJP* 17:305–319.

1017. Hyatt CW, Hopkins WD. 1995. Responses to mirrors in gibbons *(Hylobates lar)*. *AJP* 36(2):128–129.

1018. Leighton M, Setia T, Seal U, et al. 1993. Orangutan life history and vortex analysis. In Tilson R, Seal U, Soemama K, et al (eds), *Orangutan Population and Habitat Viability Analysis Final Report*, 31–42. IUCN/SSC Captive Breeding Specialist Group, Apple Valley, MN.

1019. The Population Institute. 1996. Pers com. Washington, DC, Mar.

1020. McGrew WC. 1992. *Chimpanzee Material Culture: Implications for Human Evolution.* Cambridge Univ Press, Cambridge.

Index

Page numbers for the accounts of species, including the accompanying species illustrations, are in bold. Page numbers for other illustrations are in italics. Page numbers for descriptions of taxonomic groups other than species are in bold.

Allenopithecus nigroviridis (Allen's swamp monkey), 120, **147**, 154, 167

Allen's swamp monkey. *See under* monkey

Allocebus trichotis (hairy-eared dwarf lemur), 27, **28**

Alouatta (howler monkeys), 81, 103, 158
  belzebul (red-handed howler), **107**
  caraya (black-and-gold howler), *57*, **107**
  coibensis (Coiba Island howler), **108**
  fusca (brown howler), **108–109**, 117
  guariba (=*A. fusca*), 108
  palliata (mantled howler), 107, 108, **109**, 100
  pigra (black howler), **110**
  sara (Bolivian red howler), **110**
  seniculus (red howler), 97, 107, 110, **111**, 115
    arctoidea, 111
    villosa (=*A. pigra*), 110

Alouattinae (howler monkeys), **81**, 82

angwantibo (*Arctocebus calabarensis*), **14**, 14
  golden (*Arctocebus aureus*), **14**, 14

Anthropoidea (higher primates; anthropoids; simians), 3, 8, 52, **57**, 57–233

anthropoids. *See* Anthropoidea

Aotinae (night monkeys), **81**

Aotus (night monkeys), 8, 81
  azarai (night monkey), 83
  brumbacki, 84
  hershkovitzi, 84
  infulatus, 83
  lemurinus, 84
    griseimembra, 84
  miconax, 83, 90
  nancymaae, 83
  nigriceps (southern red-necked night monkey), **83**
  trivirgatus (northern gray-necked owl monkey), **84**
  vociferans, 84

ape, Celebes black (=Celebes black macaque), 129

apes (Hominoidea), 3, 119, **207**, 208–233.
  great (Pongidae and Hominidae), v, **219**
  lesser (gibbons; *Hylobates; Hylobatidae*), **207**

Archaeoindris (extinct lemur), 2

Archaeolemur (extinct lemur), 2

Arctocebus
  aureus (golden angwantibo), **14**, 14
  calabarensis (angwantibo), **14**, 14

Ateles (spider monkeys), 7, 8, 82, *82*, 104, 105, 158
  belzebuth (white-bellied spider monkey), *82*, **112**, 114
    belzebuth, 112, *112*
    hybridus, 112, *112*

chamek (black-faced black spider monkey), **113**, 115
  fusciceps (brown-headed spider monkey), **113**
    fusciceps, 113
    robustus, 113
  geoffroyi (black-handed spider monkey), 95, **114**
  marginatus (white-whiskered spider monkey), 112, **114**
  paniscus (black spider monkey), 105, 111, 113, **115**

Atelinae (spider monkeys and woolly monkeys), **82**

Avahi
  laniger (woolly lemur), 27, 37, **47**
  occidentalis, 47

aye-aye (*Daubentonia madagascariensis*), 4, 27, **51**

aye-ayes (Daubentoniidae), **27**, 51

Babakotia (extinct lemur), 2

baboon
  chacma (*Papio hamadryas ursinus*), **140**
  gelada (*Theropithecus gelada*), 8, 120, **143**
  Guinea (*Papio hamadryas papio*), **140**
  hamadryas (*Papio hamadryas hamadryas*), 137, **138–139**
  olive (*Papio hamadryas anubis*), 119, **136–137**, 139
  yellow (*Papio hamadryas cynocephalus*), **138**

baboons (*Papio*), 120, 141, 142, 143, 147, 151

bandro (*Hapalemur griseus alaotrensis*), 45, 45–46, *46*

bonobo (*Pan paniscus*), 219, **228–229**

Brachyteles (woolly spider monkeys), 82
  arachnoides (woolly spider monkey), 7, 108, **117**
  hypoxanthus, 80, 117, *117*

bush babies (Galagonidae), 6, **13**, 18–25

bush baby
  Allen's (*Galago alleni*), 18, **19**, 22
  Demidoff's (*Galagoides demidoff*), 18, 19, 20, 21, **22**, 22, 23
  Garnett's greater (*Otolemur garnettii*), 19, 23, 24, **25**
  Matschie's (*Galago matschiei*), 17, **20**, 22, 23
  northern lesser (*Galago senegalensis*), 4, 19, 20, **21**, 23, 24
  northern needle-clawed (*Euoticus pallidus*), **18**
  rondo dwarf (*Galagoides demidoff rondoensis*), 22, *22*
  Somali (*Galago gallarum*), **19**, 21, 23, 25
  southern lesser (*Galago moholi*), 12, **20–21**, 21, 22, 23, 24
  southern needle-clawed (*Euoticus elegantulus*), **18**, 18, 19, 22
  thick-tailed greater (*Otolemur crassicaudatus*), 21, 23, **24**, 25
  Thomas's (*Galagoides thomasi*), 20, 22, **22–23**
  Zanzibar (*Galagoides zanzibaricus*), 19, 21, **23**, 24, 25

Cacajao (uacaris), 81

calvus (bald uacari), 102, **105–106**
  calvus (white uacari), 105, 106, *106*
  rubicundus (red uacari), 99, 105, *105*, 106
  melanocephalus (black-headed uacari), 93, 94, 99, **106**

Callicebinae (titi monkeys), **81**

Callicebus (titi monkeys), 6, 8, 81, 88
  brunneus (brown titi monkey), **85**, 85, 86, 87
  caligatus (chestnut-bellied titi monkey), 85, **85**, 86, 87, 92
  cinerascens (ashy titi monkey), **86**, 89
  cupreus (red titi monkey), 85, **86**, 87, 88, 92
  donacophilus (Bolivian gray titi monkey), 85, **87**
  dubius (titi monkey), 85, **87**
  hoffmannsi (Hoffmann's titi monkey), **88**
  modestus (titi monkey), **88**
  moloch (dusky titi monkey), 69, 71, 86, 88, **89**, 92
    brunneus, 89
  oenanthe (Andean titi monkey), 83, **90**
  olallae (Beni titi monkey), **90**
  personatus (masked titi monkey), 63, **91**
    nigrifrons, 91
    personatus, 91
  torquatus (collared titi monkey), 69, 85, 86, 89, **92**

Callimico, 59
  goeldii (Goeldi's monkey), **60**, 69, 72, 75

Callimiconidae, 60

Callithrix (marmosets), 6, 8
  argentata (bare-ear marmoset), **60–61**, 66, 73
    argentata, 60, *61*
    emiliae, 60, 61, *61*, 69
    leucippe, 60
    marcai, 60
    melanura, 60, *60*
    saterei, 60, *61*
  aurita (buffy tufted–eared marmoset), **61**, 64
  flaviceps (buffy-headed marmoset), *58*, **62**, 64
  geoffroyi (Geoffroy's tufted-eared marmoset), **62–63**, 64, 91
  humeralifer (tassel-eared marmoset), **63–64**, 65
    chrysoleuca, 63, *64*
    humeralifer, 63, *63*
    intermedia, 63, *63*
  jacchus (common marmoset), 61, 62, **64**, 65
  kuhlii (Wied's tufted-eared marmoset), 64, **64**, 77–78
  mauesi (Maues marmoset), 1, **65**
  nigriceps (black-headed marmoset), 1, **66**
  penicillata (black tufted–eared marmoset), 64, **66**, 77
  pygmaea (pygmy marmoset), 5, 59, **67**, 75
  saterei, 1

Callitrichidae (marmosets and tamarins), 6, 52, **59**, 60–79, 62
Callitrichinae, 60–79
capuchin
   brown. See *Cebus apella*
   Ka'apor *(Cebus kaapori)*, 1, 96, *96*
   tufted *(Cebus apella)*, 93, **94–95**, 99, 101, 104, 105, 106
   wedge-capped. See *Cebus olivaceus*
   weeper *(Cebus olivaceus)*, **96–97**, 111
   white-fronted *(Cebus albifrons)*, **93**, 94, 99, 106
   white-throated *(Cebus capucinus)*, **95–96**, 98, 114
   yellow-breasted *(Cebus apella xanthosternos)*, 94, *94*
capuchin monkeys *(Cebus)*, 81, 99
Catarrhini (Old World monkeys and apes; catarrhines), 3, 57, 59, 119–233, 148
Cebidae (cebids), 52, **81–82**, 83–117
cebids. See Cebidae
Cebinae (capuchins and squirrel monkeys), **81**
Ceboidea (New World monkeys; =Platyrrhini), 3, 57, 57–117
Cebuella (=Callithrix pygmaea), 67
Cebus (capuchin monkeys), **81**, 99
   albifrons (white-fronted capuchin), **93**, 94, 99, 106
   apella (tufted capuchin), 93, **94–95**, 99, 101, 104, 105, 106
      apella, *94*
      libidinosus, *94*
      nigritus, *94*
      xanthosternos (yellow-breasted capuchin), 94, *94*
   capucinus (white-throated capuchin), **95–96**, 98, 114
   kaapori (Ka'apor capuchin), 1, 96, *96*
   nigrivittatus (=C. olivaceus), 96
   olivaceus (weeper capuchin), **96–97**, 111
      apiculatus, *97*
Celebes black ape (=Celebes black macaque), 129
Cercocebus (mangabeys), 120, 146, 147, 153
   agilis (agile mangabey), **144**, 144
      chrysogaster, 144
      sanje, 144
   galeritus (Tana River mangabey), 144, **144**
   torquatus (white-collared mangabey), **145**, 156, 157, 161, 174
   atys (sooty mangabey), 145, **145**
Cercopithecidae (macaques, baboons, guenons, and colobines; =Cercopithecoidea), 8, **119**, 169
Cercopithecinae (cheek pouch monkeys), 8, 119, **120**, 120–167
Cercopithecoidea (Old World monkeys; =Cercopithecidae), 3, 8, 119–205, 148, 207
Cercopithecus (guenons), 120, 148, 150, **153**, 172
   ascanius (red-tailed guenon), 146, 147, 153, **154**, 160, 165, 167, 170, 171, 175
      schmidti, *154*

      whitesidei, *154*
   campbelli (Campbell's guenon), **155**, 156, 164
   cephus (mustached guenon), 145, 146, 154, **155–156**, 157, 158, 159, 163, 164, 165, 166
   diana (Diana monkey), 155, **156**, 157, 164, 174, 176
   dryas (Dryas guenon), **157**
   erythrogaster (white-throated guenon), 145, 146, *152, 157*, 161
   erythrotis (red-eared guenon), 155, 156, **158**, 166
   hamlyni (owl-faced monkey), **158**
   lhoesti (L'Hoest's monkey), **159**, 165, 167
   mitis (blue monkey), 146, 153, 154, **159–160**, 170, 175
      albogularis, 159, *160*
      kandti (golden mitis monkey), 160
      mitis, 159
      stuhlmanni, *159, 163*
   mona (mona monkey), 145, 146, 155, 157, **161**, 164, 165, 167
   neglectus (De Brazza's monkey), **162**
   nictitans (putty-nosed spot-nosed guenon), 146, 156, **163**, 165
   petaurista (lesser spot-nosed guenon), 155, 156, **164**, 176–177
   pogonias (crowned guenon), 156, 161, 163, **164–165**
      grayi, 167
      pogonias, *164*
   preussi (Preuss's monkey), 159, **165**
   roloway (=Cercopithecus diana), 156
   salongo (=Cercopithecus dryas), 157
   sclateri (Sclater's guenon), 155, 157, **166**
   solatus (sun-tailed guenon), **167**
   wolfi (Wolf's guenon), 146, 147, 154, **167**, 170
      denti, 167
Cheirogaleidae (dwarf lemurs), **27**, 28–33
Cheirogaleinae, 28–32
Cheirogaleus, 27, 28
   major (greater dwarf lemur), **28–29**
      crossleyi, 28
   medius (fat-tailed dwarf lemur), 28, **29**, 30
Chimpansee (=Pan), 228, 230
chimpanzee *(Pan troglodytes)*, v, 7, 8, 156, 176, *218*, 219, 228, **230–232**, 233
   pygmy. See *Pan paniscus*
Chiropotes (bearded sakis), 81, 105
   albinasus (white-nosed bearded saki), 94, **104**
   satanas (bearded saki), 103, **104–105**, 115
Chlorocebus, 120, 157
   aethiops (vervet), 120, **150–151**, 156
      aethiops, 150
      djamdjamensis, 151
      pygerythrus, 150, *150*
      sabaeus, 150
      tantalus, 150, *151*
Colobinae (leaf-eating monkeys, colobines), 119, **169**, 170–205

colobines. See Colobinae
Colobus (colobus monkeys), 145, 153, 169, 173
   abyssinicus (=C. guereza), 170
   angolensis (Angolan black-and-white colobus), 147, 154, 160, 167, **170**
   badius, 173
   guereza (Abyssinian black-and-white colobus), 154, **170–171**, 173, 175
   polykomos (king western black-and-white colobus), **171–172**, 173, 174, 176
      dollmani, 173
   satanas (black colobus), **172**
   vellerosus (Geoffroy's black-and-white colobus), 171, **172**
colobus
   Abyssinian black-and-white *(Colobus guereza)*, 154, **170–171**, 173, 175
   Angolan black-and-white *(Colobus angolensis)*, 147, 154, 160, 167, **170**
   black *(Colobus satanas)*, **172**
   black-and-white, 145
   eastern black-and-white. See Abyssinian black-and-white colobus
   Geoffroy's black-and-white *(Colobus vellerosus)*, 171, **172**
   guereza black-and-white. See Abyssinian black-and-white colobus
   king black-and-white *(Colobus polykomos)*, **171–172**, 173, 174, 176
   Kirk's red *(Procolobus pennantii kirkii)*, 174, *174, 175*
   olive *(Procolobus verus)*, 156, 164, 169, 172, **176–177**
   Pennant's red *(Procolobus pennantii)*, 6, 147, 171, **174–175**
   Preuss's red *(Procolobus preussi)*, **175**
   red *(Procolobus pennantii tephrosceles)*, 174, *174, 175*
   Tana River red *(Procolobus rufomitratus)*, **176**
   western black-and-white. See king black-and-white colobus
   western red *(Procolobus badius)*, 145, 156, 172, **173–174**, 175, 176, 231
   white-thighed black-and-white. See Geoffroy's black-and-white colobus
colobus monkeys *(Colobus)*, 153, 169, 173
   red *(Procolobus)*, 153, 154, 160, 169, 175
Cynocephalus sp. ("flying lemur"), *27*
Daubentonia madagascariensis (aye-aye, =Daubentoniidae), 4, 27, **51**
Daubentoniidae (aye-ayes, =Daubentonia), **27**, 51
Dermoptera ("flying lemurs"), 27
douc langur(s). See *under* langur(s)
drill *(Mandrillus leucophaeus)*, 118, **141**
dwarf lemur(s). See *under* lemur(s)

Erythrocebus patas (pàtas monkey), 120, **149–150**

Eulemur, 27, 38
   coronatus (crowned lemur), **39**
   fulvus (brown lemur), **40–41**
      albifrons (white-fronted brown lemur), *42*
      collaris (collared brown lemur), *40*
      fulvus (brown lemur), 40, *40*, 42
      rufus (red-fronted brown lemur), *41, 43*
      sanfordi, 39, 40
   macaco (black lemur), **41–42**
      flavifrons (Sclater's black lemur), 41, *41*, *42*
      macaco (black lemur), 41, *41*
   mongoz (mongoose lemur), 40, **42**
   rubriventer (red-bellied lemur), **43**

Euoticus elegantulus (southern needle-clawed bush baby), **18**, 18, 19, 22
   inustus (=*Galago matschiei*), 20
   pallidus (northern needle-clawed bush baby), **18**

Galago, 21, 24
   alleni (Allen's bush baby), 18, **19**, 22
   gallarum (Somali bush baby), **19**, 21, 23, 25
   matschiei (Matschie's bush baby), 17, **20**, 22, 23
   moholi (southern lesser bush baby), *12*, **20–21**, *21*, 22, 23, 24
   senegalensis (northern lesser bush baby), 4, 19, 20, **21**, 23, 24
      granti, 21

Galagoides
   demidoff (Demidoff's bush baby), 18, 19, 20, 21, **22**, *22*, 23
   rondoensis (rondo dwarf bush baby), 22, *22*
   thomasi (Thomas's bush baby), 20, 22, **22–23**
   zanzibaricus (Zanzibar bush baby), 19, 21, **23**, 24, 25
      granti, 23, *23*

Galagonidae (bush babies), 6, **13**, 18–25

gibbon
   agile. *See* dark-handed gibbon
   black *(Hylobates concolor)*, 215, 215, 216
   capped. *See* pileated gibbon
   Chinese white-cheeked *(Hylobates leucogenys)*, 215, **216**
   dark-handed *(Hylobates agilis)*, **209–210**, 210, 211, 213, 214, 217
   golden-cheeked *(Hylobates gabriellae)*, 215, **215**
   hoolock *(Hylobates hoolock)*, 207, **208**
   Kloss's *(Hylobates klossii)*, 182, **210–211**
   Mueller's Bornean gray *(Hylobates muelleri)*, 209, **213–214**
   pileated *(Hylobates pileatus)*, 211, **214**
   silvery Javan *(Hylobates moloch)*, **212**
   white-browed. *See* hoolock gibbon
   white-handed *(Hylobates lar)*, *6, 128, 206*, 209, 210, **211–212**, 214, 217
gibbons *(Hylobates;* Hylobatidae), 5–6, *6*, 207

Gigantopithecus blacki (orangutan ancestor), 219

Gorilla
   gorilla, 219, 224, 227
      beringei (mountain gorilla), **224–225**
      gorilla (western lowland gorilla), **226–227**
      graueri (eastern lowland gorilla), **227**

gorilla
   eastern lowland *(Gorilla gorilla graueri)*, **227**
   mountain *(Gorilla gorilla beringei)*, **224–225**
   western lowland *(Gorilla gorilla gorilla)*, **226–227**

grivet. *See* vervet

guenon
   Campbell's *(Cercopithecus campbelli)*, **155**, 156, 164
   crowned *(Cercopithecus pogonias)*, 156, 161, 163, **164**
   Dryas *(Cercopithecus dryas)*, **157**
   dwarf *(Miopithecus talapoin)*, 119, 120, **148**
   greater spot-nosed. *See* putty-nosed spot-nosed guenon
   lesser spot-nosed *(Cercopithecus petaurista)*, 155, 156, **164**, 176–177
   mustached *(Cercopithecus cephus)*, 145, 146, 154, **155–156**, 157, 158, 159, 163, 164, 165, 166
   putty-nosed spot-nosed *(Cercopithecus nictitans)*, 146, 156, **163**, 165
   red-eared *(Cercopithecus erythrotis)*, 155, 156, **158**, 166
   red-tailed *(Cercopithecus ascanius)*, 146, 147, 153, **154**, 160, 165, 167, 170, 171, 175
   Sclater's *(Cercopithecus sclateri)*, 155, 157, **166**
   sun-tailed *(Cercopithecus solatus)*, **167**
   white-throated *(Cercopithecus erythrogaster)*, 145, 146, *152*, **157**, 161
   Wolf's *(Cercopithecus wolfi)*, 146, 147, 154, **167**, 170
guenons *(Cercopithecus)*, 120, 148, 150, **153**, 172

Hadropithecus (extinct lemur), *2*
Hapalemur (bamboo lemurs), 27
   aureus (golden bamboo lemur), 1, 27, **45**, 46
   griseus (lesser bamboo lemur), 45, **45–46**, 46
      alaotrensis (bandro), 45, 45–46, *46*
   simus (greater bamboo lemur), 45, 46, **46**

Haplorrhini (tarsiers and New World monkeys), 3, 52–233
higher primates (Anthropoidea), 3, 8, 52, **57**, 57–233
Hominidae (gorillas, chimpanzees, and humans), **219**, 224–233
Hominoidea (apes, hominoids), 3, 119, **207**, 208–233
hominoids. *See* Hominoidea

Homo sapiens (human), 3, 4, 82, 219, **233**

howler
   black *(Alouatta pigra)*, **110**
   black-and-gold *(Alouatta caraya)*, 57, **107**
   Bolivian red *(Alouatta sara)*, **110**
   brown *(Alouatta fusca)*, **108–109**, 117
   Coiba Island *(Alouatta coibensis)*, **108**
   mantled *(Alouatta palliata)*, 107, 108, **109**, 100
   red *(Alouatta seniculus)*, 97, 107, 110, **111**, 115
   red-handed *(Alouatta belzebul)*, **107**
howler monkeys *(Alouatta;* Alouattinae), **81**, 82 103, 158

human *(Homo sapiens)*, 3, 4, 82, 219, **233**

Hylobates (gibbons; =Hylobatidae), 5–6, *6*, 207
   Bunopithecus
      hoolock (hoolock gibbon), 207, **208**
   Hylobates *(lar* group), 207
      agilis (dark-handed gibbon), **209–210**, 210, 211, 213, 214, 217
      albibarbis, 209
      klossii (Kloss's gibbon), 182, **210–211**
      lar (white-handed gibbon), *6*, 128, *206*, 209, 210, **211–212**, 214, 217
      moloch (silvery Javan gibbon), **212**
      muelleri (Mueller's Bornean gray gibbon), 209, **213–214**
      albibarbis, 213
      pileatus (pileated gibbon), 211, **214**
   Nomascus (concolor group), 207
      concolor (black gibbon), 215, 215, 216
      hainanus, 215
      gabriellae (golden-cheeked gibbon), 215, **215**
      leucogenys (Chinese white-cheeked gibbon), 215, **216**
   Symphalangus, 207
      syndactylus (siamang), 192, 207, 209, 211, **216–217**
Hylobatidae (lesser apes; gibbons; =*Hylobates*), **207**

Indri indri (indri), *2*, 27, **50**
indri *(Indri indri)*, *2*, 27, **50**
Indridae (indrids), **27**, 27–50
indrids. *See* Indridae

Kasi (as genus), 191, 195

Lagothrix (woolly monkeys), 82
   flavicauda (yellow-tailed woolly monkey), 90, **115**
   lagotricha (woolly monkey), 102, **116**
      cana, *116*
      lagotricha, *116*
      lugens, *116*
      poeppigii, *116*

langur
   black-shanked douc *(Pygathrix nigripes)*, 197, **198**
   Delacour's *(Trachypithecus delacouri)*, **188**
   ebony *(Trachypithecus auratus)*, **186**
   Francois's
      Trachypithecus francoisi, 188, **188–189**

francoisi, 189
golden *(Trachypithecus geei),* 122,
127, **190**
Hanuman *(Semnopithecus entellus),
half title page,* 7, 127, 169,
184–185, 191, 194, 196
Nilgiri *(Trachypithecus johnii),* 132,
185, **191**
pig-tailed *(Nasalis concolor),* 169, 182,
205
red-shanked douc *(Pygathrix
nemaeus),* 168, **197,** 198
silvered *(Trachypithecus cristatus),*
169, 186, **187**
stripe-headed black *(Trachypithecus
francoisi laotum),* 188, *189*
white-headed black *(Trachypithecus
francoisi leucocephalus),*
188, *189*
langurs
brow-ridged *(Trachypithecus),* 8, 169,
184, 195
douc *(Pygathrix Pygathrix),* 169
leaf monkey
banded *(Presbytis femoralis),* **179,** 181
capped *(Trachypithecus pileatus),* 190,
194
dusky *(Trachypithecus obscurus),* **192,**
193, 217
grizzled *(Presbytis comata),* **178,** 183
Hose's *(Presbytis hosei),* **180,** 183
maroon *(Presbytis rubicunda),* 180,
182–183
Mentawai Island *(Presbytis potenziani),*
128, 182, 205, 210
mitered *(Presbytis melalophos),* 179,
181, 192
Phayre's *(Trachypithecus phayrei),*
192, **193**
purple-faced *(Trachypithecus vetulus),*
185, 194, **195–196**
spectacled. *See* dusky leaf monkey
Thomas's *(Presbytis thomasi),* **183**
white-fronted *(Presbytis frontata),* **180**
leaf monkeys *(Presbytis),* 6, 169, 184, 186,
187, 188, 190, 191
Lemur, 27
catta (ring-tailed lemur), *26,* 27, **38–39**
lemur
black
Eulemur macaco, **41–42**
macaco, 41, *41*
black-and-white ruffed *(Varecia
variegata variegata),* 43, *43*
brown
Eulemur fulvus, **40–41**
fulvus, 40, *40,* 42
brown mouse *(Microcebus rufus),* **32**
collared brown *(Eulemur fulvus collaris),* 40
Coquerel's dwarf *(Microcebus
coquereli),* **30,** 36
crowned *(Eulemur coronatus),* **39**
fat-tailed dwarf *(Cheirogaleus medius),*
28, **29,** 30
fork-marked *(Phaner furcifer),* **33**
golden bamboo *(Hapalemur aureus),*
1, 27, **45,** 46
gray mouse *(Microcebus murinus),* 6,
31, 32

greater bamboo *(Hapalemur simus),*
45, 46, **46**
greater dwarf *(Cheirogaleus major),*
28–29
hairy-eared dwarf *(Allocebus trichotis),*
27, **28**
Lake Alaotra bamboo. *See* bandro
lesser bamboo *(Hapalemur griseus),*
45, **45–46,** 46
mongoose *(Eulemur mongoz),* 40, **42**
pygmy mouse *(Microcebus myoxinus),*
1, 4, 27, 31, **32**
red-bellied *(Eulemur rubriventer),* **43**
red-fronted brown *(Eulemur fulvus
rufus),* 41, 43
red ruffed *(Varecia variegata rubra),*
43, *44*
ring-tailed *(Lemur catta),* 26, 27, **38–39**
ruffed *(Varecia variegata),* 27, **43–44**
Sclater's black *(Eulemur macaco
flavifrons),* 41, *41, 42*
sportive. *See under* sportive lemur(s)
white-fronted brown *(Eulemur fulvus
albifrons),* 42
woolly *(Avahi laniger),* 27, 37, **47**
"lemur, flying" *(Cynocephalus* sp.), 27
Lemuridae (lemurs and bamboo lemurs),
27, 38–46
Lemuriformes (=Strepsirrhini), 3, 13–51
Lemuroidea (lemurs), 3, **26,** 26–51
lemurs. *See* Lemuridae; Lemuroidea
lemurs
bamboo *(Hapalemur),* 27
dwarf (Cheirogaleidae), **27,** 28–33
extinct, *2*
ruffed *(Varecia),* 27
sportive. *See under* sportive lemurs
"lemurs, flying" (Dermoptera), 27
Leontopithecus (lion tamarins), 59, 79
caissara (black-faced lion tamarin), 1,
77
chrysomelas (golden-headed lion
tamarin), 65, 66, **77–78**
chrysopygus (black lion tamarin), 65,
77, **78**
rosalia (golden lion tamarin), 10, **79**
Lepilemur, 45
dorsalis (gray-backed sportive lemur),
9, **34**
edwardsi (Milne-Edwards' sportive
lemur), **34**
leucopus (white-footed sportive lemur),
35
microdon (small-toothed sportive
lemur), **35**
mustelinus (weasel sportive lemur), 34,
35, **36,** 36, 37
ruficaudatus (red-tailed sportive
lemur), **36–37**
septentrionalis (northern sportive
lemur), **37**
Lepilemuridae (=Megaladapidae), 27
Lophocebus (mangabeys), 120, 153
albigena (gray-cheeked mangabey),
146, 146, 154, 156, 158,
160, 161, 163, 175
aterrimus (black mangabey), 146,
146–147, 154, 167, 170
Loridae (lorises and pottos), **13,** 13–17
Loris tardigradus (slender loris), **15**

loris
pygmy *(Nycticebus pygmaeus),* 16, **17**
slender *(Loris tardigradus),* **15**
slow *(Nycticebus coucang),* 12, **16,** 17,
221
lorises. *See* Loridae
Lorisoidea (=Loroidea), 13
Loroidea (lorises, pottos, and bush
babies), 3, **13,** 13–25
Macaca, 120
arctoides (stump-tailed macaque), **121,**
129, 134
assamensis (Assamese macaque),
title page, 119, **121,** 190
cyclopis (Formosan rock macaque),
122
cynomolgus (=*M. fascicularis),* 123
fascicularis (long-tailed macaque), **123,**
204
fuscata (Japanese macaque), 6, 7,
120, **124–125**
irus (=*M. fascicularis),* 123
maura (Celebes moor macaque), **126**
mulatta (rhesus macaque), 5, 120,
126–127, 135, 185, 190
nemestrina (pig-tailed macaque), **128,**
182, 211
pagensis (Mentawai Island
macaque), 128, *128*
nigra (Celebes black macaque), 126,
129, 130, 135
nigra (Celebes black macaque),
129
nigrescens (Dumoga-Bone
macaque), *129*
ochreata (booted macaque), **130,** 135
brunnescens, 130
radiata (bonnet macaque), **130–131,**
132
silenus (lion-tailed macaque), 131,
131–132, 191
sinica (toque macaque), **132–133**
speciosa (=*M. arctoides),* 121
sylvanus (Barbary macaque), 120,
133–134
thibetana (Tibetan macaque), **134**
tonkeana (Tonkean macaque), 130,
135
hecki, 135
macaque
Assamese *(Macaca assamensis), title
page,* 119, **121,** 190
Barbary *(Macaca sylvanus),* 120,
133–134
bonnet *(Macaca radiata),* **130–131,**
132
booted *(Macaca ochreata),* **130,** 135
Celebes black
Macaca nigra, 126, **129,** 130, 135
nigra, 129
Celebes moor *(Macaca maura),* **126**
crab-eating. *See* long-tailed macaque
crested black. *See* Celebes black
macaque
Dumoga-Bone *(Macaca nigra
nigrescens), 129*
Formosan rock *(Macaca cyclopis),* **122**
Japanese *(Macaca fuscata),* 6, 7, 120,
124–125
lion-tailed *(Macaca silenus),* 131,
131–132, 191

long-tailed *(Macaca fascicularis)*, **123**, 204

Mentawai Island *(Macaca nemestrina pagensis)*, 128, *128*

pig-tailed *(Macaca nemestrina)*, **128**, 182, 211

rhesus *(Macaca mulatta)*, 5, 120, **126–127**, 135, 185, 190

stump-tailed *(Macaca arctoides)*, **121**, 129, 134

Tibetan *(Macaca thibetana)*, **134**

Togian Island, 135

Tonkean *(Macaca tonkeana)*, 130, **135**

toque *(Macaca sinica)*, **132–133**

mandrill *(Mandrillus sphinx)*, **142**

Mandrillus, 120

leucophaeus (drill), *118*, **141**

sphinx (mandrill), **142**

mangabey

agile *(Cercocebus agilis)*, **144**, 144

black *(Lophocebus aterrimus)*, 146, **146–147**, 154, 167, 170

gray-cheeked *(Lophocebus albigena)*, **146**, 146, 154, 156, 158, 160, 161, 163, 175

sooty *(Cercocebus torquatus atys)*, 145, **145**

Tana River *(Cercocebus galeritus)*, 144, **144**

white-collared *(Cercocebus torquatus)*, **145**, 156, 157, 161, 174

mangabeys

Cercocebus, 120, 146, 147, 153

Lophocebus, 120, 153

marmoset

bare-ear *(Callithrix argentata)*, **60–61**, 66, 73

black tufted–eared *(Callithrix penicillata)*, 64, **66**, 77

black-headed *(Callithrix nigriceps)*, 1, **66**

buffy tufted–eared *(Callithrix aurita)*, **61**, 64

buffy-headed *(Callithrix flaviceps)*, *58*, **62**, 64

common *(Callithrix jacchus)*, 61, 62, **64**, 65

Geoffroy's tufted-eared *(Callithrix geoffroyi)*, **62–63**, 64, 91

Maues *(Callithrix mauesi)*, 1, **65**

pygmy *(Callithrix pygmaea)*, 5, 59, **67**, 75

tassel-eared *(Callithrix humeralifer)*, **63–64**, 65

Wied's tufted-eared *(Callithrix kuhlii)*, 64, **64**, 77–78

marmosets *(Callithrix)*, 6, 8

Megaladapidae (sportive lemurs), **27**, 34–37

Megaladapis (extinct lemur), *2*

Microcebus, 28

coquereli (Coquerel's dwarf lemur), **30**, 36

murinus (gray mouse lemur), 6, **31**, 32

myoxinus (pygmy mouse lemur), 1, 4, 27, 31, **32**

rufus (brown mouse lemur), **32**

Miopithecus talapoin (dwarf guenon), 119, 120, **148**

Mirza (=*Microcebus*), 30

monkey

Allen's swamp *(Allenopithecus nigroviridis)*, 120, **147**, 154, 167

black snub-nosed *(Pygathrix bieti)*, **200**

blue *(Cercopithecus mitis)*, 146, 153, 154, **159–160**, 170, 175

De Brazza's *(Cercopithecus neglectus)*, **162**

Diana *(Cercopithecus diana)*, 155, **156**, 157, 164, 174, 176

Goeldi's *(Callimico goeldii)*, 60, 69, 72, 75

golden mitis *(Cercopithecus mitis kandti)*, 160

green. *See* vervet

grivet. *See* vervet

Guizhou snub-nosed *(Pygathrix brelichi)*, **201**

L'Hoest's *(Cercopithecus lhoesti)*, **159**, 165, 167

mona *(Cercopithecus mona)*, 145, 146, 155, 157, **161**, 164, 165, 167

owl-faced *(Cercopithecus hamlyni)*, **158**

Preuss's *(Cercopithecus preussi)*, 159, **165**

proboscis *(Nasalis larvatus)*, 8, 123, 169, **203–204**, 222

Sichuan golden snub-nosed *(Pygathrix roxellana)*, 200, 201, **202**

southern talapoin. *See* guenon, dwarf

Tonkin snub-nosed *(Pygathrix avunculus)*, 169, **199**

vervet *(Chlorocebus aethiops)*, 120, **150–151**, 156

widow. *See* titi monkey, collared

woolly *(Lagothrix lagotricha)*, 102, **116**

Yunnan snub-nosed. *See* black snub-nosed monkey

monkeys

capuchin. *See under* capuchin; capuchin monkeys

cheek pouch (Cercopithecinae), 8, 119, **120**, 120–167

colobus. *See under* colobus; colobus monkeys

howler. *See under* howler; howler monkeys

leaf. *See under* leaf monkey(s)

leaf-eating (Colobinae), 119, **169**, 170–205

Neotropical. *See* New World monkeys

New World

Ceboidea, 3, 57, 57–117

Platyrrhini, 3, 57, **59**, 60–117, 119

night. *See under* night monkey(s)

odd-nosed

Nasalis, 169

Pygathrix, 169

Old World

Catarrhini, 3, 57, 59,119–233,148

Cercopithecoidea, 3, 8, 119–205, 148, 207

owl. *See under* owl monkey

snub-nosed *(Pygathrix Rhinopithecus)*, 169

spider. *See under* spider monkey(s)

squirrel. *See under* squirrel monkey(s)

titi. *See under* titi monkey(s)

woolly *(Lagothrix)*, 82

yellow-tailed woolly *(Lagothrix flavicauda)*, 90, **115**

mouse lemur(s). *See under* lemur(s)

muriqui. *See* spider monkey, woolly

Nasalis (odd-nosed monkeys), 169

Simias

concolor (pig-tailed langur), 169, 182, **205**

Nasalis larvatus (proboscis monkey), 8, 123, 169, **203–204**, 222

night monkey *(Aotus azarai)*, 83

southern red-necked *(Aotus nigriceps)*, 83

night monkeys

Aotinae, **81**

Aotus, 8, 81

Nycticebus

coucang (slow loris), *12*, **16**, 17, 221

intermedius, 17

pygmaeus (pygmy loris), 16, **17**

orangutan

Borneo *(Pongo pygmaeus)*, 204, 220, **222–223**

Sumatran *(Pongo abelii)*, **220–221**, 222

orangutans

Pongidae, **219**, 220–223

Pongo, 6

Otolemur

crassicaudatus (thick-tailed greater bush baby), 21, 23, **24**, 25

garnettii (Garnett's greater bush baby), 19, 23, 24, **25**

owl monkey, northern gray-necked *(Aotus trivirgatus)*, 84

owl monkeys. *See* night monkeys

Palaeopropithecus (extinct lemur), *2*

Pan

paniscus (bonobo), 219, **228–229**

troglodytes (chimpanzee), v, 7, 8, 156, 176, *218*, 219, 228, **230–232**, 233

schweinfurthii, 230, *231*

troglodytes, 230

verus, 230, *231*

Papio (baboons), 120, 141, 142, 143, 147, 151

hamadryas, 7

anubis (olive baboon), 119, **136–137**, 139

cynocephalus (yellow baboon), **138**

hamadryas (hamadryas baboon), 137, **138–139**

papio (Guinea baboon), **140**

ursinus (chacma baboon), **140**

patas monkey *(Erythrocebus patas)*, 120, **149–150**

Perodicticus potto (potto), **17**, 20

Phaner furcifer (fork-marked lemur), **33**

Phanerinae, 33

Piliocolobus (as genus), 173

Pithecia (sakis), 75, 81

aequatorialis (equatorial saki), **101**

albicans (buffy saki), 94, **101**

hirsuta (=*P. monachus*), 102

irrorata (bald-faced saki), **102**

monachus (monk saki), 75, 101, 102, **102**, 106, 116

pithecia (white-faced saki), **103**, 104, 105

Pitheciinae (sakis and uacaris), **81**
Platyrrhini (New World monkeys;
　　=Ceboidea), 3, **57**, 58–117,
　　119
Pongidae (orangutans), **219**, 220–223
Pongo (orangutans), 6
　abelii (Sumatran orangutan), **220–221**,
　　222
　pygmaeus (Borneo orangutan), 204,
　　220, **222–223**
potto *(Perodicticus potto)*, 17, 20
Presbytis (leaf monkeys), 6, 169, 184, 186,
　　187, 188, 190, 191
　aygula (=P. comata), 178, 180
　comata (grizzled leaf monkey), **178**,
　　183
　　fredericae, 178, *178*, 180
　femoralis (banded leaf monkey), **179**,
　　181
　　natunae, *179*
　　robinsoni, *179*
　frontata (white-fronted leaf monkey),
　　180
　hosei (Hose's leaf monkey), **180**, 183
　melalophos (mitered leaf monkey),
　　179, **181**, 192
　　melalophos, *181*
　　mitrata, *181*
　　nobilis, *181*
　potenziani (Mentawai Island leaf
　　monkey), 128, **182**, 205,
　　210
　rubicunda (maroon leaf monkey), 180,
　　182–183
　senex (=Trachypithecus vetulus), 195
　siamensis, 179
　thomasi (Thomas's leaf monkey), **183**
primates. *See individual taxa; see also*
　　higher primates
Procolobus (red colobus monkeys), 153,
　　154, 160, 169, 175
Procolobus
　Piliocolobus
　　badius (western red colobus), 145,
　　　156, 172, **173–174**, 175,
　　　176, 231
　　　temminckii, 173
　　pennantii (Pennant's red colobus),
　　　6, 147, 171, **174–175**
　　　gordonorum, 174
　　　kirkii (Kirk's red colobus), 174,
　　　　174, 175
　　　tephrosceles (red colobus),
　　　　174, *174, 175*
　　　tholloni, 174
　　preussi (Preuss's red colobus), **175**
　　rufomitratus (Tana River red
　　　colobus), **176**
　Procolobus
　　verus (olive colobus), 156, 164,
　　　169, 172, **176–177**
Propithecus (sifakas), 27
　diadema (diademed sifaka), **47–48**
　　candidus, 47
　　diadema (diademed sifaka), 47, *47*
　　edwardsi (Milne-Edwards' sifaka),
　　　47, *48*
　　perrieri, 47
　tattersalli (golden-crowned sifaka), 1,
　　48
　verreauxi (Verreaux's sifaka), 6, **49**

prosimians. *See* Prosimii
Prosimii (prosimians), 3, 6, 8, **13**, 13–56,
　　52
Pseudopotto martini, 13
Pygathrix (odd-nosed monkeys), 169
　Pygathrix (douc langurs), 169
　　nemaeus (red-shanked douc
　　　langur), *168*, **197**, 198
　　nigripes (black-shanked douc
　　　langur), 197, **198**
　Rhinopithecus (snub-nosed monkeys),
　　169
　　avunculus (Tonkin snub-nosed
　　　monkey), 169, **199**
　　bieti (black snub-nosed monkey),
　　　200
　　brelichi (Guizhou snub-nosed
　　　monkey), **201**
　　roxellana (Sichuan golden snub-
　　　nosed monkey), 200, 201,
　　　202
　Rhinopithecus (as genus), 199, 200, 201,
　　202
ruffed lemur(s). *See under* lemur(s)
Saguinus (tamarins), 6, 59, 60, 67, 103
　bicolor (bare-faced tamarin), **68**
　　bicolor, 68
　　martinsi, 68
　fuscicollis (saddleback tamarin), 60,
　　68–69, 71, 72, 73, 74, 75,
　　76, 92
　　weddelli, 69, 72, 89
　geoffroyi (red-crested tamarin), **70**, 75,
　　76
　imperator (emperor tamarin), 69,
　　70–71
　inustus (mottled-face tamarin), **71**
　labiatus (red-bellied tamarin), 60, 69,
　　72
　leucopus (silvery-brown bare-faced
　　tamarin), **73**
　midas (golden-handed tamarin), 61, **73**
　　midas, 73
　　niger, 73
　mystax (mustached tamarin), 69, **74**
　nigricollis (Spix's black-mantled
　　tamarin), 60, 67, 69, **75**,
　　102
　oedipus (cotton-top tamarin), 70, 73,
　　75–76
　tripartitus (golden-mantled saddleback
　　tamarin), 68, **76**
Saimiri (squirrel monkeys), 75, 81, 106,
　　148
　boliviensis (Bolivian squirrel monkey),
　　97–98, 100
　　madeirae, 100
　oerstedii (red-backed squirrel monkey),
　　98
　sciureus (common squirrel monkey),
　　93, 94, 97, 98, **99**, 100, 106
　ustus (golden-backed squirrel mon-
　　key), **100**, 100
　vanzolinii (black squirrel monkey), 100,
　　100
saki
　bald-faced *(Pithecia irrorata)*, **102**
　bearded *(Chiropotes satanas)*, 103,
　　104–105, 115
　buffy *(Pithecia albicans)*, 94, **101**
　equatorial *(Pithecia aequatorialis)*, **101**

　monk *(Pithecia monachus)*, 75, 101,
　　102, **102**, 106, 116
　white-faced *(Pithecia pithecia)*, **103**,
　　104, 105
　white-nosed bearded *(Chiropotes
　　albinasus)*, 94, **104**
sakis
　bearded *(Chiropotes)*, 81, 105
　Pithecia, 75, 81
　Pitheciinae, **81**
Scandentia (tree shrews), 13
Semnopithecus, 169, 184, 186, 187, 188,
　　190, 191
　barbei, 193
　entellus (Hanuman langur), *half title
　　page*, 7, 127, 169,
　　184–185, 191, 194, 196
shrew, common tree *(Tupaia glis)*, *13*
shrews, tree (Scandentia), 13
siamang *(Hylobates syndactylus)*, 192,
　　207, 209, 211, **216–217**
sifaka
　diademed
　　Propithecus diadema, **47–48**
　　　diadema, 47, *47*
　golden-crowned *(Propithecus
　　tattersalli)*, 1, **48**
　Milne-Edwards' *(Propithecus diadema
　　edwardsi)*, 47, *48*
　Verreaux's *(Propithecus verreauxi)*,
　　6, **49**
sifakas *(Propithecus)*, 27
simians. *See* Anthropoidea
Simias (as genus), 205
snub-nosed monkey(s). *See under*
　　monkey(s)
spider monkey
　black *(Ateles paniscus)*, 105, 111, 113,
　　115
　black-faced black *(Ateles chamek)*,
　　113, 115
　black-handed *(Ateles geoffroyi)*, 95,
　　114
　brown-headed *(Ateles fusciceps)*, **113**
　white-bellied *(Ateles belzebuth)*, *82*,
　　112, 114
　white-whiskered *(Ateles marginatus)*,
　　112, **114**
　woolly *(Brachyteles arachnoides)*, 7,
　　108, **117**
spider monkeys *(Ateles)*, 7, 8, 82, *82*, 104,
　　105, 158
　woolly *(Brachyteles)*, 82
sportive lemur
　gray-backed *(Lepilemur dorsalis)*, 9, **34**
　Milne-Edwards' *(Lepilemur edwardsi)*,
　　34
　northern *(Lepilemur septentrionalis)*,
　　37
　red-tailed *(Lepilemur ruficaudatus)*,
　　36–37
　small-toothed *(Lepilemur microdon)*,
　　35
　weasel *(Lepilemur mustelinus)*, 34, 35,
　　36, 36, 37
　white-footed *(Lepilemur leucopus)*, **35**
sportive lemurs (Megaladapidae), **27**,
　　34–37
spot-nosed guenon(s). *See under*
　　guenon(s)

squirrel monkey
> black *(Saimiri vanzolinii)*, 100, **100**
> Bolivian *(Saimiri boliviensis)*, **97–98**, 100
> common *(Saimiri sciureus)*, 93, 94, 97, 98, **99**, 100, 106
> golden-backed *(Saimiri ustus)*, **100**, 100
> red-backed *(Saimiri oerstedii)*, **98**
squirrel monkeys *(Saimiri)*, 75, 81, 106, 148
Strepsirrhini (=Lemuriformes), 3, 13–51
Symphalangus (=Hylobates syndactylus), 216
talapoin monkey, southern. *See* guenon, dwarf
tamarin
> bare-faced *(Saguinus bicolor)*, **68**
> black lion *(Leontopithecus chrysopygus)*, 65, 77, **78**
> black-faced lion *(Leontopithecus caissara)*, 1, **77**
> cotton-top *(Saguinus oedipus)*, 70, 73, **75–76**
> emperor *(Saguinus imperator)*, 69, **70–71**
> golden lion *(Leontopithecus rosalia)*, 10, **79**
> golden-handed *(Saguinus midas)*, 61, **73**
> golden-headed lion *(Leontopithecus chrysomelas)*, 65, 66, **77–78**
> golden-mantled saddleback *(Saguinus tripartitus)*, 68, **76**
> mottled-face *(Saguinus inustus)*, **71**
> mustached *(Saguinus mystax)*, 69, **74**
> red-bellied *(Saguinus labiatus)*, 60, 69, **72**
> red-crested *(Saguinus geoffroyi)*, **70**, 75, 76
> saddleback *(Saguinus fuscicollis)*, 60, **68–69**, 71, 72, 73, 74, 75, 76, 92
> silvery-brown bare-faced *(Saguinus leucopus)*, **73**
> Spix's black-mantled *(Saguinus nigricollis)*, 60, 67, 69, **75**, 102
tamarins *(Saguinus)*, 6, 59, 60, 67, 103

lion *(Leontopithecus)*, 59, 79
tarsier
> Dian's *(Tarsius dianae)*, **54**
> Philippine *(Tarsius syrichta)*, **56**
> pygmy *(Tarsius pumilus)*, **54**
> spectral *(Tarsius spectrum)*, **55**
> western *(Tarsius bancanus)*, **53**
tarsiers
> Tarsiidae, 53–56
> Tarsiiformes, 3, **52**, 52–56
> Tarsioidea, 3, **52**, 52–56
> *Tarsius*, 52
Tarsiidae (tarsiers), 53–56
Tarsiiformes (tarsiers), 3, **52**, 52–56
Tarsioidea (tarsiers), 3, **52**, 52–56
Tarsius (tarsiers), 52
> *bancanus* (western tarsier), **53**
> *dianae* (Dian's tarsier), **54**
> *pumilus* (pygmy tarsier), **54**
> *sangirensis*, 52
> *spectrum* (spectral tarsier), **55**
> > *dentatus*, 55
> > *sangirensis*, 55, *55*
> > *spectrum*, *55*
> *syrichta* (Philippine tarsier), **56**
Theropithecus gelada (gelada baboon), 8, 120, **143**
titi monkey
> Andean *(Callicebus oenanthe)*, 83, **90**
> ashy *(Callicebus cinerascens)*, 86, 89
> Beni *(Callicebus olallae)*, **90**
> Bolivian gray *(Callicebus donacophilus)*, 85, **87**
> brown *(Callicebus brunneus)*, 85, 85, 86, 87
> *Callicebus dubius*, 85, **87**
> *Callicebus modestus*, **88**
> chestnut-bellied *(Callicebus caligatus)*, 85, **85**, 86, 87, 92
> collared *(Callicebus torquatus)*, 69, 85, 86, 89, **92**
> dusky *(Callicebus moloch)*, 69, 71, 86, 88, **89**, 92
> Hoffmann's *(Callicebus hoffmannsi)*, **88**
> masked *(Callicebus personatus)*, 63, **91**
> red *(Callicebus cupreus)*, 85, **86**, 87, 88, 92
titi monkeys

Callicebinae, **81**
Callicebus, 6, 8, 81, 88
Togian Island macaque, 135
Trachypithecus (brow-ridged langurs), 8, 169, 184, 195
> *Kasi*
> > *johnii* (Nilgiri langur), 132, 185, **191**
> > *vetulus* (purple-faced leaf monkey), 185, 194, **195–196**
> *Trachypithecus*
> > *auratus* (ebony langur), **186**
> > *cristatus* (silvered langur), 169, 186, **187**
> > *delacouri* (Delacour's langur), **188**
> > *francoisi* (Francois's langur), 188, **188–189**
> > > *francoisi* (Francois's langur), *189*
> > > *hatinhensis*, 188
> > > *laotum* (stripe-headed black langur), 188, *189*
> > > *leucocephalus* (white-capped black langur), 188, *189*
> > > *poliocephalus*, 188
> > *geei* (golden langur), 122, 127, **190**
> > *obscurus* (dusky leaf monkey), **192**, 193, 217
> > *phayrei* (Phayre's leaf monkey), 192, **193**
> > *pileatus* (capped leaf monkey), 190, **194**
tree shrew(s). *See under* shrew(s)
Tupaia glis (common tree shrew), *13*
uacari
> bald *(Cacajao calvus)*, 102, **105–106**
> black-headed *(Cacajao melanocephalus)*, 93, 94, 99, **106**
> red *(Cacajao calvus rubicundus)*, 99, 105, *105*, 106
> white *(Cacajao calvus calvus)*, 105, 106, *106*
uacaris *(Cacajao)*, 81
Varecia (ruffed lemurs), 27
> *variegata* (ruffed lemur), 27, **43–44**
> > *rubra* (red ruffed lemur), 43, *44*
> > *variegata* (black-and-white ruffed lemur), 43, *43*
vervet *(Chlorocebus aethiops)*, 120, **150–151**, 156
woolly monkey. *See under* monkey

Primate
Conservation,
Inc.

1411 Shannock Rd.
Charlestown, Rhode Island 02813-3726

Primate Conservation, Inc. (PCI) provides matching grants to primatologist and conservationists for field research and conservation projects on the least known and most endangered species. PCI has funded projects on guenons and red colobus in West Africa, tarsiers in Indonesia, lemurs in Madagascar and douc langurs and the Tonkin snub-nosed monkey in Vietnam.

PCI is unable to fund many worthwhile projects due to a lack of resources. Please help us to get more conservationists into the field to secure a future for endangered primates.

Primate Conservation, Inc.
1411 Shannock Rd.
Charlestown, Rhode Island 02813-3726

Primate Conservation, Inc.
1411 Shannock Rd.
Charlestown, Rhode Island 02813-3726